Cognitive Kin, Moral Strangers?

Human-Animal Studies

VOLUME 23

The titles published in this series are listed at *brill.com/has*

Cognitive Kin, Moral Strangers?

Linking Animal Cognition, Animal Ethics & Animal Welfare

by

Judith Benz-Schwarzburg

Translated by

Mark Kanak

Edited by

Susana Monsó
Evan White
Judith Benz-Schwarzburg

TUTA SUB AEGIDE PALLAS 1683

BRILL

LEIDEN | BOSTON

Cover illustrations by Judith Benz-Schwarzburg. Used with permission.

The translation of this work was financially supported by the Cultural Department of the City of Vienna, the Messerli Foundation, and the Messerli Research Institute unit Ethics and Human-Animal Studies.

Originally published as *Verwandte im Geiste – Fremde im Recht. Sozio-kognitive Fähigkeiten bei Tieren und ihre Relevanz für Tierethik und Tierschutz*. Copyright © 2012 by Harald Fischer Verlag GmbH, Erlangen.

The Library of Congress Cataloging-in-Publication Data is available online at http://catalog.loc.gov

Typeface for the Latin, Greek, and Cyrillic scripts: "Brill". See and download: brill.com/brill-typeface.

ISSN 1573-4226
ISBN 978-90-04-35886-7 (hardback)
ISBN 978-90-04-41507-2 (e-book)

Contents

MONKEY, n. An arboreal animal which makes itself at home in genealogical trees.
MAN, n. An animal so lost in rapturous contemplation of what he thinks he is as to overlook what he indubitably ought to be.

AMBROSE BIERCE (1842–1914)
The Devil's Dictionary
Clipart courtesy FCIT

Preface and Acknowledgments

This book is an interdisciplinary study at the intersection of the humanities and natural sciences. It deals primarily with questions of animal ethics, animal welfare, and cognitive ethology, but it also includes, for example, insights from evolutionary anthropology, cognitive archeology, comparative psychology, developmental psychology, theoretical philosophy, linguistics, and veterinary medicine. Based on the results of empirical research in the area of mental ability and performance, the book is intended to explore something that has been under investigation since the introduction of Darwin's theory of evolution: namely, the possibility that the difference between people and animals is only gradual. This entails an examination of crucial aspects of the evolutionary relationship between humans and animals and their practical importance for human-animal relations today.

The book concerns itself with a research field that is presently undergoing major developments and is beginning to crystallize in its relevance for various scientific branches in international discussions. In addition to this interdisciplinary approach, the book combines theoretical and practical questions, establishing a connection that is often overlooked. For a long time, there was an unsatisfactory situation in research on the socio-cognitive abilities of animals. Studies such as the ones conducted by Irene Pepperberg on the language skills of grey parrots (Pepperberg 2002) had not been adequately analyzed and integrated into the debates for which they are relevant. The major challenge faced by the academic disciplines involved is to evaluate the research results presented in this book and other pertinent findings from recent studies and to find new, appropriate interpretations of the knowledge and skills of animals. Such integration of findings from a range of disciplines cast important light on a central aspect of the relationship between humans and animals, supplementing in a substantial way the findings of modern genetics.

My book examines from three perspectives of socio-cognitive competence (culture, language, and theory of mind), to what extent it can be assumed that nonhuman animals also have such abilities or parts and precursors of such abilities. In so doing, the book explores whether an understanding of such competences that is exclusive, monolithic,[1] and refers to healthy adults from the outset is, in and of itself, inherently problematic—not only when considering

1 "Monolithic" entities consist of only one stone, are inseparable or monumental. With regard to the definition or the conceptualization of socio-cognitive abilities such as culture, language, or theory of mind, this means understanding these abilities primarily as total

such abilities in other animals but also when an adequate understanding of these competences is sought in humans. Suggestions for how social and cognitive abilities can be described and conceptualized in animals are also presented.

Adopting the idea of continuity in the area of social and cognitive skills from an interdisciplinary perspective and on the basis of current research results also represents a gain in a very important aspect of animal ethics. The moral status of animals in ethical theories depends on the criteria for moral consideration; and these often depend on which characteristics animals possess. If the current research on the minds of animals reveals that these abilities and the needs which follow from them can be used to establish stronger protective rights, then we are faced with the task of rethinking our own exclusivity as beings who are to be morally considered in a separate way (see Ingensiep 1999, 442). If we learn more and more about animals and recognize their physical and psychological vulnerability as well as their similarity to humans, then we must above all be compelled, in addition to all the theoretical (including moral-theoretical) discussions about the human-animal relationship, to rethink our everyday practical handling of animals in the wild and in captivity. This applies to our handling of farm animals, pet animals, zoo animals, and laboratory animals alike. Unfortunately, not all these cases can be fully addressed in this book. I therefore concentrate on the two families whose species have hitherto been highlighted in discussions on animal minds, in ethical discussions on animal sentience and suffering, and in conversations on animal rights: nonhuman great apes and dolphins. With regard to these two families, examples of practical animal welfare and conservation are discussed, from the problems of habitat disturbance and destruction to the use of animals in experiments or in dolphin-assisted therapy. This discussion will reveal the threats our "cognitive relatives" are exposed to every day. What this book will thereby show in its core is the blatant discrepancy between new insights on animal minds as well as increasing animal-ethics-related demands on the one hand and the status quo of environmental, animal, and species protection on the other hand—a discrepancy which I identify as the decisive characteristic of the human-animal relationship in the twenty-first century.

From the outset of my research for this book, my focus was not only on the question of whether animals possess the socio-cognitive abilities necessary for culture, language, and theory of mind. It seemed just as important to me to ask why we should deal with this question at all. In this light, my book transcends

---

phenomena, as larger closed complexes, and to neglect that they emerge from precursors and individual building blocks.

much of the previous discussion about animal minds, both in terms of analytical philosophy of animal minds and of cognitive ethology, and follows one of my basic assumptions: relationship links with other animals who are kin to us are also always links of responsibility.

This book emerged from my dissertation, *Socio-Cognitive Abilities in Animals and Their Relevance for Animal Ethics and Animal Protection*, which I completed in 2012 at the Faculty of Arts of the Eberhard Karls University of Tübingen, Germany. It was published in book form in German 2012 under the title *Verwandte im Geiste—Fremde im Recht* by Harald Fischer Verlag in Erlangen. The present English-language edition is a significantly reduced and updated version of the German edition.

The fact that I was able to pursue my wide-ranging, practice-oriented, and interdisciplinary research interests was supported and promoted by many different sides. I was especially delighted that the Eberhard Karls University of Tübingen gave me the 2012 Doctorate Prize for my dissertation, on the recommendation of the Faculty of Philosophy. In 2013 the book was also awarded the German Study Prize (second place in the field of Humanities and Cultural Studies). My thanks go to the Körber Foundation.

I would like to thank the two supervisors of my dissertation, Eve-Marie Engels and Hanno Würbel, as well as Albert Newen for his support in the first two years of the doctorate. I would also like to thank all the colleagues who supported me in many ways in my work in Tübingen and Vienna (among them are the examiners for my doctoral thesis, the directors, faculty, and staff of the post graduate program "Bioethics" at the International Center for Ethics in the Sciences and Humanities (IZEW) of the University of Tübingen, the Forum Scientiarum Tübingen, and many individuals who have supported and continue to support my research, including Herwig Grimm and his entire team in Vienna, Arianna Ferrari, Andrew Knight, Juliane Kaminski, Miriam Haidle, Vera Marashi, Ludwig Huber, and many more). I thank Svenja Springer and Susana Monsó (whom I could trust 100%) for their support in the reduction of the manuscript, for their proofreading, and for their friendship.

I would like to thank the Deutsche Forschungsgemeinschaft (DFG) for financial support of the doctoral thesis in the form of a "printing cost stipend" and a doctoral fellowship in the post graduate program "Bioethics." I would like to thank the Cultural Department of the City of Vienna, Herwig Grimm, and the Messerli Foundation for financial support of the translation (and their patience), and especially Heinz Schweizer for his interest in my work. I also thank Gary Steiner, Gary Francione, Wendy Lochner, Volker Sommer, the anonymous reviewers, Margo de Mello, Kenneth Shapiro, and everyone else who supported the plan to translate this book. This is especially true for Claudia Schorcht from

Harald Fischer, Jennifer Pavelko, Meghan Conolly, as well as Benjamin Suchard from Brill, for Susana Monsó, Evan White, Christoph Schachenhofer, and for Mark Kanak, who embarked on this project with me. It seems to me that a book that examines the human-animal relationship in such interdisciplinarity and depth is appearing on the market at a time when science is increasingly being reduced to h-factor and peer-reviewed journal articles. Everyone involved in this book and supporting it still seem to value real books, which is great!

Without the support of my family, neither the dissertation nor the German or English editions of the book would have been possible. The fact that I embarked on this work can be traced back to my studies with a professor who brought Darwin into her philosophy seminars, but also to my childhood and my everyday life between rabbits, guinea pigs, and vegetarian-vegan cohabitants and friends. My special thanks go to my parents, Hannerose and Hartmut Benz, my sisters, Sophia and Franziska Benz, my husband, Stefan Schwarzburg, my children, Finn and Milla, and to all the animals who shared my life.

Judith Benz-Schwarzburg
Vienna, 2019

PART 1

Introduction

CHAPTER 1

Socio-Cognitive Abilities in Animals as the Object of Science—and What Has Been Neglected Thus Far

In Western philosophy, research into the abilities of animals has played a role since ancient times.[1] With the emergence of philosophical ethics among the Greeks (at the very latest), animals and children emerged as protagonists in the sphere of philosophical consideration and served either to study original human nature or to study a countercosmism that has already crystallized: the adult human being is precisely what the child and the animal cannot be or are not yet able to be. To this day, children and animals are an integral part of philosophy, and scholars refer to them as reference points to one another (Gigon 2002, 53ff.).

These days, philosophers are particularly interested in nonhuman primates and their abilities (see, for example, De Waal 2006). The discovery of great apes as "border organisms" between man and beast took also place in antiquity (Ingensiep 1999, 435ff.). In the belief structures of many non-European cultures, such as the ancient Egyptian, Asian, and African natural religions, monkeys and human beings assumed a special role between God and the world, humans and nature (see Sorenson 2009, esp. 34–69).

Some African tribes, such as the Mpiemu in today's Central African Republic or the Bangando in Cameroon, have always believed in a special link between humans and apes and a common origin for the species (Sorenson 2009, 34).[2] In mythical and religious contexts as well as in philosophical

1 Aristotle, Plutarch, Plato, Seneca, Augustine, Thomas Aquinas, Montaigne, Hobbes, Descartes, Leibniz, Hume, Reimarus, Schopenhauer, Bentham, Kant, and Heidegger, just to name a few well-known philosophers, all examined animals and their abilities. In these investigations, the human being was usually placed above the animal kingdom, as the only creature that is reasonably minded, able to form political states, or capable of exhibiting morality. For a comprehensive overview and discussions of these and similar contemporary characterizations, see Calarco (2008), Schütt (1990), Perler and Wild (2005), Wild (2007), Lurz (2009), Kalof and Fitzgerald (2007), and Kalof and Resl (2007). In the following chapters of this book, reference will be made again and again to the philosophical history of ideas regarding the human-animal relationship—in section 5.1, for example, with reference to linguistic skills, and in section 8.4.1 with reference to the concept of person.

2 Of course, fundamental differences can be used to establish the similarity between humans and animals in different cultures. While in non-Western cultures the animal (generally) and the monkey or great ape were ascribed a rather positive role, the same cannot be said for

 | DOI:10.1163/9789004415072_002

considerations, the animal's similarity to mankind has long been fascinating above all in regard to qualities attributed to humans alone, such as reason and cognition (Schütt 1990). However, only Charles Darwin founded a *systematic scientific theory* of evolution of human capacities. His evolutionary approach to behavior and cognition is surely complex, and the comparisons between the abilities of animals and humans are manifold. His work *The Origin of Species* (1859) was followed in 1871 by *The Descent of Man*, wherein he specifically addressed the question of a cognitive continuity between humans and other animals:

> If no organic being excepting man had possessed any mental power, or if his powers had been of a wholly different nature from those of the lower animals, then we should never have been able to convince ourselves that our high faculties had been gradually developed. But it can be shewn that there is no fundamental difference of this kind. We must also admit that there is a much wider interval in mental power between one of the lowest fishes, as a lamprey or lancelet, and one of the higher apes, than between an ape and man; yet this interval is filled up by numberless gradations.
>
> DARWIN [1879] 2004, 86

He clearly states that his object is "to show that there is no fundamental difference between man and the higher mammals in their mental faculties" (86).

If no animals other than humans possessed cognitive abilities, then the human being's outstanding cognitive abilities would be inexplicable for Darwin. Instead, he assumes that there are also connections between species in the mental realm. Differences are not leveled, but are indicated as only gradual. There are certainly many gradual differences between the cognitive abilities of a fish and those of a human being, which, in their totality, make a great gradual difference between these two species. Between the closely related human

---

Western, Christian societies. According to Sorenson, these are and have been dominated by anthropocentric ideas espousing the separation of humans from animals and the presumed superiority of humans over nature. For such societies, the great apes in particular present a problem because they share so many abilities and emotions with human beings. On the one hand, the similar behavior confirmed the naturalness of human behavior; on the other hand, animals were regarded as less perfected versions of humans. Therefore, western societies often reacted to the similarities between humans and great apes and other animals with derision and consternation. To this day the image of a monkey is used to caricature fellow human beings and to criticize (supposedly) lower or reprehensible behavior (see Sorenson 2009, 39–40, 47–48).

beings and apes, such deviations, on the other hand, remain more clearly visible as only gradual. Darwin does not see any fundamental differences between species. Ultimately, in Chapter 6 of the *Descent of Man*, he writes: "The mental faculties of man and the lower animals are not different in kind, although immensely in degree." He concludes: "A difference in degree, however great, does not justify us in placing man in a distinct kingdom" (2004 [1879], 173). While the principle of graduality is not at first completely incompatible with the idea of a *scala naturae*, Darwin nevertheless rejects a definite, linear understanding of such a theory if it is traced from "lower" forms of life to the human being as the "crown of creation." Despite gradual developments and gradual differences, the connection between different species and the temporal-dynamic character of evolution is emphasized. The principles of continuity and graduation are also applied to the cognitive abilities of the animals and the dogma of the separate creation of each individual species is overcome. Further, in his investigation *The Expression of the Emotions in Man and Animals* (1872), he pursued the theme of the similarity between humans and animals.

Darwin's unprecedented systematic identification of the whole human being, including her cognitive abilities, as similar to other animals allowed him to do comparative research as we do today. His theory of evolution thus certainly forms a very important background and one of the most important prerequisites for all studies of socio-cognitive abilities of animals after Darwin. These ultimately occurred in the twentieth and twenty-first centuries within disciplines such as cognitive ethology, comparative psychology, and animal philosophy[3]—separate scientific areas in natural sciences, social sciences and humanities. The study of animal behavior as a domain of modern sciences begins with ethologists, zoologists, and psychologists such as Niko Tinbergen, Konrad Lorenz, Wolfgang Köhler, and Ivan Pavlov. At the end of the 1970s, Donald Griffin finally founded cognitive ethology. His groundbreaking publication, *The Question of Animal Awareness: Evolutionary Continuity of Mental Experience* (1976) systematically and intentionally took a closer look at the complex natural behavior of various animal species and the taboo subject of consciousness in animals (see Bekoff et al. 2002, xi et seq.).

More and more representatives of philosophy of mind, epistemology, and philosophy of language have also begun expressing interest in the cognitive abilities of animals. Today it is assumed that animal philosophy or philosophy of animal mind occupy their own areas within theoretical philosophy (see Wild 2008; Lurz 2009). Philosophers of the twentieth century were increasingly

3 For the determination of the discipline of animal philosophy, see below as well as Wild (2008).

interested in the question of whether animals had concepts and beliefs. Martin Heidegger, for example, investigated whether they had the necessary foundations for these qualities; later Donald Davidson studied whether they had conceptual networks and meta-beliefs. Davidson (1982) also formulated the very strong thesis that the possession of language is a necessary precondition for the possession of thoughts.[4] Strikingly, many philosophers and scientists often refer to old research results with regard to animals and their abilities, although new, promising studies are available. Their selection of animal specimens is often inadequate or one-sided, or the intelligence of "animals" is spoken of without distinguishing between different species at all. A brief example of this is highlighted in Alasdair MacIntyre's critique of Heidegger and the opinions, for example voiced in Heidegger's lecture *The Fundamental Concepts of Metaphysics* of 1929–30. MacIntyre observes that, while Heidegger cites a great variety of animal species in his discussion—for example, domesticated dogs, bees, moths, and woodpeckers—he ignores other species such as wolves, elephants, gorillas, chimpanzees and dolphins. Thus, many species which are the focus of today's discussion of complex animal cognition are missing (see MacIntyre 1999, 44).

Heidegger's assessment of the abilities of animals, according to MacIntyre, nevertheless arrives at the idea of a qualitative difference between animals and humans and their respective experiences. According to Heidegger, animals "don't have a world", in the sense that humans do. They are "without a world" or "poor in world" (German: "weltlos" and "weltarm"), as they do not perceive the entities surrounding them *as* the specific entities they are. They cannot linguistically distance themselves from their surroundings, and their actions are driven by instincts and impulses (cf. MacIntyre 1999, 43ff.). Heidegger's claims are certainly problematic, and his concept of the qualitative difference of animal thinking is questionable. One may go so far as to counter this with regard to all animals. At the very least, the restriction of animal behavior to the purely instinctual is questionable for species that can react flexibly and complexly to problems. As early as 1917 (twelve years before Heidegger's metaphysics lecture), Wolfgang Köhler's famous intelligence tests on anthropoids had reached wide and numerous circles. MacIntyre rightly criticizes the fact that Heidegger does not treat species such as wolves, elephants, apes or dolphins, but nevertheless extends his conclusions to all nonhuman animals (MacIntyre 2001, 44). Philosophers who now rely on Heidegger's ideas must be

4 In linguistics, the thesis of the linguistic dependence of thought (or linguistic relativity) has been, and still is, most prominently represented by Benjamin Lee Whorf.

aware of the narrow range of animals he uses as test cases and the resulting one-sidedness of his claims.

More recently, since internationally renowned philosophers such as Donald Davidson and Daniel Dennett have addressed the socio-cognitive abilities of animals, a lively discussion on animal minds has begun taking place among contemporary philosophers. This conversation is for example concerned with the areas of language and beliefs, metacognition and mind reading (theory of mind),[5] consciousness in general, and many individual questions (such as a numerical understanding, logic, and memory in animals). These debates are no longer only conducted in the Anglo-Saxon and French linguistic regions, but central texts are also translated into other languages (see, for example, Perler and Wild 2005).

However, a major research gap is apparent when one examines existing work on the *ethical* implications of the cognitive abilities of animals. According to philosopher Hans-Werner Ingensiep, even modern behavioral studies are still essentially influenced by classical theoretical questions of philosophy—for instance, by the question of the difference between the essence of the human being and the existence of a language in great apes. However, with the answers offered so far, we are faced not only with the situation of having to meet the welfare needs of apes as the individuals we know they are (and this applies equally to other species). Humanity is also called upon to consider their own exclusivity when it comes to moral consideration and moral treatment (Ingensiep 1999, 442). As early as 1923, Albert Schweitzer postulated that philosophy had, to that point, found it incredibly difficult to wrestle with the question of when animals enter their thinking process. In fact, philosophy almost entirely avoided serious consideration of the issue (Schweitzer [1923] 1971, 362–63).

Animals have long since made their mark in ethics, and an entire separate area of ethics that deals with them has been established, namely animal ethics. Whether the mark they have made will become more profound in the face of our increasing knowledge of the socio-cognitive abilities of animals, or whether they will turn ethics upside down and lead to new theoretical directions, remains open to debate. But the huge upsurge of human-animal studies we see today clearly hints to the latter.

5 Having a so-called "theory of mind" means having an understanding of the inner live, desires, and beliefs of oneself and others. In many cases, the terms metacognition or mind reading are also widely used (for more detail, see Chap. 6).

CHAPTER 2

Questions and Objectives of the Book

The first half of this book summarizes and examines relevant research findings on three central areas of cognitive research—namely, on culture, language, and theory of mind in animals. A significant number of studies from different scientific disciplines are discussed (see fig. 1). In the second half of the book, these studies provide the background for moral-theoretical discussions in animal ethics.

The three areas of culture, language, and theory of mind were selected for the first half of the book because, on the one hand, they are classificatory criteria of differentiation between humans and animals—that is, they have long been (and are often still) attributed solely to humans—and because, on the other hand, they are areas of socio-cognitive abilities that are connected with one another. Culture is not just a fact or a product, for example, but also a behavior that is acquired. This has to do with social learning. Imitation is an essential component of the cultural spread of knowledge and skills. Certain forms of imitation, on the other hand, require the ability to become involved

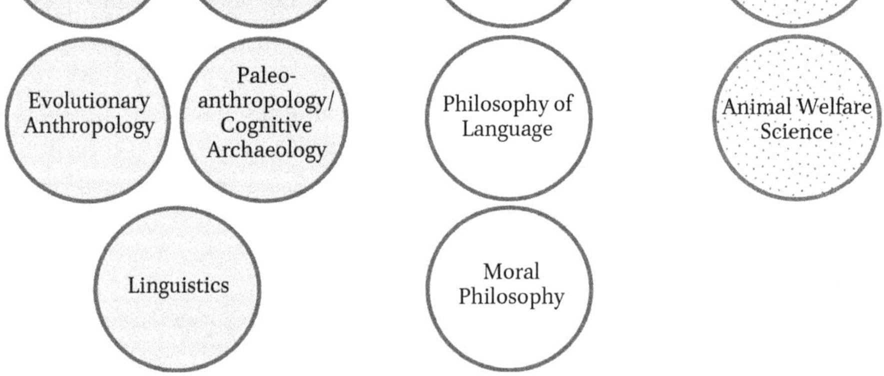

FIGURE 1 The main disciplines involved in the interdisciplinary perspectives on animals' socio-cognitive abilities provided by this book

 | DOI:10.1163/9789004415072_003

with others and their intentions, which is linked to a theory of mind. Language also has something to do with culture, since it can be regarded, among other things, as a symbolic achievement. A work that only addressed one of the three areas could not do justice to these and similar lines of connection, and thus would not suffice for any one of the three. The selection of these three areas is also due to the fact that they are complex enough to have long-term studies and extensive literature behind them, and that they occupy a central place in the debate on animal minds and animal cognition.

A large number of empirical studies on the cognitive abilities of animals are available. However, their interpretation seems difficult. Our implicit and explicit assessment criteria and the definitions used to evaluate the abilities of animals must be critically questioned with respect to their applicability to nonhuman species. It is conceivable that the empirical findings are often interpreted more appropriately when new concepts are proposed—in the case of language, for example, a modified theory of concept possession that can integrate the kinds of concepts which animals possess and make use of. The guiding question in the first half of the book is, therefore, whether animals possess culture, language, and a theory of mind and if so, in what sense this might be the case. The aim of this question is to find meaningful interpretative schemata and definitions that can include and integrate the cognitive abilities of animals. In this way, I will suggest important aspects of an appropriate theory formation, with regard to the cognitive relationship between humans and animals.

At least three approaches to the question of whether animals have culture, language, and theory of mind are conceivable, and each has its own specific value.

First, it would certainly be useful to provide a *broad overview* of these social and cognitive abilities in animals. Precisely because the reception of animal cognition has been unilateral in the sciences and humanities, especially in philosophy, and because recent studies have often been neglected, this book aims at a wide selection of animal specimens within the examples discussed.

Second, it would be useful to analyze individual cases or examples *in detail* and to present their interpretation within the debates for which they are relevant.

Third, it would be desirable to *compare* the cognitive abilities of nonhuman animals with those of humans, especially those of human children. Problems and inadequacies in the investigation of children's skills might throw light on inadequacies and their possible elimination within the study of animal cognition. In both children and animals, we face the particular challenge of finding suitable test designs that are, for example, non–language based.

None of the three approaches or viewing angles—the one comprising a broad overview on many case studies, the one providing a detailed analysis of individual examples, and the comparative approach—can be simultaneously applied. This would go beyond the scope of this book. I will, therefore, focus on one of the following pathways in each of the areas I investigate: the question of culture in animals will be addressed using a *broad overview* as much as possible, providing many *ethological examples.* This will, however, be complemented by an examination of the use of tools by chimpanzees, dolphins, and New Caledonian crows. The next two areas, language and theory of mind in animals, will be less comprehensively examined. I approach them more pointedly. On the question of language in animals, the ability to use *concepts* will become the center of my discussion. This allows me to introduce a more detailed *philosophical* interpretation which clarifies to what extent some animals also have perceptual concepts. For the discussion of further criteria for language comprehension and language skills, however, *linguistic* theories will also be addressed. On the question of the theory of mind in animals, there is lastly a special exploration into the investigation of the theory of mind *in children.* Theory of mind in small children can be observed as something that emerges as a process in ontogenesis and, on the other hand, can be recognized as a competence that is difficult to capture with previous test designs. This results in methodological implications which correlate in interesting ways with implications for our notion of theory of mind, something that is equally relevant when it comes to children as well as animals.

I have also attempted to complement insights from controlled experiments in captivity with field observations, in order to avoid the limitations of both methods. Many scholars attempt to answer the question of socio-cognitive abilities in animals with recourse to only one of the two methodological approaches, while ignoring the differences between their respective advantages and disadvantages. Rendell and Whitehead (2001) show the difference between the two methods using the example of the study of culture in animals. Controlled (laboratory) experiments tend to focus on the process of cultural dissemination—in other words, on the question of whether imitation or teaching takes place, which follows (and produces) the often assumed restriction of the phenomenon of culture to these two abilities. On the other hand, field observation focuses more on the product—that is, the cultural behaviors themselves—and is often used by those who follow a broader definition of culture (see Rendell and Whitehead 2001, 311). According to Rendell and Whitehead, experimental studies on the cognitive processes that underlie cultural dissemination are often methodologically preferred. They offer the advantage of controllable conditions, which reduces the likelihood of possible ambiguity in the interpretation of the data. On the other hand, there is the problem that what is measured may not be related to what the animals demonstrate in the wild. In contrast, field studies as a systematic collection of data on the behavior of individuals and groups often run over great time and space. The advantage of this methodological approach is that it is very closely tied to what animals are actually doing in their natural environments. Field studies, however, are often cumbersome and have the shortcoming that the results achieved can be ambiguous to a greater degree (see Rendell and Whitehead 2001, 311). The two methods have been applied cooperatively in the study of cognition in humans and animals. For a comprehensive, overall view on whether animals have culture, language, and theory of mind, meaningful data from both methods must be used. In the future, carrying out more field studies would be important not only because of the described background, but also in terms of a moral dilemma with which anyone who studies animal cognition in captivity is confronted: one would like to give credit to the abilities and capabilities of animals scientifically while at the same time accepting conditions that do not meet all their physical or psychological needs. In view of this dilemma, the ethical discussions in the second half of this book take on added importance.

After providing insight into the discussions of culture, language, and theory of mind in animals in the first half of the book, the second half examines whether aspects of socio-cognitive affinity revealed by current research shape the human-animal relationship and are therefore ethically relevant. The central question is whether animal ethics can gain more insight when social and cognitive abilities are included in ethical theory and practice. A possible continuity between humans and animals with regard to such abilities could be important, inter alia, for the question of the protection of individual animals or species. In this respect, up to now demands have been made on philosophical grounds for personhood rights for great apes and for dolphins. These approaches argue that such rights are analogous to human rights. However, socio-cognitive abilities in animals play a role not only in moral theory but in practical interaction with animals as well. This is why we are asked how we actually deal with "cognitive relatives" such as apes and dolphins, both in captivity and in the wild. In the course of this investigation, I will demonstrate the discrepancy between the well-founded demands for strong protective rights, and even rights similar to human rights, on the one hand, and the status quo of animal welfare and species conservation, on the other.

Finally, the book is devoted to discussing the advantages and disadvantages of human rights–analogous animal rights, whereby the graduality of cognitive abilities is of particular importance. Alternatively, the possible integration of social and cognitive abilities of animals within welfare ethics (as opposed to animal rights) will be investigated. Social and cognitive abilities in animals could be a decisive factor in determining the well-being and the quality of life of animals, and ignoring or underestimating these abilities could lead to their suffering. Unlike other studies, I try not to separate animal rights and animal welfare issues too much, since I see the possibility and the need to provide space for the socio-cognitive abilities of animals within both approaches.

All in all, my book will present problems and discussions at their intersection. It is, therefore, an attempt to integrate a large number of discrete subjects and questions in order to present the wide range of scientific, philosophical, and ethical questions that arise from the existence of complex socio-cognitive abilities in animals. This is the value of my book as I myself see it. The following assumptions guided my approach. On the one hand, my book is not a typical introduction to animal cognition and animal ethics written for lay people. I believe that there are enough brilliant books on the market that bring the topic to readers who are not so much interested in scientific debates but love animals. To read my book you don't necessarily have to come with a scientific or philosophical background. And loving animals will surely help you work your way through the book. But you should be keen to learn about scientific and philosophical debates and read from time to time some annoyingly detailed information. So please don't expect an average introduction. On the other hand, my book is also not a typical, in-depth philosophical or scientific investigation. If you are for example looking for an introduction to philosophy of animal minds you will surely find more suitable books for this purpose as well.

What you will find instead in this book is sort of an *interdisciplinary map* to socio-cognitive abilities in animals and their ethical relevance from the perspective of a researcher in the field. At worst you will miss many things, but at best you will be surprised and ultimately impressed by the multitude of research questions, studies, and debates that arise when you open your mind to the idea of animals as cognitive kin.

PART 2

Socio-Cognitive Abilities in Animals

CHAPTER 3

The Concept of Cognition and the Concept of Consciousness

Before the focus of this book turns to empirical studies and their interpretation, some fundamental definitions—emerging from the concepts of cognition and consciousness—need to be anticipated. Culture, language, and theory of mind are described in this book as areas of competence that require conscious action with the help of cognitive and social skills. The concepts of cognition and consciousness are, by contrast, more difficult to define because they are used very differently, especially in the humanities in relation to the natural sciences. In what follows, an understanding of both concepts in their interrelationship will be pursued, as ideas from cognitive ethology and behavioral research are brought together with central concepts from analytical philosophy of mind.

The *Oxford Dictionary of Animal Behavior* assumes that the concept of cognition refers to "mental processes that are presumed to occur within the animal, but which cannot be observed directly" (McFarland 2006, 32). This, however, is a very general definition, in which the concept of "cognitive" seems to be almost equivalent to "mental" or "psychical." While it is important to separate mental states and processes from externally visible actions, "cognitive" can mean more than just "mental" in this sense. It is possible to define cognitive states and processes as *a certain form of mental states and processes*. Differentiations may be necessary here, since, among other things, *emotional* states and processes are also mental (or internal, or psychological) phenomena, which, however, can be distinguished from cognitive phenomena. McFarland emphasizes two aspects for further differentiation following his first characterization of cognition as "mental." Cognition means the *manipulation of explicit knowledge* or in general *forms of mental abstraction*: "In the study of navigation, problem solving, social interactions, deceit, language, and thinking in animals, scientists have found it necessary to postulate cognitive processes" (32). It must be stressed that this is still a broad concept of cognition. When I use the term "cognitive" in the present book, then, unless otherwise specified I will follow the meaning suggested by McFarland. I do so because it does not simply equate "cognitive" with "inner" or "mental," but neither does it reduce cognition to pure processing abilities. My use of the concept aims to express the fact that cognitive abilities in fact deal with explicit knowledge and

 | DOI:10.1163/9789004415072_004

forms of mental abstraction. Explicit knowledge, on the other hand, is deliberately manipulated, and is also present in the form of *social* knowledge and *social* interactions—for example, when a being deliberately deceives others. Frequently, therefore, I will speak of "social and cognitive abilities" or synonymously of "socio-cognitive abilities." This is to counteract a more one-sided understanding of cognition than that of McFarland. In particular, the three areas of competence explored in this book—culture, language, and theory of mind—are characterized by cognitive processes in the sense of logical and combinatorial abilities as well as by social processes. The latter can also be addressed as forms of explicit knowledge and mental abstraction. In addition, the interactions that are possible thanks to the existence of culture, language, and theory of mind also require social competencies (e.g., social learning, imitation, shared attention, or empathy) during their acquisition. These competencies are also cognitive skills in the sense of McFarland's definition, since social knowledge can be explicit, mentally abstracted, and manipulated knowledge.

It should be clear that cognitive abilities are closely tied to the concept of consciousness, which is also difficult to define. Cognitive states and processes generate explicit knowledge, thus presupposing conscious and reasoned decisions that follow information processing through reflection, weighing and selection, and integrating new knowledge sensibly into future behavior, including social behavior. According to Donald Griffin's working definition of consciousness, an animal possesses this when it subjectively thinks about objects and events. Thinking in this sense means that there are internal, mental representations and that these are processed. These could represent situations for the animal with which she is confronted in the present, as well as memories or anticipations of future situations. Conscious thinking in this sense often includes the comparison between two or more such mental representations, as well as the choice and decision in favor of a particular behavior from which the animal assumes that she can achieve a desired goal or avoid undesirable goals. More complex forms of cognition would in principle be connected to this simple working definition (see Griffin 1991, 5, for a discussion of different definitions of consciousness see Griffin 2001a, 1–19).

From McFarland's and Griffin's definitions, it is obvious that when it comes to questions of cognition and consciousness in animals, we are in fact talking about explicit knowledge in a genuine, meaningful and not only rhetorical, pragmatic sense. What do animals know about the world around them? And on the basis of what knowledge and information-processing abilities do they act in it? In what follows, these considerations will be placed within a larger theory of cognition. For this purpose, a text by Peter Bieri is used that, from a philosophical point of view, explains some elements of the above-mentioned definitions (for example, the concept of knowledge or that of representation). I could have used many other introductory texts, some surely more up-to-date (like the ones provided in Andrews and Beck (eds.) 2018), but Bieri suffices to introduce some crucial terminology. His approach is, furthermore, classical enough to show us where orthodox theory reaches its limits when facing animal cognition. But it is also open enough to show us where animals could fit in the picture.

In his introduction, Bieri summarizes the most important elements on which epistemology builds up as an analytical philosophy of knowledge. He makes it clear that the question "What is cognition?" is often being related to the question "What is knowledge?" (Bieri 1997, 10). However, what knowledge means is difficult to assess as "knowing" comes in so many different forms. We can, for example *know or recognize something or someone* thus using the verb to know in an *object construction*. In phrases such as "It was not obvious that you wanted that," however, *propositional construction* is used. In this example, the propositional construction is formed by the "that" clause, which refers to the verb and indicates *what* could not be recognized (11–12). "Knowledge" referring to propositional content is often at the center of our attention. But, again, it comes in many different forms (e.g. to know *that* this-and-that is the case, to know *why* X has happened, to know *where* X is located, etc. (12)). In addition, one can speak of "knowing how" and either mean practical knowledge (*know how to do something*) or phenomenal knowledge (*know how it feels like to be in a specific state of experience*, for example, in love or in jealousy).

The identification of such different meanings of the verb "to know" lead Bieri to a distinction between various types of information—and, above all, between differently complex types of information *processing*. He thereby aims to distinguish what is present in some nonhuman animals and what is possibly reserved for humanity alone. At the center of cognition and knowledge, Bieri first presents the ability to discriminate: to know something means to be able to distinguish it from something else (15). However, acts of discrimination vary in their complexity: different persons might apply different kinds of criteria to base their discriminations upon, they might use different cognitive skills to perform their discriminations and they might focus on different aspects of content.

The most elementary form of discrimination is that of *sensory discrimination*. Each individual possessing organs of perception is initially provided with sensory and physiological information, which the individual's brain processes. One's ability to identify a hut on a mountain in front of a certain background means that the hut can be visually differentiated from its surroundings. What is sensitively differentiated here is not necessarily classified in any way. At this level, no belief about this hut is formed at first (whether it is big or small, belongs to the *Wanderverein* or is private property); what is expressed is simply the ability to recognize something or someone. Most animals are capable of this kind of sensory discrimination, and some of them can sense things we ourselves cannot perceive: a cat can recognize things in the night better than a human can; a dog can recognize higher-frequency sounds than a human; and sniffer dogs can recognize drugs or explosives (Bieri 1997, 15). On the level of sensory discrimination, however, the latter only means that the dogs can discriminate different odor qualities. Sensory discrimination is thus even attributable to the Geiger counter, which can distinguish radioactivity, and to the missile, which recognizes its goal with the help of heat detectors (15). But in many instances of information processing, more than mere sensory discrimination is involved. Above all, we can set apart the ability to recognize something *as something in specific*. Bieri returns to this idea later in his work.

But first, he turns to the domain of "knowing how," and explains some elementary forms of it. He starts with *knowing how (or what) it feels like* to be in a certain condition or to experience this condition. "Knowing what it feels like" in this sense is referred to as phenomenal consciousness, a form of consciousness that sentient animals are nowadays usually believed to possess. For animals too, pleasure, pain, and anxiety feel in a specific way and they feel different from one another. What is important is that animals can possess phenomenal consciousness, without being able to classifying their feelings and believing anything about them by help of second-order thoughts, that is thoughts about thoughts (Bieri 1997, 16). This assumption regarding how some animals experience emotions is central to all pathocentric approaches to animal ethics (and it is regarded as one of the classical arguments in animal ethics, discussed in Section 8.3.1. of this book). A second form of "knowing how" in animals, which is also linked to discriminatory capacities but not necessarily to intellectualized capacities like classification, is *knowing how* in the sense of *knowing how to do* something. To start with, knowing how to move for example, can be understood as the mere sensory-motor ability to select among different possible movements in a particular moment and distinguish them from those that are not part of the intended action (Bieri 1997, 16). If plants, animals, and robots "know how to move" in this sense, they may even possess this ability without a phenomenal inner aspect. Caution is necessary, however, in the case of more complex animal behavior. "Knowing how" in the sense of the process underlying a complex problem-solving behavior in flexible contexts requires the weighing, rejection, and, if necessary, the reinvention of strategies with a view to an ultimate goal. Chapter 4, on culture in animals, presents the example of the beaver's den-building as such a case.[1]

In a next step Bieri points us to various forms of information and, more importantly, *information processing*. Again, he starts with basic forms. Knowing something, or recognizing something, he continues, is to be in possession of information. Information, on the other hand, is based on the reduction of possibilities, and whoever receives information is experiencing a reduction of uncertainty. At the end of the process of acquiring knowledge, an individual possesses information and is in a condition that has informational content (Bieri 1997, 17). This informational content is then not only *stored* but also *used* in a certain way: beings or systems such as humans, animals, plants, and some might argue even robots can use information stored in them, in contrast to inanimate objects (such as stones), enabling them in some cases even to adapt to the respective situation in light of their needs, wishes, or goals. Information processing plays a crucial role on the path from sensory input to behavioral output (Bieri 1997, 17–18). At this point, inanimate objects are separated. However, different living beings can attain different levels of complexity regarding their ability to adapt. A higher degree of complexity usually goes hand in hand with a higher degree of behavioral flexibility.

In order to get a better understanding of how information processing works, Bieri introduces the *concept of representation* and defines it as follows: "Representations are bearers of information: they are those elements in a system by which the system has states of informational content" (Bieri 1997, 18, transl. MK). They form a whole system of representations or a structure of internal elements, which allow information to be recorded, stored, processed, and put to use. According to Bieri, these internal elements can in theory be described in physical or biological terms. If, however, one were to take into account the correlation to the events outside the system and their functional role in the production of behavior, then they could be regarded as representations (19).

What different types of information are now conceivable and what differently complex cognitive processing services are possible? To answer this question, Bieri summarizes what he has already said about sensory discrimination,

1 "Knowing how," in the case of complex problem solving, can be divided into individual components, which in turn can include propositional knowledge. In order to know how to crack a nut using a hammer and anvil, you have to know that some (but not all) nuts have extremely hard shells, which stone and which anvil is hard enough, which object is suitable as an anvil, at which angle and from which height the nut must be hit in order to crush the shell but not the soft core, etc. In Chapter 4, many complex examples of such behavior in the animal kingdom are discussed in detail.

and then moves on to other forms of discrimination, which ultimately correspond to acts of classification.

First, systems (from the robot to the human being) process a whole series of *physiological information*, of which they are *unaware*: a system does not always feel that it has to have and use information, and the fact that a system possesses and uses information does not necessarily mean that the system understands something. This applies to information such as respiratory movements, heart activity, and the like (Bieri 1997, 19). The use of such unconscious physical information can be distinguished from the use of *phenomenal information*, including information about colors and odors. Bieri speaks of the "quality of experience" for the receiving subject and assumes that phenomenal information is a kind of information that affects the individual and is consciously experienced (19). Bieri assigns a phenomenal consciousness in this sense to all sentient animals. However, in order to *understand* the phenomenal information, a third element would have to be added: an individual must be capable of processing *semantic information*. "It must have a system of *semantic representations*—of representations which, besides generally being bearers of information, are also bearers of *meaning*" (19–20, transl. MK). Semantic information is *generalized* information and tells an individual that something is of a particular kind or a certain type. The concepts of our natural language are exemplary for semantic representations. For instance, "red" is a kind of color for me, "cat" is an animal, and "hut" is a kind of dwelling. Our concepts are available to us in classifications and subcategories. "Red" is a visible property, while COLOR is a semantically determined, abstract category which refers to red, blue, green, etc. A classification into an upper category is thus a performance of abstraction (that is, abstraction from the currently visible object to an upper category under which the object falls).

The natural language of people is the clearest case of a semantic representation system, according to Bieri (1997, 20). It subsumes logical subjects under a predicate, sorts things under a concept, and ultimately formulates whole sentences. Moreover, semantic information is always associated with the *distinction between truth and falsehood*, since sentences usually have a truth value. In order to have a semantic consciousness in addition to a phenomenal consciousness, one must be able to distinguish between the truth or the falsehood of messages. Furthermore, semantic information is interlinked by principles, as one can move from one semantic piece of information to another by using logical rules or rules of reasoning. There is a "relationship of reasoning." *Reasons and evidence* combine semantic information with one another, and someone who speaks in a meaningful way operates in a vast *semantic network* (Bieri 1997, 20). I will further discuss some of the above-mentioned conditions for the possession of semantic information in the theory of concepts by Newen and Bartels (2007) in Chapter 5, and examples of concepts and conceptual skills in animals are presented there. It will be shown that it is indeed useful to distinguish certain forms of discrimination and assume that real concepts are somehow characterized by their position in a conceptual network. However, some criteria (like knowing about the truth value of propositions) can be conceptualized in less demanding ways than many philosophers understand them. This offers possibilities to come up with a theory of concept possession which isn't so intellectually demanding that it excludes nonhuman animals right from the beginning.

According to Bieri, forms of recognition and knowledge can each be analyzed and evaluated differently. Sensory discrimination occurs even before the performance of any classification. There are cases of sensory discrimination in which there is no phenomenal information involved and others in which it is involved. In humans, the latter is usually the case. Experiences such as our example of the perception of a hut or the perception of colors and tones usually entail sensory discrimination combined with phenomenal information and thus a conscious perception. According to Bieri, these processes occur even in animals, at least in animals of a certain degree of development or complexity: "When my cat knows/recognizes me or my car, alongside physiological information, it is also working with patterns of phenomenal information; when the sniffer dog detects marijuana, then it is due to phenomenal odor information; if the bat gains orientation due to echolocation, it operates with fields of acoustic information that have a certain quality of experience for her" (1997, 21, transl. MK). Note that at this point—well before bringing in semantic information processing—humans and sentient animals are separated from plants, less complex animals, and robots. The pattern recognition of the robot, the distinction between brightness and darkness of the plant, as well as the target detection of the missile and the Geiger counter, are not accompanied by phenomenal consciousness or qualitative experience. Now Bieri can define more clearly what it means to know how something feels: it means to be in a condition that has phenomenal information content. It represents a conscious (though not necessarily semantic-classifying) discrimination between different sensations or pieces of phenomenal information. Since sensory discrimination can take place with or without phenomenal content, sensory-motor skills are also conceivable in nonphenomenal and phenomenal variants. Plants, simple animals, and robots know how to move without a phenomenal inner aspect (and the same applies to unconscious parts of humans' adaptive movement repertoire). Much of what we and other animals can do on a sensory-motor level, however, is actually accompanied by phenomenal consciousness. For example, when we learn to tap-dance, we are "guided by patterns of phenomenal information about the position of our limbs" (Bieri 1997, 21, transl. MK). The form of sensory discrimination that is accompanied by phenomenal consciousness is already a form of explicit cognition or explicit knowledge (in Bieri's account) in the sense of conscious and purposeful behavior.

For Bieri, a stage higher than the conscious phenomenal information and its processing is propositional realization, which is dependent on the possession and use of semantic information. Real propositional knowledge in the sense of "knowing that X" and "recognizing that X" consists of being able to deal with generalized information in the form of predicates, to distinguish between the truth and the falsehood of sentences or propositions, and to classify certain propositions as reasons or evidence for other propositions. Here, too, there are forms of discrimination that go beyond sensory discrimination (which operates with physiological and possibly with phenomenal information). Meanings, truth values, and logical relations are distinguished here (Bieri 1997, 23–24). Once again, Bieri ascribes (limited) propositional knowledge to animals as well. Some animals would know *where* they are at home, *whom* they have to protect, *when* it is best to hunt. The sniffer dog knows, too, *which* luggage has the drugs. According to Bieri, we are also ready to say that the cat knows *if* it sees another cat or a dog. But the limits of the attribution of propositional knowledge to animals would normally be drawn "where we must say that animals also know *why* something happens." This capacity he only admits for certain kinds of primates (24, transl. MK). Whether this is warranted or whether there are other limits than "knowing why" with regard to propositional knowledge in animals is an open question. An antelope that sees another antelope running might in fact deduce that there is a predator approaching and thus "understand why" the other antelope tries to escape.

All in all, when talking about semantic and propositional information content, "animals are put into a real no-man's-land of uncertainty" (24, transl. MK). Our uncertainty is mainly due to the fact that we find semantic representations most clearly expressed in our linguistic terms—and animals simply do not speak like us. It is crucial, however, that Bieri considers that "semantic information can be represented in a particular being differently than by mastering a language." As long as a mental representation represents some sort of generalized information to the animal, it could be interpreted as a kind of thinking that operates against the backdrop of "*internal templates*" (24, transl. MK). Bieri explains:

> When animals recognize *that* such and such is the case, and thus *classify* something, they...measure their phenomenal information on a repertoire

> of templates, and these templates are their *concepts*. When they notice that the one template 'fits' and the other does not, it is equivalent to the fact that the one classification is 'true' and the other 'false.' If one also assumes that these templates—'concepts'—are connected by rules, one might obtain a model for a nonlinguistic way of concluding and using evidence. If one succeeded in this, one could assume that real cases of propositional knowledge are far beyond the range of speech-capable beings. (24–25, transl. MK)

Research still has problems with conceptualizing nonlinguistic concepts in animals. But evidence for simple concepts used in the principle of Bieri's ideas are actually derived from language studies, in which an animal could be taught a form of human language. Bieri's assumptions that animals can understand something about their classifications being true or false and that their concepts are interlinked by rules (and thus anchored in a conceptual network) are stressed in the analysis of these research results (see Chapter 5).

Finally, Bieri argues that at the head of the hierarchy of information types, that is, still above the level of phenomenal and semantic information itself, is the notion of opinion or belief. What is true is that some beliefs, namely those beliefs that are beliefs about beliefs (second-order beliefs), are in fact quite demanding. The same is true for an understanding of false beliefs. Whether someone has a true or false belief depends on the semantic informational condition he is in. If we gain new knowledge, we can correct an opinion, and in order to know what another means, an observer must be able to put himself into the other's state of perception, depending on his particular state of information. In principle, he must observe which information was or could be accessible to the other (possibly as opposed to himself). Since a statement does not appear in isolation, but is set in a wide-minded network, in which every single belief is a connecting point to many other beliefs, humans also ascribe whole *belief profiles* to other people, as well as associated *desire profiles*. It is an open question to what degree animals can do the same. Humans, however, make use of this capacity in order to explain the behavior of others and anticipate their behavior (Bieri 1997, 26–27). In this way, the possession of some beliefs (second-order beliefs or false-belief understanding) involves complex competencies that point to the domain of theory of mind. Bieri rightly points out that beliefs can differ in their complexity. If an individual corrects his opinion, this may include the realization that the previously held belief was wrong. This means that *a belief of one's (own) belief* is formed. In this case, we speak of *second-order mental states* (Bieri 1997, 28). Likewise, it is possible in a two-level construction to develop *a belief about the belief of another person*, i.e., to think that another individual thinks or does not think that food is hidden in a certain place. It is difficult to determine whether an animal knows or thinks *that another animal thinks or means* something specific. Therefore, in recent years, research has concentrated on first investigating whether animals *know what other animals have seen* (*or haven't seen*), whether they possess knowledge about the perceptual states of others. Such perceptual states also fall under the category of mental states and thus under the rubric of a theory of mind. In addition, there is much evidence that knowledge about the perceptions of others is also used in some animals to predict their behavior (see Chapter 6).

Bieri's account of the central concepts of knowledge theory showed us the limits of orthodox approaches to animal minds. But at the same time it revealed some possibilities how to include animals in our philosophical thinking after all. In the chapters to come we will go on discussing what animals indeed know and can do. Some ideas mentioned so far will be systematized and further specified. In Chapter 4 I will attempt to answer questions of conscious action in the sense of "knowing how' or ability in animals (especially in the context of creating and using tools). Adaptability and flexibility as well as more complex forms of social learning are also dealt with there. In Chapter 5, language in animals is then linked to reflections on semantic and propositional knowledge. In Chapter 6 (on theory of mind), considerations regarding the attribution of a belief-desire-psychology to others are discussed. All in all, these chapters should serve to provide a basic introduction to and understanding of empirical studies on the social and cognitive abilities of animals. This prepares us to turn to the ethical implications of such abilities in animals in the second half of the book.

CHAPTER 4

Culture in Animals?

4.1 Animal Tool Use and Tool Production: a Widespread Phenomenon

Culture is one of the most decisive characteristics of humans and tool culture is one of the oldest forms of culture we know about when it comes to human beings. The early use of objects has left many traces in our history, whether in the form of stoneware, wood artifacts, ceramic shards, or metal tools. The artifacts of early humans are regarded as direct testimony to the development of an important area of human thought, namely, thought directed at addressing problems with external objects (Haidle 2012, 12–3). The "symbiosis with objects" is, for modern humans, a "species-specific cognitive feature" (13), one that he has further developed in an extraordinary manner. Objects that serve as tools determine and facilitate our lives, from the lighter to the microwave, from the toothbrush to the PC.

In contrast, at first sight tool behavior in other animals seems to occur only rarely, and in a much less complex form. In order to trace it we need to be aware of what distinguishes tool use and tool manufacture as cultural behavior, what it demands of a subject and what possibilities it opens up. In the following two sections, the phenomenon of tool behavior in animals, which turns out to be widespread, will be investigated on the basis of many individual examples, and initial differentiations will be worked out before a definition of culture applicable to animals is explored in Section 4.2.4 and the sections that follow.

In modern research, tool *use* is defined as "the use of an external object as a functional extension of mouth or beak, hand or claw, in the attainment of an immediate goal" (van Lawick-Goodall 1970, 195). In more precise terms, tool use is an external employment of a freely movable object from the environment to more efficiently change the shape, position, or state of another object, of another organism, or of the user herself. The user holds or carries the tool during or immediately before use and is responsible for the correct and effective orientation of the tool (Beck 1980, 4ff.). Parker and Gibson (1977) provided the basis for this definition in their differentiation of true tool use from "proto–tool use": for the former definition to obtain, the tool or the medium must be freely movable and actively used, while in the latter case the medium is firmly connected to the environment, and only the object to be manipulated is moved

 | DOI:10.1163/9789004415072_005

(as is the case, e.g., when hard-shelled foods are hit on the ground or on a stone).

Anecdotal examples of tool use in animals have been known for centuries.[1] However, they only came to the attention of modern science as the result of the behavioral studies of chimpanzees by Wolfgang Köhler in 1917 and Jane Goodall's first systematic outdoor observations of these animals in the 1960s. Köhler ([1921] 1963) noted that chimpanzees take mental detours and use equipment in pursuit of a goal. He had confronted the animals with different experimental tasks (like getting to an object which is out of immediate reach). Although he recognized the transition between object play and tool use, he defined only the tasks he had set in his experiments as real problems, like the use of sticks as levers, spoons, probes, weapons, or projectiles. If the animals solved a problem that they themselves perceived as such, he considered their behavior to be play behavior. This includes instances in which they draped objects on the body as a kind of ornament, or painted the environment with clay moistened in the mouth (Haidle 2012, 26). Köhler understood tool *production* as the introduction of more significant detours en route to the achievement of a goal: not only a tool is introduced here as a material intermediate between problem and solution, but the actual goal is in the short term removed from the center of attention to the benefit of another problem, the manufacture of tools (Haidle 2012, 26). In his experiments, for example, Köhler was able to observe how chimpanzees assembled individual tubes to gain a longer pole to reach forage, or how they built crate towers in order to reach a hanging piece of food by standing upright on them (Köhler [1921] 1963, 74ff.). Although he had already documented the production of tools by chimpanzees in 1917, Haidle (2012, 27) reports that, up to the 1960s and 1970s, it was generally continued to assume that tool use, largely, and tool production, exclusively, are the domain of human beings. Only in humans they were considered as expressions of conceptual thinking.

Discussion of culture in animals, therefore, began with the phenomena of tool use and tool production, and to a large extent still revolves around these behaviors. Both have been observed in a variety of animals, including invertebrates. Donald Griffin dedicated a separate chapter with many examples to the subject in his 1984 book *Animal Thinking* (Griffin 1985). For him, when

1 For example, as early as the sixteenth century it was known that capuchin monkeys and baboons use hammer stones; in the seventeenth century, observations of howler monkeys (which threw branches at invaders) were already reported; and by the nineteenth century it was known that macaques can open shells and chimpanzees can crack nuts (see the references in Haidle 2012, 25–6).

considering such cases, it is crucial to pay attention to two questions. Firstly, we need to assess whether the animals *deliberately think about what they are doing*. Secondly, we need to understand whether the manufacture and maintenance of their tools indicate that they in fact *anticipate certain results in the near or distant future* (Griffin 1985, 128). Decisive differences in the tool behavior of different animals can therefore, according to Griffin's considerations, be rooted in the intentionality and goal-directedness of the behavior. That tool behavior is purposeful and deliberate seems to be a prerequisite for tool use, while other criteria can determine the complexity of tool behavior. For example, goals can be pursued at different spatial and temporal distances, or different degrees of mental detours can be applied in the employment or production of tools. The examples that follow illustrate the number and diversity of seemingly cultural behaviors in nonhuman animals. I begin with tool behavior in invertebrates before then turning to birds and mammals. I provide some theoretical considerations to distinguish between innate and acquired behavior and introduce two important criteria: that of goal-directed flexibility, and the characteristics of life in the social fabric. Against this background we will be prepared to have a more detailed look at culture in great apes, dolphins and New Caledonian crows.

4.1.1 *Tool Behavior in Invertebrates*

Many invertebrate species behave in a surprisingly complex way when gathering food, catching prey, or protecting themselves against attackers. The larvae of antlions, for example, a species of insects in the Myrmeleontidae family, create funnel-shaped traps in the sand to catch ants. They also throw sand grains at their fleeing prey (see the discussion in Griffin 2001b, 114). With respect to the marine crab *Melia tesselata*, there is an astonishing symbiosis with sea anemones, which have nettle cells that, when touched, expel a projectile with an irritant. The crab carefully removes small anemones from the underlay and holds one in each of her hind limbs. She then uses the nettle-anemone as protection against enemies: in case of threats, she directs them against the troublemaker. These are just two examples of how invertebrates use objects to achieve something specific. But how can one judge this behavior? Do these animals also have a picture or idea of what they want to do and achieve? Is it a deliberate and intentional or rather instinct-guided, inherited behavior? Is their behavior flexible and responsive to the situation and goal?

Studies of the behavior of the predator bug species *Salyavata variegata* in the tropical rain forest of Costa Rica demonstrated that we are not always facing flexible behavior when objects are used as tools to achieve a particular goal. Elizabeth McMahan (1982) studied the prey behavior of these bugs, which initially suggests an amazing level of intelligence. The animals disguise themselves with bits of a termite nest that they stick to their body. Thanks to this chemical and tactile camouflage, they can get into termite nests and thus capture a termite worker without the armed soldiers recognizing them as intruders. A predatory bug then proceeds to eat the attracted worker and then "wags" into in the opening of the nest with the remaining empty exoskeleton. Another worker termite comes to carry it away, is pulled out by the exoskeleton, and then consumed. The bug repeats this process up to thirty-one times (see McMahan 1982 and the discussion in Griffin 2001b, 134). Benjamin B. Beck (1986) also explores this bug species in his book *Tools and Intelligence*. In the introduction, he uses the description of their hunting strategy, which at first sight resembles the termite fishing of chimpanzees with the help of probes. Beck deliberately leads his readers astray in order to inform them:

> I misled you for a reason. Camouflage and baiting performed by a large-brained ape with clear morphological and behavioral affinities to humans would clearly indicate intelligence. In contrast, discovering that the hunter is a brainless insect somehow weakens the inference of intelligence. (135)

The choice would remain: either we agree to accept that insects could behave as intelligently as chimpanzees, or we discard our previous assumption of a direct link between tool use and intelligence. Beck says, "I believe we must do a bit of both" (136). If one looks at the case more closely, then one has to admit that there are limitations to the behavior of bugs. According to Beck, camouflage and attraction do not seem to be learned. First of all, the behavior was only observed in certain stages of the development of predatory bug nymphs and, second, it was observed always in the same way, and is therefore subject to an extremely stereotypical use. According to Beck, "these attributes suggest that the behavior is innate or genetically prewired" (137). Although the behavior is extremely effective in adapting to the circumstances and appears to be complex, Beck says that there is no way that this is deliberate behavior. The insects were preprogrammed like "self-guided missiles." Any hungry predatory bug would, if so disguised and placed in a termite construction, proceed in this way—even without having had the opportunity to learn this strategy (Beck 1986, 137). Beck's analysis seems to be differentiated. Not every complex behavior is evidence of insight and ingenuity in the animal kingdom. When all individuals of one species show such behavior, and especially if they do so only in one stage of their development, it can be innate rather than learned behavior.

Griffin's approach, however, was always to take into account the possibility that even insects may display behaviors beyond those that are genetically preprogrammed. One should indeed, as a heuristic assumption, take them seriously. Insects (and perhaps all nonmammals) may be underestimated. After all, we now know that bees, for example, have dialects in their dance language and can apparently learn such dialects. Although this is not a form of tool behavior, it does concern the possibility of learning ability in insects, which underlies tool behavior (and other possible forms of cultural behavior). The waggle dance of the bees contains an element that indicates the direction of flight to the food, while the duration of the dance is a measure of the distance. However, the dance duration as a distance measure, in particular, differs significantly between different bee species. Su et al. (2008) succeeded for a while in integrating Asian honeybee (*Apis cerana cerana*) and European honeybee (*Apis mellifera ligustica*) populations. It could be observed that the bees began to understand the dance dialects of the other population through a learning process. The authors titled their study *East Learns from West* because especially the Asian worker bees were able to decode the information from their European counterparts about the location and distance of feed stations and follow this information to the food sources:

> Using video recordings made at an observation hive, we first confirm that Acc [the Asian honeybee] and Aml [the European honeybee] have significantly different dance dialects, even when made to forage in identical environments. When reared in the same colony, these two species are able to communicate with each other: Acc foragers could decode the dances of Aml to successfully locate an indicated food source. We believe this is the first report of successful symbolic communication between two honeybee species; our study hints at the possibility of social learning between the two honeybee species, and at the existence of a learning component in the honeybee dance language.
>
> SU et al. 2008, 1

The authors define social learning according to Heyes (1994), Heyes and Galef (1996), Galef and Laland (2005), and Griffin (2004) as a learning process that is influenced by the observation of or interaction with other living beings (typically conspecifics) and their products. They believe that the possible chemical communication and the dance-related interaction between the Aml and Acc worker bees in the mixed colony point to this particular form of learning (Su et al. 2008, 5–6). In fact, learning processes seem to be in play here, although many of their procedures and mechanisms are still unclear. For example, further research will be needed to determine whether bees will, over time, improve their learned ability to decode and otherwise use the dances of the other species to find food. In the case of a learning process, this would be expected.

The previous examples show that complex tool behavior can already be found in invertebrates, but that it has to be differentiated on a case-by-case basis, whether it is deliberately applied, learned behavior, or innate behavior. In general, when differentiating between instinct-directed or innate behavior compared to learned behavior, it must always be borne in mind that even complex behavior patterns are usually composed of both elements. They can have learned as well as innate components—which applies to every animal, including humans. Of course, culture itself does not function directly through genetic inheritance and, therefore, does not evolve through natural selection of inherited traits. Nevertheless, the cultural *ability* of living beings, according to Bonner (1989), is a direct product of evolutionary processes. It makes it possible to pass on types of information that could not be transmitted genetically, or at least not efficiently so. Thus, the concept of natural selection can also include forms of learned behavior, and the transmission of learned behavior provides adaptive advantages: "Therefore, there has arisen a genetically determined behavioral capacity to transmit information by signs, by language, by imitation" (Bonner 1989, 3–4).

4.1.2 *Tool Behavior in Birds*

Birds, too, use tools and for decades have provided impressive examples of individual inventions. Ornithologist Gray J. Roberts in Australia observed a black kite flow to a river with chunks of bread, throw it into the water and try to catch the attracted fish and crayfish. She repeated this behavior for over an hour and displayed it in several later attempts, one of which was ultimately successful (Roberts 1980, 53). Griffin and Roberts rightly point out that this is a twofold extension of the normal behavior in food acquisition: normally, this bird would neither use the fishing technique nor eat crayfish or fish generally (Roberts 1980, 53; see also the discussion in Griffin 1985, 133). The kite might have seen people fish feeding and fishing and then taken the step to drop her own bait. Or she had already fetched food from the nearby campsite, where by chance something fell into the water. She probably recognized the connection when she saw that it attracted aquatic animals. Such explanations are in the realm of speculation—even if they seem obvious—but it is certain that this approach represents a creative change in normal behavior.

Creativity in terms of a problem-solving strategy was also shown by marsh tits (*Parus palustris*) in an experiment on their food-hiding behavior. Clayton and Jolliffe (1996) divided eleven hand-reared birds into two peer groups.

Whole seeds were offered to the first group as food. The birds, as they usually do, carried the seeds away and hid them in order to store them for future needs. The second group was offered the same feed but ground to powder, so that it was impossible to carry it away in the beak and hide it. Some of the birds in this group then began to peel off the colorful tape used to mark the food boxes, fold the pieces of tape around the ground food, and then carried the packages away and hid them. One of the animals dipped the damp side of the tape into the feed powder, then folded the tape together and hid it. A second bird in an adjacent cage adopted this unusual method (Clayton and Jolliffe 1996; see also the discussion in Griffin 2001b, 116).

There are many more examples of tool use in the context of food acquisition in other birds as well. One of the best-known and best-studied cases involves the so-called Darwin's finches of the Galapagos Islands and among them, especially, the woodpecker finches (*Camarhyncus palmitus* or *Cactospiza pallida*). All fourteen species of Darwin's finches belong to the order of passerine birds and have evolved from an ancestor coming from the mainland. Since Darwin's research, they have been regarded as a model of speciation, as they have adapted differently to the diverse conditions of life on the islands, both in physical form and especially in their diet. For Griffin, the behavior of the woodpecker finches, which spear insects with thorns or twigs, is particularly important as an example of tool use. He describes how the Galapagos finches use a cactus thorn or a small branch as an auxiliary tool and shorten it before use by shortening or breaking off protruding parts. At times, the finch retains the branch under its toes until it is reused (Griffin 1985, 132).

It is noteworthy that we are talking about tool *production* with these finches, because they adapt their material before it can be used as a skewer (Griffin 1985, 132). The fact that there are birds that not only trim sticks to suitable lengths, but also process other materials in other ways will become clearer later in the case of New Caledonian crows (see Section 4.5).

In seagulls, too, tool use occurs in food acquisition. Herring gulls in Stage Harbor—near Chatham in Cape Cod, Massachusetts—have acquired a special throwing behavior for cracking mussels and similarly difficult-to-open food (Beck 1980). Beck (1986) describes the approach of the gulls under the heading "Cognitively Complex Tool Use in Gulls and Chimpanzees." This already makes it clear that this is a more complex form of problem-solving, and is possibly even comparable to the tool use of chimpanzees. Within forty-four days, it was observed that the seagulls living on the beach of Stage Harbor threw different kinds of shells, snails, and crabs a total of more than three thousand times in the midst of flight, dropping them to the ground to crack the hard shells. Several drop points, at which the hardness of the ground varied, were used by the seagulls as a tool. The drop points were between thirty and two hundred meters from the place where the gulls pick their prey at low tide. After picking, they used a certain low and fast flight to fly straight to the drop point. This varied in distance and direction from the picking place and was often not seen from there (Beck 1986, 139–40; Beck 1980; Griffin 2001b, 55–6). The gulls' ability to flexibly use an internal map and to find their way to the drop point is therefore already remarkable (see Beck 1986).

Furthermore, the gulls very carefully considered the texture of the ground on which they dropped the prey. For example, more than 87 percent of the observed area is made up of soft dunes, but the gulls only dropped 4 percent of their catches on unsuitably soft sand. 90 percent of their throws were aimed at the rock face, which makes up only 1 percent of the area but provides the best drop surface. Gulls' knowledge of the important properties of a tool (here, the ground) suitable for their purposes implies advance planning, imagination, and the ability to strategically vary their behavior in a way that suits their respective needs per the given environmental conditions. Just as carefully, they adjusted their flight position and altitude, which ranged from four to six meters depending on the terrain. If Beck says that this procedure implies advanced planning and a strategy, then it means that this precise consideration and planning for the seagull means a saving of time and energy. For example, in order to achieve the maximum aerodynamic stability, the gulls position themselves precisely in the wind for 93 percent of their drops. Prey that had to be thrown several times, and was now covered with sand, was often carried to the water and washed off (Beck 1986, 141; Beck 1980; Griffin 2001b, 55).[2] According to Beck, there is strong evidence that all these behaviors are learned through play, practice, and observation, just as chimpanzees learn about termite fishing (Beck 1986, 143; Beck 1982). In conclusion, he says:

> These examples offer convincing evidence of such cognitively complex or intelligent processes as premeditation, formation of strategies, and the flexible use of cognitive maps to respond adaptively with regard to features that are removed from the gull's present surroundings in both space and time. (1986, 143)

For all of these reasons, for Beck, the cognitive prerequisites necessary for chimpanzee behavior are comparable to those that determine the dropping behavior of the herring gulls. This may be the case, at least with respect to a

2 A similar behavior is employed by Japanese macaques (*Macaca fuscata*) on the island of Koshima: they wash sweet potatoes to rid them of wet sand (Kawai 1963; Matsuzawa 2003).

direct comparison of termite fishing and shell-dropping. However, chimpanzees display a number of other complex acquired behaviors that cannot be considered in the context of food acquisition (this is discussed in greater detail in Section 4.3 and 4.6).

Particularly interesting for ethological research on tool behavior in birds are corvids (Corvidae), which comprise, in addition to ravens themselves, crows, jays, and magpies. Studies by Bernd Heinrich (1995, 1999) and his colleagues showed that ravens can actually develop successful and insightful strategies for getting food. Heinrich confronted five hand-raised ravens, none of which had previously dealt with the experimental device or problem, with the task of getting to a piece of food that had been tied to a cross-beam with a long rope. The line was too long to be pulled up by the animals sitting on the cross-beam. After six hours, one of the birds suddenly performed a series of complex actions: she pulled a piece of the string toward her, fixed the drawn piece with her foot, grabbing the string further down, and pulling the next piece of string upwards. Step by step, she came closer to the food without the weight of the food pulling down the string that had already been pulled up. After some time, all but one of her four conspecifics learned this strategy, each varying the method slightly. Some moved sideways as they fixed the string, others placed the string pieces already pulled up in loops, not moving from their starting position. It seems that each bird solved the problem individually. While they used different maneuvers, they might have been inspired by watching their companion obtain food in this unusual manner. All birds who had acquired the strategy of gradually pulling the string understood that their normal escape behavior, namely, to fly away with the prey in their beaks, could not work due to the food still being attached to the line. In over one hundred trials, they always dropped the food first and then flew up when frightened. The only bird who had not learned the strategy also tried unsuccessfully to fly away with the tied food in the beak (see Heinrich 1995; Heinrich 1999; and the discussion in Griffin 2001b, 117ff.). Individual inventions are rarely of interest to evolutionary biologists, who focus mainly on phenomena that occur in whole populations. However, all new inventions are likely initially limited to one individual or a few individual animals. Griffin furthermore emphasizes that individual inventions provide conclusive evidence of independent thinking in individuals, and this is noteworthy, regardless of whether it asserts itself as a behavioral strategy on the population level (Griffin 1985, 131). In general, the ambiguity arising from the possibility of innate behavior can be ruled out when a unique problem gives rise to a completely new, creative solution. Creativity is usually understood here as a problem-solving strategy and—in the case of tool behavior—as the development of a technical-mechanical solution. Creativity in birds, however, also extends beyond this and, as the last example shows, may even point to aesthetic preferences.

Male bowerbirds (Ptilonorhynchidae), native to Australia and New Guinea, are very inconspicuous, but build ornate bowers. These they decorate with colorful objects to lure females for mating. First, they carefully clear the surrounding area of vegetation to build bowers of various shapes, which are then decorated with colorful leaves, shells, pebbles, and a variety of other items that the males specifically collect for this purpose (Beck 1980, 22; Becker 1993, 57ff.; Diamond 1982, 99–100). The males compete with each other, stealing their decorative objects from each other and destroying the bowers of others—especially the most beautiful, which represent the greatest competition.

The females visit and observe several bowers before deciding to mate. There is a statistically significant tendency for more mating with males who have the most stable and conspicuous bowers. Subsequently, the male is no longer involved in nest-building and rearing of the young (see Gould and Grant Gould 1994, 124ff.; and the detailed discussion in Griffin 2001b, 93–9).

The abilities of these birds has been discussed time and again, and it has been argued that they display purely hormonally controlled enticing behaviors. According to this interpretation, the decoration of the beautiful bowers as part of their reproductive behavior does not amount to an artistic or aesthetic endeavor. Griffin emphasizes, however, that Karl von Frisch has already pointed out that the desire to impress the female is undoubtedly the motive for mankind's creative creations. The bird may well be thinking of the females and competing males when he designs and decorates the bowers (Griffin 2001b, 99). Moreover, there are good reasons to challenge the view that this behavior is exclusively hormone-controlled. Gould and Grant Gould (1994) make it clear how complex the behavior of the bowerbirds truly is. The males learn, beyond species-specific genetic information, what the females value. It has been observed that repeated trials of bower decoration take place in young birds. According to Gould and Grant Gould, for about two years the younger birds awkwardly practice building in the group, during which time they make no courtship attempts. Furthermore, they observe the construction and the finished end products of experienced males and their courtship. It might therefore be possible that not only genetic information comes into play, but that a learning process takes place as well, a process that is, perhaps, of an even more decisive role in the years of perfecting the quality of the constructions (see Gould and Grant Gould 1994, 124ff.; and the discussion in Griffin 2001b, 96ff.; 1982, 101). In addition, females conduct group visits to the bowers, where the younger ones can observe the more experienced females and learn from them which aspects of a construction are most attractive (Diamond 1987). In addition, the bird makes great efforts to keep its his bower "fresh" and to preserve its individual design (Gould and Grant Gould 1994, 125). If the design of a bower is changed due to any circumstances, the bird immediately restores its individual design (Diamond 1987). Even if, for example, a bower is raised and turned slightly askew, the returning male—unable to restore the bower to its original position—will tear it down and start rebuilding it.[3]

Ethologists even see references to culturally traditional preferences, because different species and groups of (and even individual) bowerbirds have different preferences in bower construction and decoration (see the discussion on this in Griffin 2001b, 96ff.). Bowerbirds of the species *Amblyornis inornatus* in the Wandamen Mountains in New Guinea construct particularly large, covered bowers, with a kind of "maypole" in the center, decorated in black all around. Two hundred kilometers away, in the Kumawa Mountains of New Guinea, lives another group of the same species that builds simpler, uncovered bowers that have the same "maypole," but with an all-brown decoration. In an experimental study, researchers placed colored poker chips near the bowers. The birds of Kumawa removed them immediately, while the first group from the Wandamen Mountains included the chips in their decoration. They favored blue, then (in declining order of preference) pink, orange, red, lavender, yellow, and white. Individually, these preferences and the arrangement of the poker chips differed from each other, but they followed the preferences and arrangement variants, which the animals showed otherwise with reference to their collected decoration objects (Diamond 1988).

> The difference in style between the two groups may be genetic, but given the time the birds dedicate to observation and practice, cultural transmission of "taste" seems much more likely....The ontogeny of bower-building, its flexibility, the role of individual variation, and the (very likely) cultural dimension, all suggest that this behavior goes at least one step beyond even the goal-directed building of other species.
>
> GOULD and GRANT GOULD 1994, 126

This process can no longer be simply explained by the existence of a purely mechanical, task-oriented programming. Particular individuals seem to be able to set their own goals and choose the details they focus on by picking a

3 Diamond (1982) points out that not only individual styles play a role, here; the orientation of the bower is also important. It is always positioned at such an angle to the sun that males and females can face each other during their courtship in the early morning hours, and neither need look directly into the sun when looking at each other (Diamond 1982, 99–100).

specific bower style. Gould and Grant Gould (1994) conclude: "The temptation to invoke the word 'thought' here is almost overwhelming" (126). Griffin, too, thinks that, given that young birds are observing experienced males in bower construction and decoration as well as during courtship in their arbor, it is obvious that local and individualized occurrences of styles are manifestations of what Diamond calls "culturally transmitted traits, like human art styles" (Griffin 2001b, 98; with reference to Diamond 1986, 3042).

4.1.3 *Tool Behavior in Mammals*

Birds are particularly savvy tool users and creators. But mammals also demonstrate interesting examples of similar behaviors.

Sea otters (*Enhydra lutris*) forage for crabs, sea urchins, clams, snails, and other mollusks. Such food must be removed from the seafloor with claws and teeth, and their shells must often be cracked in order to eat them. Young and old animals especially, who have problems in this regard, sometimes use tools such as stones. First, they pick a stone of suitable size and weight, clamp it under the armpit, and dive to the bottom. They then use it to loosen their prey from the bottom, emerge with it, and crush the prey while floating in the water on their back (see Hall and Schaller 1964; Houk and Geibel 1974; Riedman and Estes 1990; Barnes 2002; Griffin 2001b, 119–20; Kruuk 2006, 140–61). Sometimes several trials are necessary, during which the animals repeatedly use the same tool to reach a prey animal (Barnes 2002, 14–5). In addition to diving for food, otters use tools to get at food when out of the water. On the one hand, they use their own belly as a base to crush food on and, on the other hand, they again use stones as a hammer and anvil to open the prey (see Limbaugh 1961; Hall and Schaller 1964; Houk and Geibel 1974; Riedman and Estes 1990; Barnes 2002, 19; Kruuk 2006, 140–61; and the discussion in Griffin 2001b, 119–20). Sea otters use stones to cut and break up prey primarily in areas where they cannot obtain enough food in any other way. They also develop clear individual preferences for specific tools (Riedman and Estes 1990, 33–4; Barnes 2002, 18). In some places, for instance, they use beer bottles instead of stones. Since they float these do not need to be stowed in the armpit between two dives to use again after surfacing (Woolfenden 1985; Riedman and Estes 1990, 33).

The tool behavior of otters seems to be learned socially. Estes et al. (2003) found in long-term studies that dietary choices and dietary behavior among ten different female sea otters showed extreme interindividual variability, although all ten animals studied lived in the same place, at the same time, under the same environmental conditions. The availability of food cannot explain these differences. The observed behavioral variants also persist over time and are, according to Estes and his colleagues, "life-long characteristics." Their studies have shown that individual behavior in food acquisition is passed on matrilineally, in close bonding phases, from mothers to daughters (Estes et al. 2003, 144).

The observed tool behavior in sea otters usually occurs within direct food acquisition, but Limbaugh (1961) observed how an otter also used seagrass for fur-cleaning. Haidle (2012) cites two additional examples in which sea otters wrapped seaweed around themselves as a buoyancy aid and for stabilization during the resting phase, or drove away food-stalking seagulls with targeted water swells (307–08).

It is sometimes claimed in the literature that there are relatively few examples of tool use in mammals (especially in comparison to birds). Baber (2003, 24) indicates, for example, that he only found incidences of stone hammers in sea otters, rock-throwing bullets in polar bears, and back scratching sticks in elephants—in addition to the tools of the chimpanzees. Similarly, Griffin (2001b, 119) mentions few additional examples in mammals, apart from monkeys and great apes. In contrast, Haidle (2012), after extensive research, has published a page-long list of examples of tool behavior in animals cited and discussed in the literature, in which she mentions approximately 450 cases of tool behavior in mammals, including the giant panda, the goat, the eland antelope, Botta's pocket gopher (*Tomomys bottae*), and monkeys and great apes. Some species or subspecies are represented by several examples in this list, such as the deer, elephant, and bear. However, most of the examples are known from great apes, with bonobos (*Pan paniscus*) and chimpanzees (*Pan troglodytes*) clearly dominating the list, with approximately 140 examples.[4] At a later stage I will discuss nut-cracking, termite-fishing, and other tool behavior in chimpanzees (see Section 4.3).

In chimpanzees and sea otters—but also in herring gulls, for example—there is a special foraging behavior that involves actions that are very different from direct grasping, cracking, and eating. These species employ skills that are only indirectly related to food. Thus, the initial stages of foraging, which consist of the search for as well as the selection or processing of suitable tools, are separate from the actual consumption of food. It is extremely obvious that the animals intend to gather food and believe that the efforts at hand (such as the long transport of the tool) ultimately lead to the cessation of hunger. Here, according to Griffin (1985), it can be assumed that the animals engage in conscious thinking, in the sense of intentional behavior aimed at a specific goal (136).

---

4 This does not mean that tool use does not occur in other mammals. Chimpanzees have been studied extensively by ethologists in this respect. The fact that research on the cognition of animals is primate-centric is quite evident as we progress through this book.

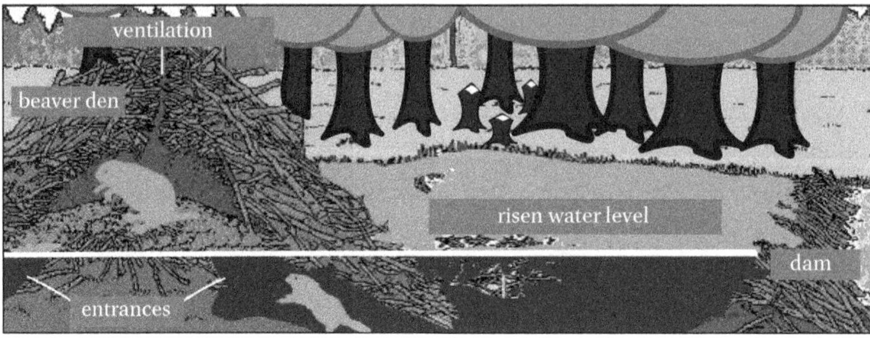

FIGURE 2 Beavers block water with the help of dams, thereby submerging the entrances to their dens.

A striking example of tool use in mammals that is not directly related to immediate food acquisition and intake provides further evidence of goal-oriented, conscious behavior in animals. Beavers (Castoridae) eat tree bark, leaves, and sprouts and set up stockpiles, which, when the waters in which they live are frozen in winter, they can reach underwater. For this reason, and as protection against enemies, the entrance to the entrance beaver den must be below a certain water level (see fig. 2). If it is too high, then the entrance may freeze over in winter. In order to be able to control the water level, beavers build dams and dam ponds.

In the process, islands are also built or channels dug for transporting tree trunks, and building materials are sometimes collected long before the construction of the nest and dams. Some dams reach record lengths of up to seven hundred meters. In a beaver castle, families often work together for a very long time, with the goal of blocking a pond and gathering and processing materials in the early stages of construction. Almost the entire construction behavior of beavers is oriented towards the later regulation of the water level. In winter, beavers can lower the water level through holes in the dike. This creates an air layer between water and ice, which allows the animals to breathe outside their cave. The dam thus has the purpose of improving the environment for the benefit of the animals at a considerable distance in time (Gould and Grant Gould 1994, 128ff.; Griffin 2001b, 99–112). Do beavers have this goal in mind, or is dam-building behavior an innate behavior? In experiments with loudspeakers it was found that beavers pay attention to the sound of running water (see Wilson 1971; Richard 1967, 1980; Griffin 2001b, 107). It must be emphasized, however, that under natural conditions they would not necessarily answer the sound of running water with a buildup of material, as if they were merely following an automatism triggered by this stimulus. Often, they also stack in calmly flowing water, while noisy streams do not necessarily instigate dam building. Most of the material is also piled upstream of the dam, not on the downstream side of the dam, where the water flows the loudest. Under certain circumstances, even behavior that is contrary to dam construction is displayed in the form of holes made in the dam to drain water. It is not very probable, according to Griffin, that the entire process of dam construction and water regulation can be explained in response to a single stimulus. This would be an oversimplification (Griffin 2001b, 107–08; see also, along the same lines, Gould and Grant Gould 1994, 130–31). Furthermore, if that were the case, then numerous genetic subroutines would have to be postulated for the various behaviors, flexibly adapted to the environmental situation, in some cases even completely contrary to dam construction. Griffin suggests that the plausibility of such subroutines established for their own subpurposes declines, the higher the necessary number and complexity of such extra tasks is. Also, their number and complexity increases the more flexible the behavior is (Griffin 2001b, 112).

Beavers actually seem to work very flexibly toward an end goal. The main objective is targeted blockage at selected locations. But beavers stop accumulating material when a certain level is reached, and do not build dams when they live in existing lakes or deep rivers, but instead dig caves into the embankment to build their lodges. They never start digging a cave in a slope that rises too slow or lies too low. This would not provide enough space for the second half of the corridor, nor for the cave to end up far enough above the water surface and remain dry. This factor, which is decisive for a useful final product, but not decisive for the first part of the tunnel, which is still below the surface of the water, seems to be taken into consideration by the beaver right from the start (Griffin 2001b, 103). The structures reach astonishing levels of perfection, and the animals work with great sophistication. They have a remarkable understanding of how water works, which they always consider when repairing and arranging building materials. Gould and Grant Gould speak of an "apparent understanding of hydrodynamics," which also helps them carry out "emergency repairs" on the dam. Beavers are thus much more quick-witted about their situation than many a person in a comparable situation (Gould and Grant Gould 1994, 130). Ryden (1989) observed the reaction of a beaver colony to a wanton destruction of its dam by humans. After the large hole in the causeway had been discovered by the patrolling beaver, it immediately began to attach branches to close the gap. Due to the heavy water flow and the few branches available nearby, this was only moderately successful, and the beaver therefore changed its tactics and began to block the gaps between the rocks under water with mud and aquatic plants. Although hardly any of the other animals had inspected the dam during the rest of the colony's observation period, several conspecifics came up and joined in: "They seemed to recognize that stick piling was ineffective and turned instead to plugging underwater gaps between the stones, even though the noise of flowing water came from the top of the rock pile" (Griffin 2001b, 109). Eventually, the runoff could at least be slowed down. The next morning, as the male left the den, he took a large branch from the structure (a behavior that is highly unusual) and carried it almost one hundred meters to the dam to attach it to the repair site, "presumably because he remembered the need for material to repair the damaged dam. Other beavers also removed sticks from the lodge when they emerged and brought them to the dam" (Griffin 2001b, 109–10).

Beavers can solve other unexpected problems. For example, they can build ramps to access human-made protective fences for valuable trees. In another example, beavers living in a human-made pond dewatered by a three-hole drain with a diameter of eight millimeters in its protective cap started to pierce and clog the holes with sharpened branches like "giant pencils" after a few weeks (Gould and Grant Gould 1994, 130; Griffin 2001b, 101).

Beavers display a much-sought-after trait: goal-directed flexibility, the ability to flexibly solve problems. They are not programmed to build; they know exactly what they need and, for example, they also use suitable structures on the banks or human-made dams that they expand. With the ability to face these immense cognitive challenges, which allows the beavers their special way of life, it is not surprising that their brain is by far the largest among the rodents in relation to their body size (Gould and Grant Gould 1994, 131).

The following theoretical considerations are worthy of continued investigation, as they are three important aspects of animal tool behavior: the condition that it is learned behavior, the condition of goal-directed flexibility, and the fact that most of the animals showing such behavior lead a life in the social fabric.

4.2 Theoretical Considerations

4.2.1 *The Concept of Learning*

After the example of the termite fishing of the predatory bugs, it became clear that there are different explanations or explanatory approaches needed for different cases of tool use. Here we have come to the point of either attributing insect cognitive abilities to a level similar to the termite fishing of great apes, or reconsidering a relative relationship between tool use and intelligence. It is the case, as we have seen, that the bug has a simple nervous system and its termite fishing is not a learned behavior. It occurs only in certain stages of development, reveals itself as such in all predatory bugs in stereotypical form, and is innate and genetically preprogrammed (Beck 1986, 137). If we go to the level of conscious, goal-oriented behavior, we are already in the realm of learning. But learning can be done in different ways. Many animals that develop novel solutions use the principle of trial-and-error learning. Beck describes this process as follows:

> First, a successful or correct response is emitted fortuitously—in play, out of frustration, or unintentionally in the course of other behaviors. When the response is followed closely by a reward, it is likely to be repeated. Such trial-and-error or "operant" learning is common in both invertebrates and vertebrates. It is certainly adaptive.... Operant learning even subserves much human behavior.
>
> BECK 1986, p. 137–38

First of all, this can take the form of a true chance discovery: an animal randomly opens a locked box and finds food in it, without even wanting to. In a learning process, what happened in this instance (the movement that led to the opening of the box) is remembered and applied in a later, similar situation. Second, trial-and-error might include situations in which an animal wants to open the box but does not deliberately apply a strategy, instead doing all sorts of more-or-less obvious actions that could lead to the goal of opening. As Beck rightly says, there is an adaptive learning component in a trial-and-error procedure, in which success reinforces the strategy applied—when the problem arises for the second time, the animal can apply a matching solution, however it was discovered. Still, this can be differentiated from real insight into a problem and its suitable solution.

Real insight, unlike the two variants of trial-and-error described above, mean seeing a problem and knowing what the solution is. Whereas trial-and-error only leads to a solution by chance, real insight involves a conceptual analysis of the problem that purposefully leads to a solution. However, contrary to what one might think first, this understanding of "pure insight" does not emerge out of thin air either; it also has some requirements. In a trial-and-error process, playful recourse is made to all possible modes of action from one's own repertoire of action, and such modes are attempted one by one. Equally, in the mental analysis of a problem, a problem is detected (such as a locked box in which food could be stored), a goal is identified (namely, to open the box), and possible solutions are selected with reference to prior experience. In both cases, learned knowledge is applied to a new context. Even if a strategy is deliberately applied through real reflection, it is usually not completely different from the other behavioral repertoire of an animal (or human being). Conceptual analysis often consists of identifying components of the problem that are already familiar from other contexts and thus identifying which parts of an already acquired strategy of action that succeeds in similar contexts might be helpful.

Arzt and Birmelin (1993) regret that the application of a strategy to a new context is often underestimated. They go into more detail on the previously mentioned experiments by Wolfgang Köhler, in which chimpanzees managed to reach bananas that were hanging too high with the help of a box. A criticism often expressed in these experiments degrades the apes' alleged intellectual achievement to a simple application in the new context, since the animals may already have learned how to deal with crates or the like (Arzt and Birmelin 1993, 120). Arzt and Birmelin (1993) balk at the unjustified disregard of the application of already acquired knowledge to a new context. They are of the opinion that the application of experience in a new context led to further great inventions, especially in humans (120). They point out, for instance, that in 1879 Thomas Alva Edison would never have invented the electric bulb without the familiar experience that electric current makes a wire glow. The same principle applies to many other examples, such as the invention of the telegraph in 1838 or the first microscope. The series of such examples can be extended as desired, and it should not be forgotten that our so-called creativity is never creation out of nothing (Arzt and Birmelin 1993, 121). Should it not be conceded that chimpanzees and other animals would not be able to solve such a problem without previous experience with movable objects?

The really interesting thing about Arzt and Birmelin's text is the following report on a Köhler banana test, which was repeated by scientists at Harvard University in an original way (Epstein et al. 1984): it concerned the question of what previous experience is actually necessary to solve the problem of reaching bananas that are hanging too high with the help of a sliding box. Instead of chimpanzees, the scientists used animals that had no experience with moving crates, namely pigeons (Columbidae). Different things were taught to the birds, who were initially helpless in the situation: first, they learned to stand on a box and pick the hung banana (1). In a separate tutorial, they were taught how to use the beak to move a box (2). The third step included instructions on how to move the box to a green point on the ground (3). Neither learning experience (1) nor step (2) alone resulted in the pigeons pushing the box under the banana to be able to pick from it. Only when they had learned step (3) could they solve the task. After looking back and forth between banana and box, the pigeons pushed the box purposefully under the banana, without losing sight of it. According to Arzt and Birmelin, the pigeons examined the situation and then acted (and obviously intentionally) on a plan in which two new steps, based on no previous experience, had to be performed. Separate instructions had to be combined—moving and climbing the box as well as replacing the green target point with the target point banana (Arzt and Birmelin 1993, 123). This experiment shows that previous experiences determine later possibilities of action. However, if suitable strategies were developed in other contexts, then it is also an act of conscious analytical thinking to try or apply them in similarly recognized contexts. The realization then consists of combining individual steps that have already been learned and, if necessary, modifying them so that they make sense in the new context with regard to an identified goal. Anyone who solves a problem with a solution strategy that he has taken from another context has not invented the solution strategy itself. However, he has at least identified a problem, identified a relative similarity of his current problem to a previous one and, if the strategy is appropriate, correctly assessed the mode of action of the strategy with regard to solving the new problem in its specific context. Gould and Grant Gould (1994) are also unwilling to dismiss Köhler's chimpanzees as "only" demonstrating operational behavior that is within innate limits:

> Chimpanzees may indeed have insight…indeed, having the wit to realize that something done in play—as one of many possible behavioral sequences, explored apparently just for the fun of it—might solve a novel problem probably *does* qualify as insight. (81)

After all, what we see here is a "low-level, everyday sort of insight," according to Gould and Grant Gould. But the strict requirement to demand nothing new from scratch, to allow conscious inspiration, cannot be desired. Like Arzt and Birmelin, Gould and Grant Gould (1994) point out that "by that standard few humans would get passing marks for cognitive prowess" (81).

Where tool use occurs in animals, it may involve inborn behaviors (as in the case of the predatory bugs) or various forms of learned behavior. Even tried and tested solutions represent a transfer of (previously learned) behaviors to new contexts. However, in the previous examples there were already indications that animals can use a solution strategy unerringly and without much trial and error. In Bernd Heinrich's controlled experiments with ravens, for example, one of the animals, after only a very short time and apparently without any immediate effect, executed the correct solution in all complex individual steps, drawing the piece of meat along the line closer and closer and holding the string repeatedly fixed with the foot. In the following chapters, further laboratory experiments will be discussed, which, thanks to controlled conditions, also document the emergence of intelligent behavior in the sense of a purposeful transfer of knowledge to a new context.

For the moment, it should not be forgotten that intelligent behavior can, of course, also exist where no use of tools takes place. This is a very important point:

> Indeed, some tool behavior, like that of the chimpanzees, is cognitively complex, but some, like that of the assassin bug…are either innate or based on simple learning processes. Likewise, behavior not involving tools can be cognitively complex or simple. Tool use may be among the agents selecting for intelligence, but it is neither necessary nor sufficient for intelligence to involve.
>
> BECK 1986, 134

However, tool use clearly helps animals to solve problems and deal with challenges from the environment. Being able to do so with respect to a distant goal in space and time (be it in the context of tool use or otherwise), seems a good indicator for intelligence. As we have seen, intelligent solutions often require modifications of behavior or the possibility of deliberately choosing between innate behaviors to solve problems.

4.2.2 *Goal-directed Flexibility*

Only when an animal modifies innate or learned behaviors or chooses between them to solve problems does she show what we can call goal-directed flexibility. This flexibility seems to require that the animal understand the overall goal of an action and is able to modify her strategy to reach it which leads to flexible problem-solving (Gould and Grant Gould 1994, 127ff.). One approach to capture what's going on here is assuming that such goals are represented in the brains of the respective animals (cf. Beck 1986, p. 144). Beck writes: "Whatever their nature, the use of internal representations or images of environmental features appears to be a common element of many behaviors that we deem intelligent" (144). These mental representations are stored as knowledge in the memory. In addition, they are certainly not limited to the examples that have been discussed already, such as so-called mental maps. They may also be of the nature of a relationship between individuals (e.g., between mother and child), a principle ("Lightning follows thunder"), or a prediction ("If I attack the alpha male, he will strike back"). It is important to note—without wishing to elaborate on the role of language as a sign of intelligence—that in this context Beck sees evidence for these kinds of concepts even in animals that have no human language and no linguistic understanding: "The representations need not be linguistic" (Beck 1986, 144). Here he agrees with the explanations given by Peter Bieri, who emphasized mental representations, too, but also assumed that linguistic ability is not a prerequisite for the representation of semantic information (see Bieri 1997, 24; and the explanations in Chapter 3 of this book). Later, when it comes to the question of language in animals, I will speak in more detail of a theory of mental representation (see Section 5.2.3).

As we have seen, some animals, such as the beaver, clearly perform complex, adaptive, and flexible actions associated with a very distant ultimate goal. Some animals make their own tools, collect material long before use, and adapt it to a later destination. Problem and solution occur here temporally and spatially apart to a significant degree. Some tools are carried far to their place of use and used again and again. Such use shows flexibility in different specializations (see the case of the sea otters, for example). Such specializations are passed on through social learning, observation, and imitation. Especially in social animals, goal-directed flexibility seems to be a main determinant of behavior.

4.2.3 *The Meaning of Life in the Social Fabric*

Intelligence can be seen as an adaptive strategy of living beings, helping them to cope with the demands of life in a broad ecological and social niche and with an ever-changing environment. This definition can be found in Beck (1986, 138–39), but we have also encountered it in the example of the beavers, in Gould and Grant Gould's discussion of anthropologists agreeing that the challenges of our own niche as hunters and gatherers led to the extraordinary increase in human brain volume. This assumption was transferred to the beaver, stating that, analogously, it is therefore very probable that the additional brain mass of the beaver developed because it had to meet the enormous cognitive challenges of its own niche (Griffin 1985, 153).

About 2.5 million years ago, humans started to use tools. Significant increases in our brains' size and complexity began at the same time. But both, according to Beck (1986), are probably due to a third development from this time:

> The proliferation of stone tool culture and the growth of the human brain and intelligence were probably both propelled by a third factor: evolution of an advanced capacity to form and manipulate internal representations. We infer this capacity from reconstructions of the lifeways of early hominids. (145)

Life in the social community, in the niche of hunters and gatherers, brought new demands. It had to be planned and communicated, with planning and communication also referring to events in the future. Knowledge about one's own living space had to be acquired. Demographic and social knowledge thus gained in importance. People lived longer and had a longer childhood to learn all this. Beck mentions the importance of cooperative hunting and of sharing and mutual care. Such interactions would have caused a selection that targeted memory capacity and the ability to use abstract social information and concepts. In this context, Beck is referring to abstract social information such as friendship, reliability, the competence of cooperation partners, or the meaning of mutual obligations. He adds that even tool-making at the stage of selection of raw material requires the ability of abstract conceptualization, here with a view to a functional end product (Beck 1986, 146).

To summarize, such interactions that bring about life in the social fabric play a very important role in the development of intelligence. It is therefore easy to explain why intelligent behavior and conscious thinking are also often found in social animals. The way our ancestors lived 2–2.5 million years ago forced them to acquire, store, manipulate and retrieve knowledge when needed. According to Beck (1986), this knowledge was stored in the form of mental representations of their external social and physical environment (146). Preforms of such abilities could go back beyond the last common ancestor of ape and human, back to the animal kingdom. Animals, too, may have already possessed the basic ability to store and use mental representations.

Against the background that complex tool behavior is an expression of basal competences that are developed and passed on in different variants and specializations, especially under social conditions, one justifiably refrains from the assumed simple or unique causal connection between tool use and intelligence. To this day, it is still not clear why humans started walking upright. This process hardly required, according to Richard E. Leaky and Roger Lewin, any major physical changes from an evolutionary point of view. It was enough for the arms to be longer, the feet bigger, and the pelvis wider. But why these physical changes took place, according to the two scientists, still cannot be said with certainty. Probably several factors came together so that this ability could arise, among which the authors consider small carnassial teeth, increased socialization through prolonged childhood, the need for a changed protective behavior, and the use of tools (Leaky and Lewin 1996, 84).

Here, too, we find the connection between tool use and social structures. Subsequently, Leaky and Lewin make the important statement that none of these factors alone would have been sufficient to initiate this evolutionary effect (Leaky and Lewin 1996, 84).

4.2.4 *A Definition of Culture Applicable to Animals*

A definition of culture that is applicable to animals and does not exclude them right from the beginning should be based on the previous reflections on the diversity of tool use and tool production in the animal kingdom, on some of the conceivable underlying learning mechanisms, and on the significance of such behavior in the developmental history of living beings. Against this background, further selected research results on culture in animals will be discussed and the following question will guide this discussion: is tool behavior in animals an actual instance of animal culture?

Definitions of culture have been attempted in the history of philosophy and science in numerous ways from within a wide range of disciplines. Kroeber and Kluckhohn (1978, 41–79) present six important sets of definitions of culture according to their different priorities: (1) descriptive definitions (broad definitions that enumerate various contents of culture), (2) historical definitions (which emphasize the social transmission of lifestyles and traditions), (3) normative definitions (emphasizing the transmission of rules, ideals, and values), (4) psychological definitions (focusing on culture as a problem-solving strategy, with emphasis on learning, habits, and psychological concepts), (5) structural definitions (which place special emphasis on the patterns and organization of culture), and (6) genetic definitions (which focus on those processes that have allowed the emergence of culture and see culture as a product or artefact; in this sense, they emphasize the transfer of ideas and symbols or understand culture as a remainder). The authors cite over 150 different definitions, from Bain (1942) ("Culture is all behavior mediated by symbols"), to Murray (1943) ("The various industries of a people, as well as art, burial customs, etc., which shed light upon their life and thought"), to Becker (1950) ("Culture is the relatively constant nonmaterial *content* transmitted in a society by means of processes of sociation") (see Kroeber and Kluckhohn 1978, 39–79). It can thus be concluded that there is no single, generally accepted definition across all disciplines.

Even within the scientific disciplines that deal with the phenomenon of culture, what constitutes culture in the abstract sense is conceived quite differently. Thus, in cultural anthropology, it is sometimes conceived in analogy to natural evolution, so the aspect of adapting to environments is emphasized, which points in the direction of functionalism and to the field of cultural ecology.[5] Alternatively, culture is seen from the inside perspective, as its own system of symbolic forms, as is the case in Cassirer's cultural philosophy, for example. Or culture, as in cognitive anthropology, is understood as a code of thought. However, almost all of these approaches, according to Streck (2000a), operate in an exclusively human field that seems rather disconnected from the laws of nature that might be effective in the animal kingdom (141).

5 This focuses on adaptive strategies and describes the way in which people set up in certain areas (cf. Streck 2000b).

A low-threshold definition of culture that is less anthropocentric, and therefore applicable to animals, seems to be difficult to find.

This also applies to less ethnological but more philosophically oriented definitions. In addition to a French tradition, one can find a German and an English tradition, each with a different focus, and with reference to authors of the eighteenth and nineteenth centuries. Last but not least, such thinkers differ in whether, in the tradition of the Enlightenment, they understand culture in the sense of a civilization going back to the universal rationality of all human beings or, instead, make spiritual (and also religious) values, the individual genius, his or her work and emotions the focal point (Kuper 1999, 5ff.). Anthropology as a separate philosophical discipline has been spoken of since the publication of Herder's *Ideas for the Philosophy of the History of Humanity* (1784–1794) and Kant's *Anthropology from a Pragmatic Point of View* (1798). At the turn of the twentieth century, philosophical anthropology experienced a significant upswing with the writings of Max Scheler, Helmuth Plessner, and Arnold Gehlen. All three philosophers claim a special position for humans in relation to other living beings (Bohlken 2009).

In Gehlen's approach, the important role of culture in philosophical anthropology becomes particularly visible. The idea of the human being as "deficient" (German: Mängelwesen)—an idea that is already found in Herder and Nietzsche—is here combined with a strong culturalism. Because humans, unlike animals, are poorly equipped with regard to instincts, sensory organs, and endogenous protective mechanisms, they have to change the world around them to ensure their survival. Therefore, through culture, humans create a "substitute world" that becomes their "second nature" (Gehlen [1942] 1998, 240). So, the real nature of humans is culture. The emergence of cultural techniques is thereby explained, according to Bohlken, by the fact that they are relieved of defined stimulus-response schemes and thus open up new possibilities for action. Along Gehlen's lines, language is to be understood as another cultural substitute. It establishes an intermediate level between perception and action that enables foresight and planning. As a mediator in this sense, it breaks the "spell of the immediate" (Bohlken 2009; with reference to Gehlen [1942] 1998, 240, 246). The human being is not imprisoned in the immediate, in the here and now, but is rather, in every respect, "cosmopolitan" and an "unconfirmed being" (Gehlen [1940] 1993, 18–9; with reference to the works of Scheler)—thanks to his culture. In contrast to animals, he can set up in many niches and, based on accidental circumstances, he develops a cultural sphere, which at this point replaces the environment. Gehlen therefore describes "culture" as an anthropo-biological concept (Gehlen [1940] 1993, 88).

Apart from the fact that the above-mentioned abilities of planning, thinking with foresight, and flexibility have already been encountered in some of the animal examples mentioned so far, the thesis of human beings as deficient is also refuted today (Bohlken 2009). It is precisely the specialization of mankind to specialize in so many ways, to adapt to all sorts of niches and to be able to acquire many and varied competencies (Bohlken 2009).

In spite of recognizable patterns and common themes that can be anchored within individual subdisciplines (such as the idea of culture as substitute action in Gehlen's philosophical anthropology), to this day neither in philosophy nor in other sciences nor in everyday understanding is there a unified and universally valid concept of culture. Kuper (1999) states: "Culture means something rather different to market researchers in London, a Japanese mogul, New Guinean villagers, and a radical clergyman in Teheran, not to mention Samuel Huntington[6]" (3). When it comes to culture in animals, then, a relatively nonconditional concept of culture would have to be applied.

Two ways of starting to engage with the dazzling phenomenon of culture seem possible: first, there is the possibility of starting from an all-encompassing definition that allows as many ideas of culture or components of culture as possible to be summarized. Second, one could pursue a simple and basic definition that is interested in common elements in all cultures and overlaps within all notions of culture.

A starting point for the first way was provided by the English anthropologist Edward Burnett Tylor, who in his work *Primitive Culture*, of 1871, defined culture as follows: "Culture or Civilization, taken in its wide ethnographic sense, is that complex whole which includes knowledge, belief, art, morals, law, custom, and any other capabilities and habits acquired by man as a member of society" (Tylor 1971, I, 1). However, the question of what exactly falls under the "complex whole" of Tylor's areas of knowledge, belief, art, morality, and custom, as well as the law, remains open. A withdrawal to Tylor's determination, therefore, only seemingly helps, and this is primarily because his definition is simply "squishy" (Sommer 2007, 14). However, scientific history has repeatedly used individual features from Tylor's catalog as a litmus test for cultural property or has made any other individual characteristic of humans the central criterion. However, according to Sommer, it seems difficult to categorically exclude

6 In his highly contested book *The Clash of Civilizations* (1996), Samuel P. Huntington (1927–2008) turned against the notion of a universal world culture and instead assumed that civilizations—such as Islamic and Western civilizations—are in a constant state of clash and conflict. Instead of a human rights policy, Huntington called for a geopolitics of power, led by the United States, and grounded in a strengthened Western identity.

animals: we could also use tool use, choral singing, or political action as exclusion criteria and would promptly be confronted with the tool behavior of many animals, the songs of the humpback whales, or alliance formation in monkeys (Sommer 2007, 14).

Tylor's definition is monolithic;[7] culture becomes a self-contained complex, the "great whole." It also seems that Tylor's monolithic definition or similar anthropocentric and monolithic definitions do not make human culture one *special expression* and *special shape* of culture, but unjustifiably elevate it to a condition for the existence of culture in general. The suggestion, however, that one can only speak of culture if exactly these components are pronounced, or even if all these components are present, appears arbitrary. For instance, cultures without faith, at least without monotheistic or generally religious beliefs, would be conceivable.

Tylor's definition is less Eurocentric and less influenced by colonial, elitist, and discriminatory concepts of superiority than other definitions. However, it is still anthropocentric. It clearly makes skills and habits *that are suited to humans as members of society* a prerequisite for us to be able to speak of culture. To a certain extent, one could argue that Tylor has—*per definitionem* and in a circular way—only defined *human* culture. Anthropocentric-monolithic definitions that presuppose the totality of human culture as the baseline for *culture in general* are completely inapplicable from the outset if one turns to the phenomenon of culture in other animals. Of course some people might believe, according to Sommer, that only a Swiss Army knife is a real tool, only the performance of a Mozart mass is singing, and only Bismarck is a politician. But that only describes *some elements* of culture of *some humans* and doesn't help at all to find out what culture in humans (let alone in other animals) is in general (Sommer 2007, 14).

It should be clear that Tylor's "big picture" cannot always be found in its totality, neither in every human society nor in other animals. That is to say, we would still speak of culture among human societies that previously lived or are still living without written law, classical science, and largely without the technological achievements of the Western world. In fact, in an evolutionary understanding of human descent, we assume that even the ancestors of *Homo sapiens* were cultural beings. The oldest records of tool use in human history can be seen in the 2.5-million-year-old ancient stone tools discovered in the Oldovai Gorge in Tanzania (see Semaw 2000). The earliest known works of human art, small ivory sculptures from the caves of the Swabian Alb, are more than thirty thousand years old (Conard 2007). If one considers these achievements

7 "Monolithic" means that something consists of only one stone or component, is inseparable or monumental.

as the beginnings and basic forms of the cultures that modern humans have developed today, then there is no sharp divide between ourselves, our early ancestors in the genus *Homo*, the last common ancestor of humans and apes, and ultimately, perhaps, to the great apes still alive today. While we lack any evidence of the early cultures of hominins until archaeological finds transport them to the surface of the earth, the behavior of living apes (including their culture-suspicious behavior) is luckily directly observable.

Given these difficulties when it comes to definitions of culture, we now turn to the second path. Rather than following a definition that, like Tylor's catalog, makes the great totality of human culture the starting point of a general definition of culture, we shall instead pursue inquiries into basic and common similarities underlying all cultures. This is not to deny that our definition preserves the Tylorian character of a collective term. Culture could consist in an accumulation of individual building blocks—as long as the required building blocks are not oriented to humans from the outset. Cultural zoologists who embark on the path of finding a simple and broad definition see themselves, according to Sommer, as universalists who understand culture as an evolutionary continuum, with fluid transitions between species and groups of forms (Sommer 2007, 14). Carel van Schaik and many of his colleagues in primatology and ethology note that there is a definition of culture that is particularly well suited to comparative research. It states: "Culture is a system of socially transmitted behavior" (van Schaik et al. 2003, 102; for similar views see Matsuzawa 1999; and Whiten et al. 1999, 682). Thus, according to this simultaneously very broad and simple definition, which is widely accepted by biologists, culture is the social transmission of behavior (Sommer 2003, 128ff.; Bonner 1989, 9).[8] Our first examples have already shown that, even when it comes to animals, some behaviors are passed on from generation to generation and thus become traditions in this sense. But when is it really justified to actually speak of "culture"? On the one hand, this requires a more precise definition of what such traditions are and, on the other hand, it must be clear how they relate to the broader concept of culture.

Here, too, I will attempt to simplify things and thus follow Whiten's (2005) definition of tradition as far as possible: traditions are clearly distinguishable behaviors that are shared by two or more individuals in a social group and persist over time. They are acquired, at least in part, by socially-assisted learning from subsequent individuals who reintegrate them into their behavioral

8 The definition can be ascribed to Boyd and Richerson (1996, 1985). Some experts use an even narrower definition and restrict the required social transfer of knowledge to imitation and teaching, a view that is not without a degree of controversy (see also the discussions in Rendell and Whitehead 2001, 310ff.; and Sargeant and Mann 2008).

repertoire (Whiten 2005, 53; with recourse to Fragaszy and Perry 2003). Behavior is traditional when it is customary or habitual in at least one group; i.e., the behavioral variant is shown by all, most, or at least some relevant individuals, while being missing in at least one ecologically comparable position (van Schaik et al. 2003, 102; Whiten 2005, 52).[9] For a definition of culture based on such traditions, Whiten (2005) believes that "the culture of a community is made up of its own unique array of traditions" (52).

At least four characteristics of traditions or culture can be fundamentally determined: (1) If it is culture, then, in different populations of a species or subspecies, *there normally are clearly observable, distinguishable behaviors* for one and the same task (for example, nutcracking). (2) There is also a *characteristic distribution pattern that cannot be explained by ecological or genetic reasons*. It should be emphasized that this concerns the comparison between different behavioral variants and an explanation of *these variants*, not why a behavior occurs in general. Of course, the fact that it occurs and makes sense is usually also due to ecological factors. For example, nutcracking can only occur if there are nuts.[10] In what variant the nutcracking occurs, however, is not determined just by the presence of nuts. Two groups of a subspecies, although sharing the same genes and living in the same environmental conditions, can nevertheless show different variations in nutcracking behavior, which can only be explained culturally. Hard-shelled nuts often exist in the habitats of many chimpanzee groups. Some groups crack these, but some do not. In addition, those groups displaying this behavior use different techniques (one group uses,

---

9 For a specification of these criteria for identifying traditions in animals see Whiten et al. (1999: 682).

10 However, this is a very trivial basic requirement. In general, actions are also reactions to external circumstances, or adaptations to them. In cultural anthropology, human culture is assumed to be dependent on climate, soil conditions, and the like. For example, different human cultures have evolved different dwellings or different types of clothing to adapt to these parameters—from the bamboo hut to the igloo, from the fur hat to the straw hat. However, below this level, finer variants such as the style of painting of front doors or the colorfulness of costumes are no longer explainable solely by the necessity of walls and doors or the usefulness of using warm woolen cloth. They are often based on the traditional inventions and preferences of a group of people in a particular region. The perspective of cultural ecology on adapting to environmental conditions is, moreover, only *one* possible anthropological (and rather "functionalist") perspective on the phenomenon of culture. In contrast, other perspectives always have autonomy, which refers beyond the mere reaction to natural conditions, as the center of culture (Streck 2000a, 141; expanded treatment given in Streck 2000b, 145). Also, ecological reasons in the trivial sense are always present in the behavior of animals. But here, too, the comparison between different groups is about the social transfer of group-specific variants that can only be explained to a limited extent ecologically.

for example, a stone as a hammer, while another uses thick branches; one group uses a wedge to stabilize the anvil, but the other does not, etc.). (3) As already mentioned, central *social learning mechanisms* do indeed play a role in the acquisition of traditions. The respective variants of behavior are passed on to the next generations through observation and imitation, and are thus handed down. Ecological barriers, such as large riverbeds that animals cannot cross, are often the key obstacle in preventing different traditions from being observed and adopted by neighboring groups. (4) In order to speak not only of individual traditions but also, in fact, of culture, a whole, unique *set of such different traditions* must be present in a community.

Some behaviors in animals are in fact extremely complex: for example, those involving both technological and social elements based on complex and also social learning mechanisms and, moreover, appearing within a whole accumulation of other traditions. In what follows, this will be illustrated by means of some examples, which are also intensively discussed in current research.

4.3 Culture in Chimpanzees

4.3.1 *Nutcracking and Termite Fishing*

The traditions of chimpanzees seem to fulfill all of the stated conditions and, moreover, they have a diversity and complexity that has not been observed in other species, at least so far. Chimpanzees also have the two most well-known examples of tool use in the animal kingdom: the use of a hammer and anvil for cracking nuts and the use of sticks as probes to catch insects.

In the wild, chimpanzees in some areas are confronted with having to crack hard-shelled nuts in order to access their contents. They use stones or hard woods as an anvil and suitable stones as hammers (Boesch and Boesch 1983; Boesch and Boesch 1990; and fig. 3 below). In a study of a prehistoric site in Ivory Coast, Mercader et al. (2007) learned that chimpanzees in West Africa

FIGURE 3
Wedge, anvil, and hammer used by chimpanzees for cracking nuts (Drawing after Matsuzawa 1996, 201, fig. 15.4.)

cracked nuts with stones at least 4,300 years ago, long before people started farming in the area. Their findings suggest that chimpanzees have developed this approach independently, or that the use of stone as a tool was already a feature of the last common ancestor of apes and humans. The ability of chimpanzees to crack nuts with exactly the right amount of movement, at exactly the right impact strength and angle, while not grinding the contents of the nut is astounding and requires sophisticated sensory-motor and coordinative skills (Boesch and Boesch 1993). Boesch and Boesch (1983, 1990) observed that different stones are used for different nuts and that the animals remember where to find nuts and stones. Suitable stones and combinations of anvils and hammers are used again and again, and suitable places for nutcracking are repeatedly visited (Boesch and Boesch 1984; Carvalho et al. 2009).[11]

In addition to the nutcracking tools, chimpanzees use grasses, branches, and sticks as "probes" to poke into termite mounds and ant nests.[12] The goal of this behavior is to reach the protein-rich insects that crawl up or cling to the probes. Insects that bite or run upward on the probe are carefully pulled out of the insect structure with the probe and stripped off directly with the lips, or first stripped off by hand and then pushed into the mouth. To make a probe, the animals select a branch, strip off leaves and branches, break the stick to the correct length, eventually form its tip into a kind of brush, and carry the probe—sometimes over long distances—to the termite mound. They poke around in the opening and try out several holes until one proves to be productive (Sanz et al. 2004, 2009, 2010; McGrew 1994, 113, 120–21; Boesch and Boesch 1990; Goodall 1964). It is by no means a stereotypical use of the probes. This procedure requires a lot of practice, and the young ones observe the adults for several years working with the tools before they manage this skill (Griffin 2001b, 121; Boesch 1991; Lonsdorf 2006). The tools used, their processing and use, as well as the techniques for eating the prey differ significantly between individual chimpanzee groups (Whiten et al. 1999).

In addition, some chimpanzees and other great apes demonstrate nutcracking and termite fishing skills in zoos when they are given the opportunities or the necessary basic material for such behavior (Celli et al. 2003; Calcutt et al. 2009; Hirata et al. 2009).

11 Mulcahy and Call (2006) were also able to prove in the case of bonobos and orangutans in captivity that the animals used tools that they had selected, brought in, and kept with them up to fourteen hours later to solve a problem. The authors see this as an indication that great apes plan and prepare for future acts.

12 They each display different behaviors when it comes to aboveground and underground terrestrial nests (Sanz et al. 2004). Chimpanzees display a similar behavior when dealing with beehives (Bermejo and German 1999).

4.3.2 *The Complexity of Chimpanzee Traditions*

The use and manufacture of tools on behalf of chimpanzees goes far beyond the simple technical use of stones and sticks. It has already been said that these animals not only employ ready-to-use objects for problem-solving, but also specifically modify objects with a view to problem-solving. However, chimpanzees can not only use a tool to solve a problem but also use *multi-part tool sets* differentially adapted to different types of termite and ant nests (see. Sanz et al. 2004, 2009, 2010; Suzuki et al. 1995). For example, chimpanzees in the Republic of Congo use tree saplings to break up the nests of army ants. In a second step, thinner branches are used as a probe to pick up the ants. Although these migratory ants are also found in the habitats of other chimpanzees, they are not hunted there with the help of multi-part tool sets. It is a specialized and very effective form of hunting (Sanz et al. 2010).

Similarly, chimpanzees also have a sense of suitable tool sets when cracking nuts. Carvalho et al. (2009) come to the conclusion that the animals repeatedly use certain combinations of hammers and anvils. This could mean that they see such tool sets as a single tool that consists of two parts, but only has a functional value as a nutcracker in that specific combination. The signs of wear are deepened on the tool set with increasing, repeated use in the same way and in the same place, and with each new stroke. Using the same set again and again increases the effectiveness in nutcracking (Carvalho et al. 2009, 103).

Meanwhile, chimpanzees living in the savannah of Ugalla, Tanzania, have been reported to use tools to harvest the underground storage organs of plants (Hernandez-Aguilar et al. 2007). Such behavior is believed to have played a role for different features of hominin evolution, such as for the conquest and colonization of savanna habitats, the development of the skull and tooth morphology of the genus *Australopithecus*, as well as for the evolution of the genus *Homo*. This is due to the fact that plants' subterranean storage organs could be used in times of drought and food shortages (cf. Hernandez-Aguilar et al. 2007, 19210). Living in the extreme conditions of savannah habitats has evidently led to an adaptive behavior in chimpanzees that is similar to the early hominins. Their similarities to our human ancestors in terms of environmental challenges, genetic makeup, encephalization, and body size make these animals suitable models for paleoanthropology and archeology as well. The hypothesis that the excavation of plant caudices and their consumption is a key factor in the morphological and technological evolution of humans can be further explored (Hernandez-Aguilar et al. 2007, 19210). This could also be the case for one of the aforementioned key factors in human evolution, namely, hunting with weapons. Another chimpanzee population, in Fongoli, Senegal, also living in the savannah, uses small wooden spears to kill small nocturnal primates,

Galagonidae. According to researchers, the production of these spears consists of up to five individual steps and includes the sharpening of the branches. The sharpened branches are then used like spears, not like other probes and digging tools (Pruetz and Bertolani 2007, 412). It may not be clear whether this behavior represents something genuinely new in the chimpanzee behavioral record, as compared to poking for insects, but the authors seem to suggest this. They conclude:

> Such tool-assisted behavioral patterns—both considered to be relevant to the evolution of humans—are unique among chimpanzees and emphasize the importance of conserving and studying rare savanna populations.
>
> HERNANDEZ-AGUILAR et al. 2007, 19213

The high level of complexity in the tool behavior of chimpanzees is thus reflected in the production and specification of suitable tools as well as in a (spatially and/or temporally distant) goal-oriented planning of the actions and even in the use of multi-part tool sets. Such complex behavior was certainly also occurring at the beginning of the tool culture of our human ancestors.

Human cultures, however, are highly cumulative, despite the diversity and complexity of their individual traditions; that is to say, cultural advancement relies on the achievements of entire predecessor generations; these advancements are then increasingly refined. A similar significant accumulation and an associated ratchet effect,[13] however, has yet to be discovered and observed in animals. There are only isolated indications of cumulation within traditions in animals. For chimpanzees, we know that some use wedges to support anvils in order to set them (see fig. 3).

This represents a further development in the sense of an additional improvement in the technology of cracking nuts with the aid of a hammer and anvil. Aside from chimpanzees, there are, for example, also initial signs of cumulative technology evolution in the case of New Caledonian crows (see Section 4.5).

4.3.3 *Underlying Learning Mechanisms*

For cultural behavior, social learning is required. This also occurs in animals. For instance, young chimpanzees obviously learn by the observation of older conspecifics. Matsuzawa et al. (2001) coined the term "master apprenticeship"

---

13 In cultural development, the ratchet effect refers to the accumulation of certain modifications over time. We do not need to start from scratch but can build on the conserved achievements of predecessors (Tennie et al. 2009; Tomasello 1999).

to characterize this cultural learning process in chimpanzees. The most patient and tolerant masters of tool use, most often the mothers, do usually not teach the novice learners to imitate them directly, for example, by guiding their hand or clearly encouraging them. However, they nevertheless show exceptionally high tolerance levels in the interaction that they do not show in other circumstances, thus giving the novice learners the opportunity to learn from them through repeated, long-term observation and through access to the tools (Hirata 2006, 202).

Still, in the comparative assessment of tool behavior in humans and other animals, it does not seem possible, to withdraw to the position that there is generally no direct influence from "teachers" in nonhuman animals. Watching chimpanzees at Tai National Park, Ivory Coast, Boesch (1991) observed three forms of social influence associated with nutcracking (excluding the widespread sharing of cracked nuts): (1) the encouragement to use tools, (2) the facilitation of cracking, and even (3) active instruction (Boesch 1991, 530). Mother animals can stimulate the nutcracking behavior of their young by leaving hammers or nuts on the anvil. Normally, adult animals take their hammers with them when they are looking for new nuts in the area, since there is a risk of loss of the tool to other chimpanzees (Boesch 1991, 530). Leaving the hammer behind allows the youngsters to try it themselves, which the younger ones among them do in about half of the cases observed. When it came to chimpanzees without juveniles, in contrast to the four mothers observed, they were never observed leaving intact nuts on the anvil. In five cases, according to Boesch, the mothers even placed the nuts precisely on the anvil before moving away (Boesch 1991, 530). In addition to encouraging nutcracking, the mothers directly influence and improve the effectiveness of their young ones' tool use by providing them with particularly suitable hammers and nuts. They do so even though they personally have fewer nuts and even though they can crack fewer nuts than usual with the other, less well-suited hammers while their juveniles practice (Boesch 1991, 530). In some cases, where the juveniles had difficulty cracking nuts, it was even observed that mothers taught them how to handle them properly. Boesch writes: "We observed two cases in which the mother, noticing the infants' difficulties, made a clear demonstration of how to solve them" (Boesch 1991, 530–31). For example, some nuts consist of three separate compartments with cores. One has to turn the nut slightly between the strokes and reposition it in order to hit the next compartment with the notches.

> After successfully opening a nut, Sartre (a young ape) replaced it haphazardly on the anvil in order to attempt access to the second kernel. But before he pounded it, Salome (his mother) took it in her hand, cleaned

> the anvil and replaced the piece carefully in the correct position. Then, with Salome observing him, Sartre successfully opened it and ate the second kernel. Here, the mother demonstrated the correct positioning of the nut.
>
> BOESCH 1991, 531

The second example of such direct instruction went as follows:

Ricci's daughter, 5-year-old Nina, tried to open nuts with the only available hammer, which was of an irregular shape. As she struggled unsuccessfully with this tool, alternately changing her posture, hammer grip and the position of the nut, Ricci was resting. Eventually, after 8 min of this struggle, Ricci joined her and Nina immediately gave her the hammer. Then, with Nina sitting in front of her, Ricci, in a very deliberate manner, slowly rotated the hammer into the best position with which to pound the nut effectively. As if to emphasize the meaning of this movement, it took her a full minute to perform this simple rotation. With Nina watching, she then proceeded to use the hammer to crack 10 nuts (of which Nina recovered six entire kernels and a portion of the other four). Then Ricci left and Nina resumed cracking. Now, by adopting the same hammer grip as her mother, she succeeded in opening four nuts in 15 min. Although she still had difficulties and regularly changed her posture (18 times), she always maintained the hammer in the same position as did her mother. She whimpered whenever encountering difficulties, to attract her mother, but Ricci did not return. (Boesch 1991, 532)

Boesch concludes that in this second example, the mother animal corrects the error in the handling of the younger animal, and that the latter one has obviously understood and accepted this correction. Nina exactly copied the hand position her mother had demonstrated to her.

All in all, according to Boesch, Tai chimpanzee mothers are interested in the behavior of their young in doing this task and influence their attempts, although at the expense of their own, consequently less effective success rate in cracking nuts. They seem to have the ability to compare the young ones' behavior to their own idea of cracking nuts, and to be able to anticipate the impact of their own actions on their behavior. Important functions of instruction as we know them in humans are thus fulfilled: "Recruitment, direction maintenance and marking critical features are apparent in the stimulations, frustration control and scaffolding in the facilitations as well as in teaching alongside with demonstration. These acts are adjusted to the level of skill attained by the infant" (Boesch 1991, 532). The drive to use tools intensifies and reaches its peak when the juveniles are three years old and learn basic nutcracking skills. Assistance in the sense of the described handling facilitation also begins when the young chimpanzees are about three years old and is observed up to a maximum of five years of age. This may be due to the fact that it takes about ten years for the mothers to actually master nutcracking themselves. Other cultural behaviors in chimpanzees, according to Boesch, require less intense practice and a shorter learning period. Mothers are also increasingly disadvantaged because their own nutcracking becomes inefficient by supporting that of the young. There is increasing pressure to accelerate the acquisition of behavior in young animals before the next infant is born. The mothers approach this by taking direct influence on the learning process of their young (Boesch 1991, 532).

Experimental evidence for the transmission of traditions and social learning in chimpanzees has also been produced in captivity (Whiten et al. 2005, 2007). Japanese researchers, for example, have shown in their experiments that in rare cases, the tool is even given to the learner, and that young chimpanzees increasingly use the very same tools that the adults they have been observing have used. It is therefore probable that tool selectivity will be passed on (Hirata 2006, 211). Hirata et al. (2009) were able to show through which individual steps tool behavior is passed on in captivity by observing a trained male chimpanzee in a nutcracking situation with novice conspecifics. Although there was no control group without a trained animal, and it cannot therefore be said whether and how they could have solved the problem, for the observed group the ability appeared to be passed on by observing the model:

> The results showed that the process of acquisition could be broken down into several steps, including *recognition of applying pressure* to the nut, emergence of the *use of a combination of three objects*, emergence of the *hitting action*, using a *tool for hitting*, and *hitting the nut*.
>
> HIRATA et al. 2009, 1; italics added

The chimpanzees recognized these different components and practiced them one after the other. They would have combined these factors in their actions step by step, which would have led to their first success. Their strategy would not have improved clearly and immediately after the observation of an experienced conspecific. However, the mutual observation seems to have a gradual, long-term influence on the acquisition of nutcracking by naive chimpanzees (Hirata et al. 2009, 1).

4.3.4 *The Variety of Traditions*

Termite fishing and nutcracking are well-researched examples of traditions in chimpanzees that are passed on socially and differ from population to population; their occurrence in some but not all populations cannot be explained solely on the basis of genetic or ecological mechanisms and factors, and they include a high level of complexity, including cumulative development. But they are just two of many examples of chimpanzee traditions that, taken as a whole, justify our speaking of culture in these animals.

Whiten et al. (1999) created a synopsis of a total of thirty-nine traditional chimpanzee behaviors, based on decades of work by researchers in the six best-studied populations. The recorded traditional behaviors include very different things, from rain dancing and the use of leaf wads as sponges to hand clasping, a characteristic hand-holding position during grooming. The latter represents the first documented *social* tradition in chimpanzees (McGrew and Tutin 1978). The traditions of chimpanzees are thus clearly not limited to tool use, and they do not focus solely on the context of food acquisition. In chimpanzees, for example, traditional self-medication behaviors can be found through the consumption of certain plants. This way, they have developed a method to rid themselves of parasites. For this purpose, rough, sharp-edged leaves of certain plants are torn off, carefully folded up and swallowed whole. Fecal examinations have revealed that the leaves are excreted undigested, together with numerous intestinal parasites. It is believed that the rough leaf structure scrapes the worms from the intestinal walls. Also, tannins contained in various leaves have a laxative effect, such as in the case of the bitter marrow of the bush *Vernonia amygdalina*, which grows in the west of Tanzania and is eaten by chimpanzees (Koshimizu et al. 2003, 1209). Often, the plants used for self-medication by animals are also used by the human population living in these areas. It is expected that the use of such plants by animals will provide insights into as-yet undiscovered active substances that can also be used by humans, such as for migraines and malaria. Self-medication is also not an innate but an acquired behavior that is only handed down by some populations. Migrating animals learn the behavior of conspecifics (Koshimizu et al. 2003; Huffman 2003; Huffman et al. 1997; Huffman and Wrangham 1993; Schuster et al. 2007, 56, 66).

Regarding the variety and complexity of traditions in chimpanzees, Whiten et al. (1999) summarize the meaning of their composition as such:

> This comprehensive analysis reveals patterns of variation that are far more extensive than have previously been documented for any animal species except humans….We find that 39 different behaviour patterns, including tool usage, grooming and courtship behaviours, are customary or habitual in some communities but are absent in others where ecological explanations have been discounted. Among mammalian and avian

> species, cultural variation has previously been identified only for single behaviour patterns, such as the local dialects of song-birds....The extensive, multiple variations now documented for chimpanzees are thus without parallel. Moreover, the combined repertoire of these behaviour patterns in each chimpanzee community is itself highly distinctive, a phenomenon characteristic of human cultures...but previously unrecognised in non-human species. (682)

Traditions—including forms of self-medication—also occur in gorillas and orangutans, other primates, birds, as well as in many other animals (Lozano 1998). However, in most cases it is unclear whether these traditions fit into a whole set of other (and sometimes different) traditions in these species, or whether they are rather sporadic and limited to specific contexts (such as food acquisition). Even as such, they would be remarkable examples of socio-cognitive abilities, especially if they are actually passed on socially and are individually varied. However, the predicate "cultural" also requires complexity and diversity within a set of individual traditions.

4.4 Culture in Orangutans, Gorillas, and Dolphins

Although the traditions of chimpanzees have so far been the most intensively researched and, because of the multitude of such behaviors, culture is most clearly spoken of here, one should not overlook the fact that similar behaviors occur in other great apes—in the case of orangutans, even similarly complex and diverse behavior. In this instance, field observations provided evidence of at least nineteen innovations that remain stable over time, are highly complex, and also serve different purposes in the contexts of survival, the improvement of living conditions, and social communication (van Schaik et al. 2006). For example, orangutans use small sticks that they hold between the lips to release the seeds of Neesia fruit from their capsules (van Schaik et al. 2006, 2003). Use of tools in the wild has also been observed in gorillas (Breuer et al. 2005). In 2005, for the first time, Breuer et al. confirmed tool use in a free-living population in northern Congo. The gorillas used sticks, branches, and small tree trunks for different purposes. Sticks were used as walking sticks and to test the depth of water during upright walking through the swamp. Branches and small tree trunks served as support and to better maintain balance when collecting food, or as bridges when crossing water surfaces (Breuer et al. 2005). Here again it is very clear that, among other things, the habitat type can cause innovations in behavior. Overall, however, gorillas are difficult to observe in the wild as they live in hard-to-reach areas in central Africa, some of which are in the midst of war zones.

Behavioral science has also achieved interesting new results in recent years in the case of whales and dolphins. Australian dolphins (Tursiops) in Shark Bay, for example, use marine sponges as nasal protection while they survey the bottom for food. Animals that do this are called "spongers." This behavior is passed on matrilineally by mothers to daughters and presumably began with an ingenious female. Environmental factors do not appear to be responsible for this behavior, as the use of sponges occurs in deep water canals in the region, where nonspongers also look for food. In addition, similar biotopes elsewhere do not produce spongers. Again, this is not genetically determined information; the offspring learn this characteristic behavior by observation. Krützen et al. (2005) performed DNA tests and examined the genetic material of 185 dolphins (thirteen of which were spongers) with regard to ten different inheritance mechanisms. A study was carried out on twelve hypervariable gene loci (the contents of which are inherited from both parents) as well as a study of mitochondrial DNA (which is only passed on by the mother). The result was that, with the exception of one male, only some females use the sponges. These animals are related and the genetic lineage from a single common ancestor emerged only recently (as indicated by haplotype H of mitochondrial DNA). In one case, however, the behavior had just been taken over by a young male offspring of one of the female spongers. The analysis of the mechanisms of inheritance revealed that the use of sponges is unlikely to be genetically coded. Possible inheritance mechanisms would not have been an explanation for the distribution. The authors point out that observing the social learning of dolphins in this area is very difficult due to the presence of sharks and so far had not been recorded. Dolphins in captivity, however, would show social learning behavior and the ability to imitate behavior. However, it would be possible in principle for the sponging behavior to be learned by all juveniles (female and male), but then unlearned by some (especially by males). Possible reasons for this include differences in social behavior that do not allow males the time to use sponges (Krützen et al. 2005).

It is important that the behavior occurs in a nutritional context but is not directly linked to a nutritional problem-solving. For the complexity of the current thinking process, this means that the cognitive distance between the basic problem and the solution is significantly extended, whereby it comes to a delay in fulfilling a fundamental need, by observing and pursuing a less significant subordinate problem (Haidle 2012, 197–99).

Dolphins also demonstrate many other complex behaviors that may be present in different traditional forms in individual populations, subpopulations or groups of animals. This is also the case with hunting strategies other than that of using sponges. In interaction with humans or independently of them, individual groups of dolphins have evolved social and individual, site- and group-specific, savvy and unusual hunting methods by following fishing boats to obtain discarded fish, visiting provisioning locations, and catching fish cooperatively with net fishers. They use their rostra to dig into the substrate, smack their tails on the water surface over shallow seagrass beds to disturb prey, whack fish with their tails, and stir up sediment to trap fish (Sargeant et al. 2005, 1401). Some dolphins also throw themselves in sync onto shallow stretches of beach in order to catch the fish that have been driven in front of them on their own bow wave (so-called hydroplaning). Preliminary research seems to support the thesis that group-specific hunting strategies for dolphins are actually largely socially transmitted (Sargeant and Mann 2009).

Other whales also show a high variance of specific hunting methods: Recorded foraging techniques include lobtail and bubble net feeding in humpback whales, cooperative hunting and beach feeding by killer whales, and bird-associated foraging and lung feeding by minke whales (Sargeant et al. 2005, 1400–1401). In the case of dolphins and other whales, therefore, there is research in relation to such (even traditional) differences—as in the case of chimpanzees and orangutans—which also speaks to the idea of "culture." In this instance, Rendell and Whitehead (2001) presented a compilation of cultural behaviors that, in addition to the above-mentioned hunting strategies, emphasizes, above all, the individual songs of the whales.

4.5 Culture in New Caledonian Crows

In recent years, however, behavioral research has gained new insights that point to complex traditions not only in mammals. Tool use has also been increasingly investigated in the case of hook probes used by New Caledonian crows in the laboratory and in the wild. Weir et al. (2002) observed, for example, a New Caledonian crow that bent hooks out of wire pieces in a controlled experiment. The crow used the curved hooks to pull small food pails up the bottom of a glass tube by its handle. It displayed this behavior without having previously gained experience with the situation and the materials.[14]

---

14 The crow Betty had never come across such pieces of wire before, but she already had experience with bendable material, since one year before the experiment she was allowed to work freely with pipe cleaners for about an hour (Weir et al. 2002).

Even in the wild, New Caledonian crows use spikes and hook probes to access maggots in tree trunks. Hunt and Gray (2003) noted that the crows they observed in the wild in New Caledonia changed and cumulatively improved the design of specially adapted pandanus leaves. From straight spears, over time they developed more efficient, multistage skewers over several intermediate stages (Hunt and Gray 2003, 867, 872–73). All skewers are made by tearing off a strip from the leaf, with the hooks of the leaf located on the outside of the leaf later being used to skewer, hold, and pull out the food. First, with their beak, the crows tear the leaf, then grasp the torn piece and tear it down lengthwise as a strip. With this "cut and rip" technique, they have developed their design from the wide tools (which are stable but too wide for narrow holes below) to narrow tools (which are narrow enough at the tip but elusive and no longer stable enough) to one-step tools and multistep tools. The latter have one step or even multiple steps, so they are always narrower at the top, but still wide at the end of the grip and stable in themselves (Hunt and Gray 2003, especially fig. 1 on p. 868).

Hunt and Gray (2003) assume that the designs have emerged from each other. As important characteristics of cumulative change, they assume that a fanning out or diversification of the original tool design is present, that an increasing change in the design is identifiable, and that there is a precision to the social transfer of a design. The later designs must also be significantly more efficient than the previous ones, and earlier designs gradually becoming uncommon in favor of later, improved versions. All of this, the authors say, would affect the pandanus skewers (Hunt and Gray 2003, 867, 872–73). Furthermore, purely ecological explanations for the three designs can be ruled out; i.e., the tool designs do not correspond in each case to different places of use or other circumstances. Considered factors such as raw material variation, altitude above sea level, or rainfall are not significant causal factors for the different geographic distribution of designs. Also, the physical properties of the leaves (their length, thickness, and width) do not affect the design, and even in most cases several designs have been made from one leaf. It seems unlikely to researchers that the different designs emerged independently. The similarities in the method of production suggested that they had gone through a process of cumulative change. The high accuracy of the design reproduction is a given in the respective locations: In eighteen of twenty-one locations, 70 percent of the tools were of this design type that had apparently prevailed in the majority of cases (Hunt and Gray 2003, 870ff.).

It is certain that the ability to make and use tools is at least partially innate when it comes to crows. But social input still plays an important role in communicating and developing specific techniques and designs. In tool-use experiments, for example, young crows preferred to use objects whose use could be observed by another subject (Kenward et al. 2006, 1340; Kenward et al. 2005).

Furthermore, New Caledonian crows are not only capable of complex tool use but also of sequential tool use or meta tool use. This means the use of a tool *on another tool* in order to achieve a goal. Crows solve recursive tasks such as using a short stick to gain a long stick so they can push it out of a box (Taylor et al. 2007).[15] In order to test this, Taylor et al. (2007) constructed a test setup in which meat was hidden in a horizontal hole fifteen centimeters deep. It was impossible to reach only with the beak or with a short stick. Two identical wooden boxes, 1.75 meters apart, were provided with vertical bars on their front sides, between which a beak (but no head) fits. In box 1 was a stick four centimeters behind the bars. In box 2 was a stone (unusable for solving the problem) to exclude trial and error. In front of the two boxes was a short stick. The question was whether the crows would use the short stick to fish the long pole in box 1 in order to fly to the feeding hole to extract the food. All seven crows showed the required sequential tool behavior, six of them attempting to reach the long stick with the help of the short one on the first test run. Their method was not trial and error but an analogous conclusion. Three of the animals already succeeded in the first attempt and only one crow tried to pull out the useless stone (Taylor et al. 2007, 1504ff.). In control experiments, the positions of long and short stick were reversed. Although all the crows still put the long stick in the box first but usually did not touch the short stick, only one crow pulled it out; it did not, however, then carry it to the food. Taylor et al. (2007) think it is possible that the insertion of the long stick was due to habit, as this behavior had been rewarded by reaching the food. However, the immediate, spontaneous, and rapid correction of errors in an adequate, goal-oriented manner shows a causal understanding of the relationship between tools and the ultimate goal (Taylor et al. 2007, 1506).

Even though, in the case of New Caledonian crows, complex, at least partially socially-transmitted behavior is present, it must be clarified that, at the level of the current state of research, such tool behavior seems to be limited to the context of food acquisition. A variety of different behaviors, even behaviors outside the context of food acquisition, as in the case of the traditions of the chimpanzees, orangutans, and whales, does not yet seem to be documented to the same extent. While this justifies the talk of individual traditions, the

15 For a follow-up study on sequential tool use in New Caledonian crows, and for proposals concerning new terminology in the debate on the use of meta-tools, see Wimpenny et al. (2009).

"culture" of the crow we know so far seems to be primarily a food culture or tool culture in terms of diet. It should not be forgotten, however, that many other socio-cognitive activities have been observed in corvids. In Chapter 6, it will be argued that some of these—whether in the context of food acquisition or not—indicate that they have social knowledge that in some ways may be comparable to that of chimpanzees. Birds in general have long been underestimated for their cognitive abilities because they have no neocortex similar to that of the human brain. Only in recent years have new brain models been developed that can better explain homologies between the brains of birds and the human brain. In 2005, the Avian Brain Nomenclature Consortium proposed a rethink of the nomenclature and theory of the traditionally prevalent model of the brains of birds. According to the prevailing model, stated the researchers, the brain of birds is believed to consist almost exclusively of basal ganglia associated with instinctive behavior, while lacking a neocortex, which is required for the flexible behavior of mammals. However, toward the end of the twentieth century it became increasingly clear that this simple assignment is not correct and that the brain of birds can nevertheless control similarly complex behavior via structures other than a neocortex (see Jarvis et al. 2005, Güntürkün and Bugnyar 2016). It would not be surprising if in the next few years further interesting research results were obtained on the complex socio-cognitive abilities of corvids.

4.6 Can We Speak Meaningfully of "Culture" in Animals?

Overall, many species of animals display tool behavior. Among the vertebrates, according to Haidle (2012), only reptiles and amphibians, for which to date no such behavior has been documented, constitute an exception. The functions of tool use across species are also diverse: they encompass for example food acquisition; breeding or raising of young; threatening, defensive, and protective behavior; masking, personal care, increasing one's own well-being; and sexual stimulation. Animals use tools to stabilize their bodies, straighten themselves up, and expand their mobility. Whether they decorate themselves with objects, which could be the case when it comes to the antlers of different types of deer or the draping of apes, for example, says Haidle, is unclear. However, van Leeuwen et al. (2014) found evidence for a spontaneously emerged tradition with presumably no adaptive value in chimpanzees.

Tool behavior occurs in the context of solitary and social play, where tools are also used as a social stimulus (to establish contact, as an aggression buffer, or to attract attention) (Haidle 2012, 165). Thus, there are actually many examples of tool use in animals. Other culture-related behaviors not related to tools, which could not be discussed in detail in this book, include group-specific preferences for colors and shapes (referred to in the case of the bower birds) or group-specific chants (e.g., in humpback whales).[16]

In summary, on the question of culture in animals, it can be said that many nonhuman animals—and not only mammals—have individual, *clearly distinguishable, socially transmitted* behaviors between different populations, subpopulations, or groups within species. In this sense, it can actually make sense to speak of *traditions* as clearly defined and empirically supported in some animals. It is a matter of *complex* behaviors that are clearly directed at one *ultimate goal* and are followed *flexibly*. Genetically preprogrammed, stereotypical behavior must be excluded. In some species—according to current research, especially in chimpanzees, orangutans, dolphins, and other cetaceans—many different such individual traditions are now known. These are not only limited to a domain such as tool use or a context such as food acquisition, but also concern health-care and self-medication or social interaction (as in the case of handholding positions, welcome rituals, or hierarchy demonstrations). In addition to the complexity of individual traditions, there is also a certain diversity present in the set of traditions that, taken together in their unique composition, can form the culture of a community. In some species, therefore, culture is now spoken of in this sense (Whiten et al. 1999; Matsuzawa et al. 2001; Boesch 1996; van Schaik et al. 2003; Rendell and Whitehead 2001). Thus, a distinction can be made between species that possess a set of diverse traditions in the sense of culture and species in which we have hitherto only referred to individual traditions in individual domains (such as traditions in the use of tools) or in individual contexts (e.g., traditions in diet).

Differences between species seem fewer with regard to the complexity of individual traditions, but do seem to exist with regard to the *diversity* of existing traditions within a species. The question has to be left open, for example, as to whether or not New Caledonian crows have "culture" in the sense of diverse, different traditions or only when it comes to individual complex traditions. The impression that the latter is the case could be due solely to the still less-than-ample state of research. On the other hand, there are no binding criteria for when traditions are *varied enough* to be able to form the required unique culture set. However, it should be understood that at least two, if not more, different domains and contexts should be involved, and that not all culturally suspicious behaviors should be considered to be in the domain of tool use or in the context of food acquisition. At the very least, "the big picture" of

16 For a detailed discussion of the latter, see Rendell and Whitehead (2001).

Tylor's cultural definition can in a sense be translated as a criterion for nonhuman animals only if one requires more than *one* tradition for culture. It would not be important, then, that animals display the big picture of *human culture* (from the outset, Tylor's definition was oriented to human beings and required *their* combined abilities and habits), but that when it comes to culture, the matter always concerns a *complex of different cultural realms*. The notion that, in order to be able to speak of culture, this concerns precisely those realms mentioned by Tylor and also the total of all of these realms can certainly—even if it concerns humans—be contested. It could also be contested that a *qualitative difference* exists between two species, such as between the chimpanzee and the New Caledonian crow, of which one only has traditions known to us in one context (such as food acquisition), or primarily in one domain (namely, the use of tools), and the other features multiple traditions in different contexts and domains. The argument for the lack of diversity of traditions, which can be attributed to traditions but not yet culture, seems more to designate a *quantitative* difference. Finally, in some species, such as the New Caledonian crow, the intensity and complexity of handling tools such as skewers and probes, as well as existing signs of cumulation and sequential tool use, may justify applying a limited concept of culture. It is true that one cannot speak of culture in the same broad sense as in chimpanzees and cetaceans, but perhaps at least of a *tool culture* in the sense of a *food acquisition culture*.

However, apart from the lack of criteria for a sufficient variety of traditions, interpretation schemes are still lacking with regard to the necessary complexity. Still, initial systematic considerations have been made. In order to assess what an animal does when it comes to complex tool behavior, Haidle (2012) developed a scheme that reveals the stages of complexity—in comparison between today's living animals and the extinct ancestors of the genus *Homo*. The cognitive complexity of some tool behaviors becomes evident only, so her assumption goes, if one describes the creation and use of tools in its smallest units of action. Also, we need to consider the perception of needs and problems that precedes any action as well as the subject's changing focus of attention during the action (human or animal). This needs to be integrated into a coherent and sufficient description or schematic representation of tool use behavior (see Haidle 2012, 143ff.). Only then does a representation of mere action sequences become a real *thought-process diagram*, or *"cognigram"* as she calls it, with several individual phases. This can then satisfy a multiply-nested process such as the production of an Oldowan chopper.[17] In the schematic

17 In archeology, choppers are usually designated as round or oval pebble tools of the early Paleolithic, whose cutting edge was produced by working an edge on one side. Choppers are among the oldest stone tools of humanity (Semaw 2000).

schemes Haidle proposes, it becomes clear which problem level or which focus of attention is open in a certain action stage[18] or moves to the center of attention. Different focal points of attention may be directed to, for example, the acting subject itself, the tool (e.g., a hammer or an anvil, which may also be focused as a location), or the object to be processed (e.g., a nut) (Haidle 2012, 162–63).

Tool behavior can then be reflected in the number of active and passive attention foci[19] and the successive linking of the same. In comparison, for the case of the complexity of nutcracking in chimpanzees and the production of an Oldovan chopper, this means that during nut-cracking, only two foci are active, i.e., have to be controlled by the subject with respect to their modifying effect: the subject itself and the hammer as a tool. While anvil and nut are allocated their own foci of attention, they remain passive; none of these action elements effect any change (nut) or the change cannot be controlled by the subject (anvil). While the hammer affects the nut in a controlled manner, the anvil remains unaffected. By contrast, the production and use of a chopper requires three active foci of attention: the subject, the chopper as a tool 1, and the percussion stone to manufacture the chopper as a tool 2. This simple example already demonstrates the difference between a completely unmodified, used-as-found tool, such as the hammer or percussion stone, and a modified, manufactured tool, such a chopper, a flake, or a twig probe. (Haidle 2012, 162)

However, this does not mean that animals are not able to master thinking processes that require multiple active foci and involve tools that are controlled by the subject in their effect. While hammering and anvil nutcracking is only a controlled effect on the nut, in the manufacture and use of a probe there are already two foci, which as a consequence have a changing influence on the process (the subject on the probe and the probe on the termite nest). Possible feedback, resumption of a previously manifested tool in the chain of action, and the course of further problem perception outside of the primary chain of action can make a cognigram even more complex. For example, according to Haidle's analysis scheme, the use of a termite tool extraction set, as observed

18 Phases are defined as process stages, which can be considered as a combination of closely related individual actions with an intermediate result, such as the search for raw material (e.g., a suitable hammer), the use of the tool (e.g., on a nut) or the satisfaction of needs (e.g., the eating of the beaten nut) (Haidle 2012, 158ff.).

19 A focus here is an attention focus of the subject. An A-focus (active focus) is *controlled actively*; it may refer to a physical or psychic need of the subject or to a tool or various variables of the same. In contrast, a P-focus (passive focus) cannot be actively controlled and may be a place or an object (Haidle 2012, 164, fig. 22). For real tool behavior in contrast to proto-tool behavior, the active effect of media and their control was required (see Parker and Gibson 1977; Section 4.1 above).

by Sanz et al. (2004), is a rather complex example of tool behavior in animals (see fig. 4) (Haidle 2012, 211ff.). Here, she says, in addition to the subject focus and the object focus, two tool foci were already opened in parallel at the beginning of the process: that the animals sometimes bring both tools to the termite nest verifies "the foresight to recognize the need for both action components, perforation and extraction." However, in the case of the nasal protector used by bottlenose dolphins and the support used when harvesting aquatic plants by Western lowland gorillas, four simultaneously opened focal points of attention could be determined, but there, two would be sub–focal points of the same individual, namely the acting subject. In the case of termite extraction sets, the individual for the first time involves "three external objects in parallel in a single action phase". This, together with the anticipation of several elements of action, represents a "significant expansion" of the distance between problem and solution (Haidle 2012, 213). The increasing distancing of problem and solution in more complex thought processes is crucial for Haidle. In fact, the perception of one or more needs and problems and their solution with the help of a tool can sometimes be very far away in terms of time. Especially when problem-solving is not based on a ready-to-use object but rather on an object to be

Use of a tool-set to extract terminates by *Pan troglodytes* (after Sanz et al. 2004)

0. Perception of basic need: feeding
0a. Perception of sub-problem 1: open termite nest/exract termites
0b. Perception of sub-problem 2: tool necessary to open nest
0c. Perception of sub-problem 3: tool necessary for probing

PHASE I: manufacture of probe I
1. Search for appropriate twig

PHASE II: manufacture of probe II
2. Detaching the twig
3. Shortening/removal of leaves/fraying of brushtip

PHASE III: transport of probe
4. Transport of probe to termite nest

PHASE IV: search of chisel
5. Selection of chisel on site

PHASE V: opening the termite nest
6. Pushing with chisel (several times)
7. Inspection of chisel

PHASE VI: probing for termites
8. Extraction of termites with probe

PHASE VII: satisfaction of need
9. Consumption

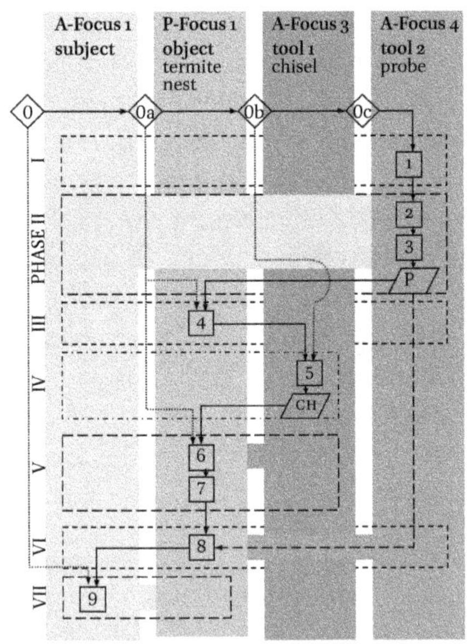

FIGURE 4 Cognigram for the use of a termite extraction set by chimpanzees
© M. HAIDLE. SOURCE: HAIDLE 2012, 212

worked on, or if a behavior refers not only to one context but to several contexts; some case studies therefore reach higher levels of complexity. In the case of dolphins using sponges as nose protection, the behavior does indeed occur in a food acquisition context, but it is not directly linked to the solution of a food problem (see fig. 5). Haidle states: "A tool is not necessary to satisfy the basic need for the object 'food'... however, the additional need to protect the skin during the search for food is satisfied using a medium, for which a fourth focus of attention must be activated." She rightly notes that the cognitive distance between the basic problem and the solution is clearly broadened and that the main need satisfaction is significantly delayed (Haidle 2012, 199).

The number of attention foci and the number of intermediate steps, as well as their distance to the end goal, thus represent the mental and practical detour between problem and solution. In this case, direct answers to the main goal (namely to reach a piece of food) must first be suppressed. The animal must specifically address subproblems (such as the finding or production of a meta-tool or skin protection) and only reaches the solution of her problem by means of interim results with a time delay. This requires the ability to apply hierarchically organized thinking and action. Taylor et al. (2007), in their study on meta-tool use in New Caledonian crows, also think that in this instance, a newly introduced behavior (the application of a first tool to a second tool) is being flexibly integrated into an established behavior as a sub-goal in order

Use of a sponge as nose protection by *Tursiops sp.* (after Krützen et al. 2005)

0. Perception of basic need: feeding
0a. Perception of sub-problem 1: access to food
0b. Perception of sub-problem 2: sediment of sea floor hurts nose
0c. Perception of sub-problem 3: protection necessary

PHASE I: search for tool

1. Search for appropriate sponge

PHASE II: production of tool

2. Detaching the sponge
3. Putting it over the nose/protection

PHASE III: scaring up prey/use of tool

3. Putting it over the nose/protection
4. Scaring up prey from ocean floor
5. Catching the food

PHASE IV: satisfaction of need

6. Consumption

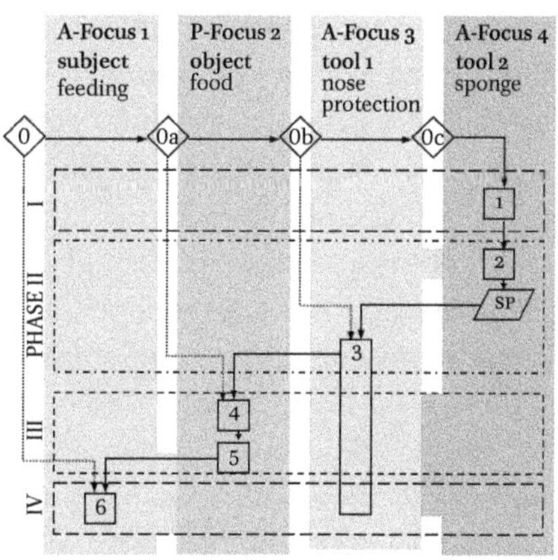

FIGURE 5 Cognigram for the use of a sponge as nose protection during the removal of food by dolphins

to reach the main objective (the application of the second tool to get the food). According to Taylor et al. (2007), structurally this is about producing a recursive pattern, which requires cognitive processing similar to that used in speech production (Taylor et al. 2007, 1504). In principle, behaviors that are fundamentally based on food acquisition can—when analyzed as thought processes in all their details—refer beyond this context and should not be underestimated.

Haidle (2012) sees a difference between the tool behavior of animals and that of human ancestors, especially in the cumulative development of tool designs. The tool behavior and traditions of the latter have rapidly gained not only diversity but also complexity. The number of simultaneously opened focal points of attention and the number of individual phases involved (and thus the complexity of the cognigrams) increases significantly in the production of spears, for example. In this instance, one can even speak of a decoupling of a tool from the original need, since one and the same tool is perfected and is used over and over again, even in completely different contexts—and this rarely occurs with animals. Human beings have created "living tools" that are kept, repaired, and optimized for later use. The decoupling of concrete needs and satisfaction, moreover, would have made room (for example) for the development of aesthetic and symbolic devices or for the development of novel composition devices, e.g., spearheads that were used at the same time as weapons and as cutting devices (Haidle 2012, 219ff., 259ff., 279ff.).

The increasing and impressive complexity of tool behavior in animals and its further improvement and propagation in the production of human stone tools, hand axes, and hunting equipment is clearly and demonstrably shown by Haidle on the basis of her cognigrams. However, major cognitive innovations like systematic devices (such as bows and arrows and spear-throwing devices)[20] did not appear until eighteen thousand years ago—and thus not at the dividing line between animals and human beings (Haidle 2012, 282). Above all, however, basic components of complex structures of human tool use are already present in the cognigrams of some nonhuman species. This can be seen in their broad problem perception, in their insight into problem-solving, in their flexibility and innovative ability, in the number of action steps and the number of phases of the action processes, in the activation of several passive and active attention foci, in the number of interacting attention foci, and, finally, in the distance between problem and solution that can be ascertained in

20 Such tool complexes could, according to Haidle, be regarded as an extension of compositional devices, their individual elements being coordinated and suitable only in combination (Haidle 2012, 282).

the sense of mental and practical detours. Although Haidle (2012) concludes that in human evolution the problem-solution distance has widened to a complete "decoupling of a specific need and a tool for its satisfaction" (277), rather gradual steps and gradations seem to emerge.

It is an open question which processes accelerated the transmission of increasingly complex and diverse traditions in human beings. Certainly, more social interaction–oriented abilities play a role (their presence in nonhuman animals is examined in Chapter 6). Shared attention, mutual support, direct social guidance, quasi-unconditional imitation, and certainly the development of effective language communication all seem to be important here. In fact, all these abilities are basically also found in animals. As in the case of human children, the first acquisition of complex abilities takes place in chimpanzees in certain sensitive childhood and adolescent phases and in close mother-child interactions (Lonsdorf 2006). Social animals learn in part over many postnatal years from parents and other conspecifics. Imitation also seems to play a significant role in learning (for example, in the case of chimpanzees, see Boesch 1991; Whiten et al. 1999). However, there are differences here: for instance, human children seemingly put greater trust in imitation and cultural conventions than chimpanzees—which suggests the depth of the cultural dimension of human social life. Even in an attempt in which the copying of a proposed act 1, as opposed to an alternative, unannounced act 2, is obviously meaningless, children tend to copy. In this instance, apes swing from imitative behavior in a pragmatic manner to emulative[21] behavior (Horner and Whiten 2005). Here, too, the difference seems to lie in the intensity of the imitation, not in the fact that imitation basically only occurs in humans and not in other species.

Cumulation, multistage and sequential tool use, concurrent focus of attention, social learning, and even active teaching and imitation are also found in some animals, and despite the visible differences in the intensity or complexity of cultural behaviors, the impression is that at least some animals have culture. Existing differences with the ancestors of humans and with *Homo sapiens*, especially with the human child, seem to be more gradual—although increasing at certain points in cultural evolution.

Certainly the cultural dimension of human social life in many aspects has an enormous depth and complexity that goes beyond what we have seen in animals so far. Nevertheless, the distinction "cultural" cannot be limited exclusively to achievements such as the construction of Notre-Dame Cathedral, the

21 In emulation versus imitation, learning is focused on what the observed model is aiming for as a result of action, not on how exactly that result is achieved.

composition of Mozart's Requiem, or the invention of the PC as the greatest achievements of human cultural evolution (for a similar argument, see Sommer 2007). This would then have to exclude many, many people who can neither conceive of nor construct buildings, nor compose or play music, nor develop or use a PC from the realm of living beings who have or possess culture—not to speak of early forms of culture during the time of hominization. We clearly speak of culture when talking about some of the earliest humans we know of, who produced the ivory sculptures found in the Swabian Alb in the early and middle Pleistocene over thirty thousand years ago (Conard 2007) or left the first stone tools in the Great Rift Valley up to 2.5 million years ago (Semaw 2000). Such early cultures are considered to be the cradle of humanity, laying the foundation for modern forms of technology and art that have evolved today.

If the *human being* is to be designated a cultural being, the specializations and excellence of individuals cannot be singled out as a yardstick and condition for cultural ability, nor can modern humans be considered in isolation from their own evolution.

In addition, culture is not a form of conscious behavior that human beings simply moved toward in the course of the evolution of *Homo sapiens* but it is also to be regarded on the ontological level. Culture is taught and passed on to individuals on their life paths from childhood to adult, step by step. The ontogenetic development is also culturally influenced and shaped (Rogoff 2003). In the process, the development of culture is embedded in and guided by the social environment—children go through all stages of cultural learning: from the barely cultured newborn to adolescent and adult representatives of a culture they see as their own, with all its specific components and traditions.

Just as one can ask when culture begins on the timeline of human evolution, one can also ask when it begins on the timeline of individual ontogenesis. The answers to both questions can only point to the first recognizable, fundamental components of culture, and they should always focus on the gradual course of the emergence and acquisition of culture; they should have the lowest common denominator in view, as well as appreciating and valuing early and strange forms of culture that are foreign to us. The same is true if one goes a step "farther" in the evolution of humans and moves beyond the last common ancestor of chimpanzees and humans to the culture of other species.

To define culture very simply and at the same time very broadly as socially transmitted behavior, and to view it from an evolutionary and ontogenetic point of view, may therefore neglect some modern cultural achievements of *Homo sapiens*. But it does justice to culture as a gradually emerging and socially shared asset, thus bridging a misleading separation of culture and nature, as well as a decoupling of human beings from other "natural beings" as supposedly the only "cultural being." In the history of philosophy and (philosophical) anthropology, "nature" and "culture" were usually used to describe two very different spheres. The culture of humans was regarded as distinct from nature and understood as something that does not reveal itself in the behavior of other social animals. We need to recognize culture as something that is part of our biology. Considering culture in the less-cultivated ancestors of *Homo sapiens* and even in animals (as paradigmatic "natural beings") might help us on this way.

CHAPTER 5

Language in Animals?

5.1 Language: the Ultimate *Differentia Specifica*?

The ability to speak has probably been the most prominent differentiating criterion between humans and animals since the relationship between the two begun to be considered. In the following, I will briefly introduce the most important basic aspects and functions of language, provide insight into the philosophical research on them, and present their role as differentiating criterion between humans and animals.

The central function of language, the *naming of things,* was already emphasized in the most ancient, mythical-epic texts, such as in Homer's and in Hesiod's writings—in conjunction with the skeptical question of whether the names we give things even truly capture their true nature. The reflection on the meaning of names as well as the reflection on lies as a form of false statement continued to be of interest throughout the ages. These issues have been explored in poetry and tragic drama (such as that of Aeschylus and Euripides) and by philosophers such as Xenophanes, Democritus, Parmenides, and Heraclitus.

Finally, in sophistic linguistic criticism, the attention extends from the names of things to what we would today call a "word," and the *correspondence of name and thing* is clearly required (see Borsche 1995, 1438ff.). These days, one would say that the sounds we use to designate things are *arbitrary*. They do not correspond in any way to the things described, apart from nouns like "cuckoo." For example, a large object may be referred to with a short word and a small object with a long one. The correspondence of word and thing is nevertheless *conventionally determined* and is shared by a linguistic community as knowledge. No one expects that a speaker would suddenly refer to an object that everyone knows as a "plate" with the word "fork." This would lead to confusion and would fail as a communicative process. Normally, however, every speaker would want to share a language in meaningful, understandable, and therefore successful linguistic interaction.[1] The *tool-like character* of language

1 Of course, for reasons of rhetoric, facts can also be deliberately veiled in language and not be named as what they really are.

 | DOI:10.1163/9789004415072_006

as a means of communication was also emphasized in ancient times, such as in Plato's *Cratylus*, the first surviving coherent linguistic-philosophical text in Greek literature. Plato tries to define language according to its function. Speaking means designating. Naming in this sense, then, constitutes a tool for teaching others and at the same time for distinguishing the essence of things (see Plato [after 399 BC Chr.] 1986 and Borsche 1995, 1444–45). Centuries later, in his Organon model, Karl Bühler resorted to such a functional determination of language as a tool (Bühler 1934).[2] Our speech as a tool of intraspecific communication includes all actions that express or want to communicate something, and is usually addressed to one or more listeners. In so doing, verbal language is a very effective means (though it is only one among of several) for complex, intraspecific communication. There are also *nonverbal languages* (such as sign languages) or *forms of communication that are nonverbal*, such as those involving individual gestures or odors.

But in strong distinction to individual gestures, languages are *whole systems of meaning* that require the listener to understand longer, combined meanings.[3] A modern understanding of language describes language as consisting of a multitude of interconnected discrete and distinct sounds, so-called *phonemes*, which *distinguish meaning*. These form (combined) units known as *morphemes*, which *carry meaning* and fulfill grammatical functions; a morpheme can, for example, indicate a plural.[4] As the smallest meaning-bearing units, composite morphemes again form whole words, which then have their own meaning or *semantics*.[5] With the help of grammatical rules (*syntax*), the individual words are combined into longer statements or sentences, which in turn have their own more complex meaning. In theory, words can become

2 Bühler's theory and its applicability to animals is discussed in greater detail in Section 5.3.

3 According to Borsche, in order to define words as signs that are meaningful and sometimes need to be interpreted, and thus to define language as a string in need of interpretation, the perspective is shifted away from the speaker, toward the listener. This shift becomes the starting point for the language concept of Augustine (Borsche 1995, 1446).

4 The word "games" is composed of two morphemes, the root word "game" and the ending "s," the ending having the grammatical meaning or function of the plural. However, the grammatical morpheme that expresses the meaning "plural" can have different forms, for example also the form "es" in "theses."

5 For example, the words "rat," "mat," and "fat" have completely different meanings, even though they differ only according to their (in-and-of-itself meaningless) distinguishing initial sound.

infinitely connected according to grammatical rules[6] and exist in associative or hierarchical relationships[7] to each other.

Ferdinand de Saussure, who cofounded modern linguistics at the beginning of the twentieth century, conceived the linguistic character as a fixed connection of a particular phonetic pattern that signifies a state of affairs (French *signifiant* = "the signifier") and a certain idea that is designated (French *signifié* = "the signified") (de Saussure [1916] 1967). This implies that language is the meaningful expression of an idea and is thus not only related to the things and circumstances surrounding us, but also to something mental. Language *represents thoughts*, and speaking and thinking affect each other. This definition can be found in many variations in the history of philosophy as well (see Borsche 1995; Schneider 1995; Majetschak 1995).[8] Between the signified and the signifier stands the notion or concept that we make of something. This is called a *semiotic triangle*, consisting of a thing (what is the case), a concept (what one means) and a symbol (what one says).[9] Edward Sapir's 1921 definition of language is exemplary for highlighting language as a *symbolic sign system for conveying mental ideas*: "Speech is a purely human and non-instinctive method of communicating ideas, emotions and desires by means of a system of voluntarily

6 Theoretically speaking, we can build and understand an unlimited amount of grammatically correct sentences. Practically speaking, however, we face limits when it comes to concentration, short-term memory, and our general understanding of a linguistic context. Theoretically, therefore, an infinitely long grammatical sentence can be expressed, but it will be understood only to a certain degree of complexity. Nevertheless, since Noam Chomsky at the latest, the *creativity* of language has been emphasized as an infinite use of finite means, and language, understood as such, is being considered an important feature of human verbal language. Chomsky also combined the idea of creativity with the concept of *recursiveness*: a noun phrase can be embedded in a noun phrase, a sentence in a sentence, etc. (see Hauser et al. 2002a; Chomsky 2002). Hoffmann (2007, 29) notes, however, that such an embedding usually has grammatical marks. These control understanding operatively—understanding would otherwise run up against the rather limited nature of our short-term memory.

7 For example, we associate the word "mother" with a whole array of associations, such as "father," "family," "child," "caring," "role model," "giving birth," "becoming," "single parent," "love," etc. The significance of the conceptuality of a term and the multiple ideas that accompany it will reappear in the context of animal concepts.

8 The relationship between thought and speech was even so pointedly formulated that the former is sometimes conceived of as impossible without the latter. The thesis of the language-dependence of thought has already been mentioned in Davidson and Whorf's formulation in the introduction to this book (see Chap. 1). Today, there are some critics of the language-dependence of thought, such as José Luis Bermúdez, who considers thoughts without language possible even in animals (Bermúdez 2003).

9 Such a triadic sign system has often been traced to Ogden and Richards (1923), but has many forerunners in philosophy and linguistics, including Aristotle. For a variation of this idea, see the volume "Metamorphoses of the Semiotic Triangle" in the *Journal of Semiotics* 10/3 (1988).

produced symbols" (Sapir [1921] 2006, 4). But according to Borsche, Augustine already considers human words the "clothing" (in the sense of lies: "disguises") of thinking, and (human) thinking "the imagery of ideas" (see Borsche 1995, 1446; Borsche 1990, 142ff.). In addition, Aristotle already addresses language as a symbolic system in which the semanticity of one's own and foreign sounds as a symptom of the soul's upheaval is abolished. It is instead being replaced by an arbitrary composition of elements that is to be collectively recognized and constitutes a new "symbolic" meaning. Symbolic meaning, according to Borsche, here has a compound meaning, and in substance a *conventional meaning determination* is meant, which refers to arbitrary sounds which a speech community resorts to, understands, and agrees upon. The ability to articulate, which Aristotle describes in the context of his teaching of perception and here in the system of the soul as part of natural philosophy, is something of which humans (along with other species—birds, for instance) are inherently capable. According to Borsche, the human power of voice and articulation is, for Aristotle, only a matter of speech and thought. *Logos* in Greek means both *ratio* (reason) and *oratio* (speech). Aristotle characterizes the human being as *zoon logon echon*, an animal who possesses reason and language alike. He furthermore makes it clear that humans share all parts of their soul with animals except the rational part. Against this background, human speech becomes an expression of reason, something that is always meaningful: language is nothing "natural" but is, instead, "symbolic"—by composition and convention (see Borsche 1995, 1442).[10]

The idea that words represent thoughts retains its power, past and present, right up to modern times. Bacon calls words traces of reason (Bacon [1623] 1857, 654), whereas Hobbes says that words are signs of thoughts,[11] and for most philosophers, sentient beings that have reason and thought are still exclusively human beings. Descartes makes language as the expression of thought the most important distinguishing feature between humans, on the one hand, and animals and inanimate automatons, on the other: animals, he writes in his *Discours de la Méthode* (Part 5, Section 10), could never combine words or other

10 The determination of humans as *zoon logon echon* and the system of the soul can be found in Aristotle's *Nicomachean Ethics*, EN I.7 and 13, as well as EN X.7, 1178a 6–11 (see Aristotle et al. 2009, 18ff., 173ff). Borsche also refers to text passages from *De anima*, *De generatione animalium*, the *Historia animalium* and the *Politeia*.

11 If human sounds are being connected to form signs for thoughts, then they are called language, and their individual parts are names (Hobbes [1655] 1967, 15). For a discussion of the role of language in Bacon, Hobbes, and other philosophers of the modern age, see Majetschak (1995).

signs to use them, as we do, to share our thoughts with others (Descartes [1637] 2001, 105).

According to Majetschak, in English Empiricism language is no longer understood as a representation of things, but increasingly as a sign for ideas or combinations of ideas (Majetschak 1995 1472, i.a. with ref. to Locke [1690] 1975). Spoken and only written language is in this sense in the history of philosophy understood as a *medium of worldview of a language community*. In particular, the French Enlightenment (for example, in the works of Condillac and Rousseau) addresses the social, geographical, and political circumstances that influence individual languages. In addition, consideration is given to the emergence of language from gestures, from mutual assistance, and from the communal nature of human beings (see Majetschak 1995, 1476–77).

With the emergence of philosophy of language in the second half of the eighteenth century, language became regarded less and less as a medium of communication and increasingly as a *constitutional medium of thought* (see Majetschak 1995, 1479). Anthropological foundations for the ability to speak also became a focus of attention, for example in Herder's work. His treatise on the origin of language begins with the statement that humans and beasts have the "language of sensation" in common. Animals, too, produce wild, inarticulate sounds—these, however, only accompany their sensations and are therefore described by Herder as mere sounds of sensations (see Herman 2002, 5–6, 16). All animals, except for the dumb frog, he says, sound out their sensations. However, this would not mean that any animal has itself the actual beginning of a humanlike language (Herder [1772] 2002, 16). Central to human language is its intentional use (Herder [1772] 2002, 17) and connection to reflection and reason: humans, as enlightened beings, are freely active in this enlightened, reflexive state of being and have invented language (Herder [1772] 2002, 31). The sphere of the animal is instinct-bound. According to Herder, each animal has her circle of inheritance, in which she lives and dies, in which she stays for her entire life, and the smaller the sphere of animals, the less they need language. Animals thus have "little or no language." Humans, by contrast, have language because they have no such uniform and narrow sphere (Herder [1772] 2002, 21–22). For Herder, language becomes the actual distinguishing character of our species from the outside, whereas reason is from within (Herder [1772] 2002, 43). Humans, whose instincts are weak, have defects (something other anthropologists following in Herder's footsteps agree on, see Sect. 4.2.4). Humans developed the capacity of language by virtue of their reason, as compensation for their weak instincts, while the instinct-guided animal, in its environmental bondage and fixation, adheres to fixed-action patterns belonging to another, non-rational sphere. Herder's views were very influential but today they seem incompatible with current knowledge about animal communication.

Still, language skills remained insolubly intertwined with the ability to reason, with consciousness, and finally—in the linguistic analysis of the twentieth century—with logic. As such, language remains something profoundly and specifically human in most of the philosophers who were contemporaries or successors of Herder (from Kant to Fichte and Schelling, and later to Frege and Wittgenstein). At the beginning of the twentieth century, Sapir's already cited definition stated, for example, that language, as a decidedly noninstinctive expressive ability, is distinguished from the communicative, instinctively controlled vocalizations of animals. He adds: "Animal communication, if 'communication' it may be called ... is brought about by involuntary, instinctive cries." These vocalizations are not, he reasons, "in our sense, language at all" (Sapir [1921] 2006, 4). Significantly, this description of animals as trapped within their instinctive barriers is still present, and even intensified, in the philosophy of the twentieth and twenty-first centuries, although since the middle of the twentieth century, research into animal abilities by other scientific disciplines increasingly describes animal abilities that go far beyond instinct-led behavior. Since the emergence of philosophy of language (at the latest), the world of humans has become a world formed by language, and it seems that the philosophical theories of the twentieth century, which increasingly emphasized the centrality of language for humans, found it difficult to set limits to the importance of language. Philosophy goes through a linguistic turn in a broad sense. It sees people no longer as substantially *thinking* or *imaginating* individuals, but rather as substantially *speaking* beings, whose peculiar being in the world can only be understood in relation to their linguistic abilities (see Lauer 2010, 311). Language becomes the *world-constituting moment*—this is seen, for example, in Heidegger's work. Here, language is no longer viewed as an expression of prelinguistic abilities; it is no longer understood in terms of its symbolic, expressive character, its communicative function, or its signifying function, but as an existential-ontological foundation (*existenzial-ontologisches Fundament*). It expresses the meaning of the whole (*das Bedeutungsganze*) (Heidegger [1927] 1967, 161). Language announces the arrival of the being itself, at the same time enlightening and concealing it. Language is the "house of being"—and "in its dwelling dwells man" (Heidegger [1946] 1976, 326, 313; transl. MK). Only humans have a linguistically opened horizon of being (Majetschak 1995, 1493); the animal does not have this horizon of being.[12] In the

12 On Heidegger's view that only the human being (above all bound to his ability to speak) has a world, see Chapter 1.

Tractatus Logico-Philosophicus (theses 5.6 and 4.002), Ludwig Wittgenstein also equates the limits of our language to the limits of our world and emphasizes humans as the living beings with the capacity to "build languages" (Wittgenstein 1922, 74, 38).

Language ability has persisted over the centuries as a philosophical criterion of difference between humans and animals. This remains the case, although in recent decades more and more interesting research results have been published on complex communication in animals, some of which I will explain in the following sections. These results seemed to find their way into philosophical debates only sporadically. The field of linguistics, as well, has only considered these results in a limited way. This becomes quite clear when one considers the repeatedly re-edited introductory work by Grewendorf et al. (1999), which is frequently used in the academic training of linguists. The first edition of this book was published in 1987. Under the heading "Basic Questions in Linguistics," the authors also discuss the topic of language in animals, citing, of course, the literature of the 1970s and 1980s. No revision of the content occurred between the third edition of the book in 1989 and the eleventh edition in 1999. Although revision may not have been required for linguistic issues, with respect to the chapter on animal language it would have been desirable. The position of the authors is therefore very striking and dramatically abbreviates the debate about language ability in animals:

> The interesting thing about language is that only the human being has this special means of communication, so that a study of this particular means of communication is a study of specific human characteristics, perhaps even of *the* differentia specifica.
>
> GREWENDORF et al. 1999, 15; transl. MK

A linguist, however, can take a very different position in terms of the meaning of language for humans, its acquisition, and its interaction with other cognitive and noncognitive abilities. Some scholars, therefore, place language in the human cognitive system differently and thus leave significant space for the question of individual components of language or prelinguistic abilities:

> There is no reason at all, and certainly no evidence, to conceptualize language as something completely different from other mental activities.… On the contrary, almost all higher cognitive processes are involved in language activities and also in sensory and motor activities. Therefore, one can assume that language and the acquisition of language are controlled

> by the same information-processing mechanisms as other cognitive activities.
>
> SZAGUN 2000, 4–5

Interestingly enough, an attempt is made here to understand language in the context of a strong connection with other cognitive and noncognitive processes in a living being and not as something detached from them. It is perhaps not surprising that the author of the preceding extract deals in her research with language acquisition in children. When it comes to children or animals, who do not speak like healthy human adults do, it might be very illuminating to inquire into the origins and individual components of language and to pursue a more functional notion of language that is less intellectualistic than most of the orthodox approaches I have introduced so far. As in the case of culture, I will try to sketch what language in animals could mean without pre-supposing an anthropocentric definition of the respective ability in the first place. A less comprehensive review of examples in animals is given here than in the previous chapter. Instead, the focus is first on a specific area of linguistic ability, namely the question of concept possession in animals. It will be shown how, with the help of one specific philosophical interpretation, concepts in animals can be understood and assessed. Afterward, functional and intentional aspects of animal communication will be examined before I come to a concluding treatment of the complex, sentence-like communication of language-trained apes. In terms of content, I will take the following characteristics of language in animals into consideration:

1. The naming of objects (in the sense of naming and categorizing individual items)
2. Language as an expression of thoughts/concepts, as mental representations, organized in semantic nets
3. Language as a tool of successful communication within the social group (which exerts demands on both senders and recipients of semantic and pragmatic abilities and points beyond the expression of instincts and sensations)
4. The semantic function of complex speech characters and their decryption by the recipient
5. The intentionality of the sender
6. The linking of characters into a sentence-like piece of communication (which points to the ability to address things or events in the past or future that are not yet or not anymore present)
7. Structuring and hierarchization through composition rules (which require at least basic grammatical skills)

As in the case of culture, it is not likely that we will discover animals whose forms of communication are exactly equivalent to human language. But this is also not necessary to speak about language in animals in the first place. We can refrain from investigating language in animals according to human language as a total phenomenon. Instead, what we know about our language can be broken down into its significant individual building blocks. These smaller units of language as well as their functions can then be researched in animals. For example, it may be asked whether some species, while not using our vocal apparatus, still use vocalizations or even nonverbal channels to communicate messages with a semantic content. Or it may be asked what the "concepts" used mean for the particular animal and what function they fulfill. Every single characteristic of human language, from its vocal-auditory production to its grammaticality, can be studied individually in animals as well. Certain important features of language will certainly be found in the communication of animals, while others may not.

It is not assumed that the discussions that follow can provide a simple yes-or-no answer to the age-old philosophical question of whether animals have language. Here, however, some important points of discussion can be compiled and weighed to give us an impression of how—if at all—one can meaningfully speak about language in animals. Some conceptions of language and some aspects of human language are not yet under discussion in the case of animals, and perhaps some of these concepts and aspects can never be meaningfully discussed with regard to animals. In the same way, it is theoretically possible that language in animals is differently organized and structured than in humans, and that it therefore includes characteristics that in principle remain inaccessible to us. And last but not least, when talking about animal language, one should not forget that we cannot simply talk about "the" animal, but rather about many different species of animals and their specific and widely varying forms of communication.

As far as language ability is concerned, there are essentially two windows into the minds of animals: a first method examines the diverse natural communication systems of animals. To do this, researchers ideally set out into the field or observe the communication behavior of intact social groups in captivity, where they research the communicative abilities of animals, as they occur under the most natural conditions possible. A second method aims to teach verbal or nonverbal human languages to animals such as parrots, apes or dolphins, and to observe their capacities to master these languages. Therefore, in the field of language in animals, there is again a methodological parallelism in research, as was the case in the research of culture in animals. An overall impression of whether animals have language or important aspects of language abilities can only be gained by discussing the results of both research methods.

5.2 Concept Possession as Prerequisite for Language

The main function of language is to communicate. In order to be able to communicate and to be understood by the members of a language community, language has to operate in terms whose meaning is clear to the speakers and listeners. But not only must the respective meaning of a word be clear; its relation to other words must not be arbitrary. Words are grouped and arranged in categories. For example, banana and apple both belong to the category FRUIT; they can at the same time belong to different subcategories of colors, such as the subcategory YELLOW and the subcategory GREEN. Our language takes account of logical principles in its conceptual categorization, because our statements incorporate causal and relational principles, such as cause and effect or equality and difference. A sentence about an apple and a banana—for example, "These two fruits are similar with respect to their color"—says something about the conceptual categorization ability of the speaker, who has recognized that both objects are close to each other with regard to the category FRUIT and also to another category, namely that of COLOR.[13] As far as these principles are concerned, it is important that we normally require agreement among communication partners. Even if a listener in the aforementioned example may doubt that the apple *is just as* yellow as the banana, she understands, nevertheless, what the speaker meant when he said that both fruits have a similar color. In the following analysis, the focus will be upon the question of whether animals have concepts as a main component of language ability. I will examine this by focusing on the cognitive capacities involved. In the case of tool use, the thought processes underlying a specific behavior were already of interest. But there, concrete materials and objects were involved. In the case of the language behavior, we are operating in a much more intangible area. Nevertheless, here again one can search, for example, for *flexible* linguistic behavior in contrast to

13 In this case, a speaker may well have the concept RED and assess and name red things as equal or different, without there being any certainty that he is therefore also operating with the abstract concept of EQUAL or DIFFERENT. However, this does not mean, as I have said, that one can deny access to the perception-based concept RED (cf. Newen and Bartels 2007, esp. 300, regarding the limits of their interpretation of the concepts of the grey parrot Alex).

rigorous stimulus-response processes.[14] It has already been shown that language is understood essentially as being connected to thought structures and that concepts—as the basic building blocks of language—represent ideas of things, their nature, and their relation to one another. In its function, therefore, language is extremely comprehensible through a model of mental representation, on which I will establish my thoughts below.

In so doing, I largely refer back to Bartels (2005) and Newen and Bartels (2007), who understand concept possession as the possession of *conceptual mental representations*.[15]

I chose their account as it is being developed against the backdrop of and in critical engagement with the most important philosophical contributions to the debate (such as Davidson 1999, Glock 2000, or Allen 1999). However, as they find other proposals unjustifiably demanding, they aim for a de-intellectualized account. This leads them to a proposal that actually bares the potential to describe concept possession in animals as it becomes visible in empirical studies. By closely referring back to them I aim to show how a philosophical interpretation of empirical data can help to get a clearer understanding of what an animal can or can't do. This requires us to have a more detailed look than in the case of the culture chapter on one specific (and unfortunately quite technical) debate. But in the end this will enable us to see to what extent we can actually talk *in a philosophically meaningful and empirically informed way* about concepts in animals.

Let's start with a simple assumption. We can initially differentiate the external world and its facts from the interior mental world of a speaker and her thoughts. What is represented mentally in thought can be present either in

14 Newen and Bartels (2007) also cite flexibility as a minimal criterion for having concepts and try to distinguish cognitive from noncognitive and conceptual from nonconceptual capacities. For cognitive behavior, it is above all important that a stable mental representation be developed and not simply that a rigid stimulus-response routine be executed. After all, even complex routines remain routines. On the other hand, as will be explained below, we are looking for behavioral responses that can be modified in light of new or additional information (cf. Newen and Bartels 2007, esp. 286ff. and 294–95).

15 The authors describe the foundations of their theory as a "representational theory of concepts distinguishing between types and tokens: Tokens of concepts are mental representations. Concepts as types of mental representations are individuated by their content. The content of a mental representation is determined by the mental capacities of the cognitive system having the representation, and can be spelled out in terms of functional roles." To prevent misunderstandings, they clarify that they "are not claiming that concepts are epistemic capacities but that they are individuated by them. Concepts are types of mental representations constituted by epistemic capacities" (Newen and Bartels 2007, 284).

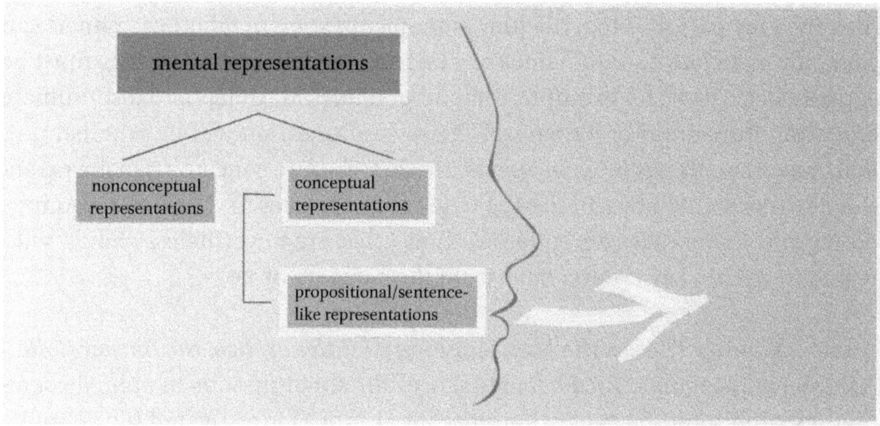

FIGURE 6 Mental representations may be categorized as conceptual and nonconceptual (e.g., in the more pictorial form of mental maps). Propositional representations require the combination of conceptual representations. Illustration based on analyses in Bartels (2005) and Newen and Bartels (2007)

nonconceptual (e.g., figurative) or in conceptual form (see fig. 6). Later, following Newen and Bartels (2007), I will discuss a third type of mental representation, so-called propositional representations.

What exactly does it mean for an individual to have conceptual representations? When a speaker uses the perception-based concept RED one could at first say she knows, if indeed possessing this concept, what is meant by the term, given that at least three important things are clear to her. First, (a) she knows that red is neither blue nor yellow because she knows that red is different from other color properties. In addition, she knows that there is a wide variety of red things, such as red balls, red cars, and so forth. Therefore, (b) she has a stable representation and recognizes it in very different objects, even in completely new objects. Finally, (c) her use of the term "red" also implies that red is an occurrence of the dimension COLOR, and that the category COLOR has different subcategories, all of which are themselves different colors. Real (minimally defined) comprehension is likely to occur only if all three of these criteria are met. Nevertheless, criteria (a) and (b) already describe key tasks that constitute the possession of concepts.[16]

The following two sections turn first to criteria (a) and (b), introducing relevant case studies. They will not, for the moment, treat the issue of perception-based concepts, such as, for example RED (as seen later on in the example of

16 See the remarks of Newen and Bartels (2007, esp. 296–97) on their criteria C1–C4. Elements of these criteria have already been addressed or will be discussed in more detail later in this chapter.

Alex the grey parrot). First, the fundamental question of whether animals can generally associate certain names or designations with certain things must be addressed. Do dogs, for example, remember names for things and discriminate one thing from another by name? The second question will be whether animals can group things into categories. Do they capture concepts that are extensionally represented by the class of objects that applies to them? For example, do pigeons know that very different things that are nevertheless similar with regard to certain key criteria fall within the category of TREE?

5.2.1 *Naming Things: the Fast-Mapping Abilities of Rico, the Border Collie*

A border collie named Rico became one of the shooting stars in animal cognition research some years ago. Kaminski et al. (2004) investigated Rico's ability to name more than two hundred individual things: at the command "Bring the sun!" the dog ran into an adjoining room where a random selection of his toys had been arbitrarily placed by the researchers. Completely on his own, he briefly got his bearings, then unerringly selected the stuffed sun and brought it to the test room, where the researcher and owner were waiting for him.[17] The fact that Rico can distinguish between all these things and their respective names is in itself an amazing achievement of learning and memory. But he can also connect a new name with an unknown object using the exclusion principle. This ability was at the very center of the experiment. If you asked Rico to bring the "sock," although he had never seen it before and had never heard this designation, he hesitated for a moment, searched among the objects, then selected accurately and brought the new object immediately and on the first try.[18] He thus was able to exclude all those objects whose names he knew, and to refer the new name to the new object.

In developmental psychology, it is assumed that this strategy is based on a process called "fast mapping." During this process, a living being, after being exposed to certain information, forms a rough new concept or hypothesis very quickly. Carey and Bartlett (1978), who coined this term, assumed that fast mapping plays a decisive role in language acquisition in children. Even though fast mapping alone is not enough to explain the rapidly increasing number of newly acquired words by toddlers, it is one of several operational principles for

17 Rico was almost always successful. He correctly chose thirty-seven out of forty items in the control experiment and brought them to the people in the next room (Kaminski et al. 2004, 1682).

18 In these identification experiments, which introduced a total of ten new items in ten sessions, with each session consisting of several rounds of tests, Rico was successful in seven of ten tries (Kaminski et al. 2004, 1682).

learning new words.[19] Fast mapping by way of the exclusion principle helps to explain how children tap into the correct meaning of a word and anchor it in their own lexicon, despite inadequate information about the subject and without seeing it at the first hearing of its name. They nevertheless create an entry for the new word in their lexicon and keep it in mind. The ability to do fast mapping is not only evident in children, but also in Rico. This border collie even proved capable of remembering a new word four weeks after the first experience of it and bringing the correct object, which he now had to select from a group of objects, four of which were unknown and four of which he already knew.[20] Kaminski et al. (2004) conclude:

> We assume that Rico's performance can be decomposed into a set of simpler mechanisms. These consist of (i) his acquisition of the principle that objects have labels; (ii) a general learning mechanism, namely learning by exclusion (emergent matching) (19); and (iii) the ability to store that knowledge in memory. Therefore, our results strongly support the view that a seemingly complex linguistic skill previously described only in human children may be mediated by simpler cognitive building blocks that are also present in another species. (1683)

They add, however, that it would be a matter for future research to determine whether or not Rico's ability to establish a relationship between a designation and an object actually corresponds to the knowledge of names of things as seen in toddlers (1683). Moreover, for the existence of real concepts with a specific meaning, additional skills might be required (see the aforementioned criterion [c]) that are not apparent in Kaminski et al.: whether Rico would apply the learned names to other objects of the same class, for example, is not clear here. Faced with a very different looking sock, would he bring that sock as well, even though it only remotely resembles the familiar sock? If two socks were before him, would he bring both socks when told, "Bring the sock"? Would he know that he would have to bring the yellow stuffed sun and the yellow ball if he received the order to "bring all yellow toys"? Would it be clear to him that the yellow sun is a toy amongst other toys? None of this can be gleaned from the experiments conducted by Kaminski et al. We do not know if Rico knows

19 More detailed information on the learning of words and the principles that apply here can be found in Bloom (2000).

20 The researchers write: "Four weeks after the identification task, he correctly retrieved the target item in 3 out of 6 sessions ($P < 0.1$). This retrieval rate is comparable to the performance of 3-year-old toddlers" (Kaminski et al. 2004, 1683).

how his "conception" of the individual things whose names he knows compares to other "concepts" he may have. Rico knows names. He learns new names using the exclusion principle. How much he knows about the designated things, their nature, and their relations to each other, and whether he possesses concepts on the basis of such knowledge, is not yet clear here. Newen and Bartels (2007) similarly reason: "Naming individual objects presupposes the identification and re-identification of objects on the basis of multiple stimulus generalizations, especially of smell and appearance." If a cognitive system such as the one exemplified by this dog were capable of dealing with acoustic signals such as names, then it would be expected that the dog would represent and understand these signals not only for individual objects but also for classes of objects. In the case of Rico, it has not been investigated whether he does just that. Newen and Bartels therefore continue: "Identification and re-identification of single objects presupposes multiple stimulus generalization, but it need not include the competence to identify and re-identify classes of objects." But this, they explain, "is exactly the core competence that justifies the ascription of conceptual representations" (Newen and Bartels 2007, 288). The ability of stimulus generalization, and why it may not be sufficient to provide real concepts, will now be discussed in more detail by way of another example.

5.2.2 *Extensionally Characterized Concepts in Pigeons*

Other attempts at concept formation and concept recognition in animals reveal more about the interconnectedness of different ideas than the experiments with Rico. At the outset, it was stated that through our language, we order and classify our world in mental categories. We group together what belongs together semantically and separate what is different. Rico's stuffed sun and his ball are both toys and clearly nothing to eat. In this sense, having concepts refers to having broader mental representations in which particular things and their properties have their place. Do animals know that different individual things can fall under a common generic term, into a common class of objects?

In the 1970s, Richard Herrnstein was already posing the question of whether, on the one hand, pigeons (*Columba livia* f. *domestica*) *recognize individual trees* and, on the other, whether they have an *abstract mental category* or the concept TREE. To test this, in training phases he used a food reward to encourage their pecking on slides that showed a picture of a certain stimulus, such as a tree. If the slides did not show this stimulus, the pecking was not reinforced. During the experiments the pigeons learned to distinguish pictures with trees from pictures without trees. This learning was accomplished despite the very large variety of trees on the slides. They differed in terms of appearance, shape, size, color, and location. Even if additional stimuli such as people, buildings, cars, and clouds were in the picture, the pigeons were not confused. The same was also the case with pictures of trees in which the tree could only be seen in the background or at the edge of the picture (Herrnstein 1979).

Two possible explanations for the performance of this task are conceivable. First, the pigeons may have remembered every single slide where their pecking had been rewarded. Second, it is possible that they have deduced what the slides have in common. The experiments showed that their reactions when dealing with new slides with or without trees were first of all correct and secondly not accidental. They also guessed correctly in those cases in which they hadn't seen the tree or the picture before. Furthermore, a criticism to the effect that pigeons could simply have an innate idea of trees does not apply. They demonstrated the same good performances on a range of other stimuli for which it is unlikely that evolution could have promoted the recognition of these stimuli in pigeons: alphanumeric numbers, cars, or fish—natural stimuli other than trees which were not taken from the natural environment of pigeons and certainly do not hold the same importance for their natural behavior and survival (Herrnstein and De Villiers 1980). There is a great deal of evidence that the pigeons indeed chose and derived the second solution strategy. They understood what was common to the pictures that presented such different trees to them—some with green leaves and others with red, some with needles and some without, some bald and some leafy, some small and some large, some growing high out of the water and others only showing the tip of a branch above the water line. Pigeons have a "concept" of trees in that they recognize what belongs to the category TREE. The animals in the experiments probably oriented themselves to the presence or absence of several features, such as leaves and/or green color. They generalized these stimuli to other pictures and recognized them on the individual slides, and in this sense we can speak of stimulus generalization. If a slide did not illustrate the features, it was not pecked on. They chose using a so-called match-to-sample or non-match-to-sample method.

The more complex the stimuli are, however, the more difficult it is to attribute the behavior to a *simple* or *sole* match-to-sample criterion. There is a whole set of criteria that needs to be developed, which is why *multiple* stimulus generalization is present here. The continuation of the studies with stimuli that are more complex and more unnatural for the animals is especially interesting here. In later stages, pigeons were even presented with pictures by Monet and Picasso (Watanabe et al. 1995, 165). The animals also successfully learned to recognize color slides of these paintings. After the training sessions, they also correctly distinguished previously unseen images by the same painters. In addition, they were able to recognize the respective style of painting in other images of similar styles: "They showed generalization from Monet's to Cezanne's and Renoir's paintings or from Picasso's to Braque's and Matisse's paintings" (Watanabe et al. 1995, 165). However, precisely because this concerns animals' ability to display *unnatural concepts*, the scope of the results of these experiments is also controversial. The animals certainly did not understand what differentiates impressionists from cubists in an *art-history* sense. Their achievement, however, is filtering out a set of criteria of both painting styles and adhere to those differences in their discrimination. These criteria could arise from the sharpness of the contours, the choice of color, and the depicted objects and their arrangement. However, none of these factors alone could be isolated as the key differentiator. The researchers suggested that the pigeons should be considered to have developed a *polymorphic concept* (Watanabe et al. 1995, 173).

The requirement of differentiating stimuli in terms of classes like FISH vs. NOT FISH pointed to the pigeons' use not of a natural but rather a synthetic concept. It is controversial to invoke synthetic concepts in such experiments, which is clear in the debate about the Monet vs. Picasso experiments and their significance. The use of an art-historical distinction may, with respect to the pigeons, mean the establishment of mere pseudo-classes, whose distinguishing features make sense to the experimenter but which must be considered completely arbitrary for the animals. "For pigeons," according to Monen et al. (1998, 224), "the proposed categorization into cubist and impressionist art … can best be considered synthetic. The actual distinction, as described by art historians, is based on the fact that although both impressionist and cubist artists took visual reality as their starting point, the impressionists tried to catch the image of the fleeting moment, whereas the cubists tried to catch the image of all moments." This is certainly not the content of the pigeons' concept MONET or PICASSO. Instead, they recognized a number of differences in the images themselves. These are clear: "Impressionist paintings are light, with loose brush strokes in pastels, leaving much of the white underground painting visible. The impressionist paintings chosen by Watanabe et al. (1995) are mainly greenish and bluish, and the objects have blurred outlines. They are mainly landscape scenes and still lifes. The objects are clearly recognizable." Picasso's pictures, on the other hand, have a very different character: "Cubists show the object as seen from different positions in a single painting. This results in the characteristic fragmentation. In typical cubist paintings from around 1912 … used in Watanabe et al. (1995), color contrast is reduced to a minimum. They are mainly in brownish and grey colors and have clear outlines. The brush strokes are not very striking. The paintings are figures and still lifes but are heavily deformed" (Monen et al. 1998, 224).

Many (probably even most) lay observers seem to distinguish pictures by Monet and Picasso in exactly the same way, without knowing and/or understanding the art-historical assessment of the painting styles. Interpretations that "tie the competence criteria for cognitive capacities to an exaggerated sense of typical human performance" hint at a specific "bias that loads the deck against animal mentality" (Buckner 2013). Buckner calls this the error of "anthropofabulation" (the combination of anthropocentrism with confabulation about our own prowess). If we don't want to fall prey to this errors, we should, as Monen et al. (1998) do in the end, concede that the pigeons' rapid and safe discrimination on the basis of the identified characteristics suggests that they somehow categorized these characteristics. The discrimination performance then illustrates their ability to recognize the similarity of stimuli and the differences between individual stimuli associated with different classes (Monen et al. 1998, 224).

TABLE 1 Extensional and intensional features of concepts

Herrnstein's pigeons and Rico, the dog	**What else we might be looking for**
Discriminating/being able to recognize trees from nontrees, recognizing what "sun" is	Recognizing what the relationship between a green pencil and a green wooden cube is
Holding a stable representation and showing a correct reaction to a feature (or a set of features) or to a command with a name → relatively strong stimulus dependence	Holding a stable representation and showing a correct reaction to different commands, each of which addresses different characteristics of the same object → stimulus independence
Learning by fast mapping/(multiple) stimulus generalization or (non)match-to-sample strategy → successful stimulus discrimination	Representing different objects as belonging to the same class with the same characteristics
Capturing a concept that is represented in an extensional way by the class of objects to which it applies	Capturing a concept that is characterized extensionally (by the class of objects to which it applies) and intensionally (by its position in a network of other concepts)

As an interim conclusion, we can say that even for successful discrimination of this kind, a stable mental representation of the sought-after feature or property is needed for multiple stimulus generalization and (non) matching-to-sample strategies. We are certainly dealing with a cognitive effort here. But it is clear that for the pigeons, for example, the concept TREE is *extensionally characterized by the class of objects that apply to it.* The animals recognize what is a tree and what is not a tree. Although categorization is a necessary feature of human language, human language is obviously not necessary for categorization in the sense of extensionally characterized concepts. Even animals have such concepts at their disposal. In this sense, being able to form concepts of living beings and objects in the environment is certainly a prerequisite for orientation and the identification of food or predators for many animals. Natural selection presumably favors this capacity for concept formation and discrimination under certain conditions.

Table 1 summarizes what we have found so far with respect to categorization/discrimination in pigeons and in Rico and what could be asked in addition for ascribing concept possession. We continue to search for conceptual knowledge in animals that goes beyond the recognition of particulars and cannot be explained by stimulus-response and stimulus-generalization alone. Let us remember that when it comes to complex conceptual mental representations, there should certainly be flexible behavior. This might mean different things in the case of concept formation. Newen and Bartels (2007, 287) think that rigid behavior is often caused by simple stimulus-response mechanisms based on a single type of stimulus. By contrast, however, they observe that "concepts necessarily presuppose an integration of more than one source of information in order to modify the behavior which is the response to a given stimulus." Real concepts should therefore be represented independently of stimuli and should not only be available in a one-to-one association of stimulus and behavioral response. The relation between stimulus and reaction, according to Newen and Bartels (287)—with reference to Pylyshyn (1990)—should be "many to one." This means that one and the same behavior (such as an escape behavior) can be initiated due to many different stimuli (and their combination), such as the smell of smoke, the sight of fire, the hearing of burning, and so on. The type of response to perceptual input also depends on multiple sources of information and is modifiable by them. If one smells smoke coming from the yard next door but knows that the neighbor wanted to do a controlled burn of timber, then at first this is no reason to panic. It is precisely this ability to modify the reaction due to the integration of additional information that is characteristic of complex concept formation (Newen and Bartels 2007, 287). When new information becomes available, behavior, including language behavior, must be modifiable rather than remaining in a routine.

Furthermore, it is not sufficient to have at one's disposal terms that can be used to discriminate individual things. For true categorization, not only must red be recognized, but it must also be understood that redness is a color property. Red not only has to be differentiated from yellow (at the level of individual colors), but red must also be understood as different from other properties, such as the property of being wooden (a differentiation at the higher level of the categories COLOR and MATERIAL). Depending on which property of an object is required (color or material), a conceptually controlled reaction would have to be different. Finally, concept formation often has to do with the recognition of different features in one and the same item. In a red ball, its redness and its roundness are both relevant. Red as a color property and round as a form property must also be recognized as such in other things to which these properties also belong in order to be available as abstract concepts.[21]

Thus, we are looking for a response that is not the result of rigid stimulus-dependent causes, but comes about through a more complex evaluation process. This evaluation process extracts from the perceptual data classes that are oriented towards certain properties that belong to things (Newen and Bartels 2007, esp. 291). Overall, a sufficient network of properties of an object must be distinguished in order to have a true concept of that object. Is there such a form of conceptual ability in an animal?

5.2.3 *Concepts in Alex the Grey Parrot*

There was indeed an animal that probably possessed the abilities defined above with regard to terms or more complex concepts: a grey parrot (*Psittacus erithacus*) named Alex. Unfortunately, Alex died unexpectedly in September 2007 before he could be observed in further language studies. However, the existing results are presented below and continue to be discussed based on the philosophical interpretations in Bartels (2005) and Newen and Bartels (2007).

5.2.3.1 Training Method and Vocabulary

Irene Pepperberg worked with Alex for more than twenty years and has continued to research the cognitive and communicative abilities of other African grey parrots after his death (Pepperberg 2002, 2008). She uses a training method

21 This means that "the animal is able to represent the same property while dealing with different objects and is also able to represent different properties while having only one and the same visual input of an object. The animal is able to deal with a systematic object-property-representation" (Newen and Bartels 2007, 293).

that is particularly suitable for these animals, the so-called model-rival training system, in which one person acts as a trainer while a second person serves as a model for the animal's behavior and a rival for the trainer's attention. The two persons begin to have a conversation during which, for example, a new item is named. At first, the parrot just sits and listens. However, out of curiosity and jealousy, he interferes with the speakers and begins to name the item as well. If he does so, the trainer's praise and attention are directed towards him (Pepperberg 2002, 25ff.). In the documented training interview, as illustrated in table 2, Irene Pepperberg (trainer) and Bruce Rosen (model/rival) attempt to improve Alex's pronunciation of the word "five." In this way a very flexible

TABLE 2 Training talk with Alex the grey parrot. Adapted from Pepperberg 2002 (table 2.1, 27)

TRAINER AND MODEL/RIVAL USE THE NEW WORD "FIVE":
Irene Pepperberg (in the trainer role): "Bruce, what's this?"
Bruce Rosen (in the model/rival role): "*Five* wood."
Irene Pepperberg: "That's right, *five* wood. Here you are ... *five* wood." (Pepperberg gives Rosen five wooden sticks; he starts breaking one, which Alex also would do if left with the sticks.)
ALEX JOINS IN:
Alex (joining the conversation out of curiosity and to call attention to himself): "ii wood."
Bruce Rosen (now in the trainer role, swapping the broken stick quickly and presenting the five sticks to Alex): "Better..." (Rosen turns away and repositions himself to establish eye contact with Alex)"...how many?"
Alex: "No!"
THE RESEARCHERS CONTINUE TO ENGAGE WITH EACH OTHER:
Bruce Rosen (turning away from Alex and instead establishing eye contact with Pepperberg): "Irene, what's this?" (Rosen shows Pepperberg the sticks.)
Irene Pepperberg (now in the model/rival role): "ii wood."
Bruce Rosen: "Better..." (Rosen turns away from Pepperberg and then makes eye contact with her again): "... how many?"
Irene Pepperberg: "FIVE wood." (Pepperberg takes the sticks) "*Five* wood."
ALEX SPELLS THE WORD CORRECTLY:
Now Pepperberg (takes over the trainer role, turning her gaze to Alex and presenting the sticks) "...How many wood?"
Alex: "Fife wood."
Irene Pepperberg: "OK, Alex, close enough ...*fivvvvve* wood ... here's *five* wood." (She puts one of the sticks in Alex's beak and the others in his grasp.)

---

learning situation was created, one in which the roles of the questioner and the responder were repeatedly reversed between the trainer and the model/rival. It was not always the trainer who asked the questions, and it was not always the model/rival person who answered. The animal was also integrated into the communication from time to time. So Alex did not simply hear a "language duet" in which the roles were fixed and unilaterally assigned, but rather, Pepperberg emphasizes, a reciprocal, communicative process. He learned that communication is a mutual game with words, "a two-way street in that one person is not always the questioner and the other always the respondent"—that is, it is a process designed to influence and change the environment (Pepperberg 2002, 26ff.). In this way, Alex learned the names for different objects, colors, materials, and numbers. Even short utterances (e.g., "Wanna go") and introductions to questions were part of his vocabulary. He also knew the words "color," "matter," "same," and "different."

5.2.3.2 Alex's Perception-Based Concepts and Their Interpretation

In Pepperberg's experiments, Alex saw one or more items presented to him on a tray, and he was asked about their characteristics. Thanks to his learned vocabulary, he could answer these questions. In most experiments he was right in 80 percent or more of his answers, but he achieved the best results in trials in which he was confronted with new things and situations.[22] Alex's abilities will be clarified with some examples. For example, if Alex was confronted with a red key and a yellow wooden cube, he was able to say which of the items fell into a particular queried category. In response to the question "What toy red?," his answer was "Key!" But Alex could also gauge the items in terms of their belonging to different categories. When shown the yellow wooden cube and asked "What color?," he answered "Yellow." When asked "What matter?," his answer was "Wood." Even more impressive was his ability to classify objects as equal or different in terms of different categories. If he was confronted with a yellow wooden cube and a yellow key and asked "What's same?," he answered "Color"—even if he had never seen that color before. To the question "What's different?," his answer would be "Matter" (Pepperberg 2002; Bartels 2005, 156–86, esp. 175–81).

According to Newen and Bartels (2007) and Bartels (2005), with this behavior Alex shows that he can identify and re-recognize objects and characteristics, that this ability has relative stimulus-independence, and that his classification

22 For Alex's categorization regarding "color" and "shape," see Pepperberg 2002, 58; on the discrimination of "same" and "different," including its transfer performance for new objects, (73). For his understanding of the concepts and relational concepts, see 125–67.

TABLE 3 Newen and Bartels's (2007) criteria for having concepts at one's disposal

Criterion 1:	Having a stable representation (for example, of a property A_1) in very different objects or events
Criterion 2:	Having a representation of at least one other characteristic (A_2 ... A_n) of the same object or event
Criterion 3:	Having relative stimulus independence
Criterion 4:	Being able to locally represent the characteristics in minimal semantic networks

performance involves an adequate level of abstraction. It goes beyond stimulus generalization, and involves organizing things into classes. He situates his concepts in minimal semantic networks (Newen and Bartels 2007, 295ff.). This behavior is close enough, according to Newen and Bartels, to having real concepts at one's disposal, based on the four criteria summarized in table 3:

Regarding criteria 1 and 2:[23] Alex has a stable representation of the characteristic of being red while bypassing very different red objects. Remember that even when it comes to items he has never seen before, he can say that they are red when asked about their color. Furthermore, with respect to an object, he can represent not only the property of being red, but also other characteristics (such as the fact that the cube is made of wood). Newen and Bartels (2007) clarify:

> If a system can only represent the redness of one kind of object—e.g., of cars but not of tables, chairs, houses, etc.—then it is not able to classify the property of being red, and can only discriminate red-car situations. Furthermore, if the cognitive system can only discriminate the property of redness of a car but cannot distinguish some other property of the same object—e.g., having wheels, having a special form, being (mainly)

23 Newen and Bartels (2007, 296, 305) formulate criterion C1 as follows: "The system has a stable representation of A_1 while dealing with very different objects (or events)." Criterion C2 states: "The system represents not only the property A1 but also some (but at least one) other properties, A2, ..., An, of the same object (or event)—while the other properties are not the dimension or general categories under which the property A1 can be subsumed."

> made of metal—then it makes no sense to claim that the system is able to classify an object. A fortiori it cannot represent red as the property of that object but only the appearance of a red-situation. (296)

Alex can clearly discriminate red in different objects and can, according to different characteristics of the queried objects, also name these objects correctly.

Regarding criteria 3 and 4: Criteria 1 and 2 are certainly necessary conditions for having concepts, but they do not seem sufficient. They do not yet fulfill some conditions that were required after the discussion about Rico and the pigeons. For example, they do not yet cover the need for concepts to be independent of stimuli and anchored in conceptual nets. Additionally, there are also criteria 3 and 4, which Alex also fulfills. According to these criteria, genuine concepts exist if the following applies:

> C3. If we have relative stimulus independence such that it depends on some additional mechanism—which detects and weighs stimuli other than the key stimulus of redness—to determine that the system focuses on redness while perceiving a red square, in contrast to some other property; a relevant independent stimulus could be, e.g., a learned acoustic signal which demands focusing on the color instead of the shape of the object; and
>
> C4. If the property of being red is represented in a minimal semantic net: the property red is represented as an instance of the dimension color—not as an instance of an incorrect dimension (e.g., danger).
>
> NEWEN and BARTELS 2007, 296, 305

In fact, in the experiments Alex saw the same visual stimulus three times (i.e., the same item on the tablet three times) and heard three *additional, neutral and acoustic stimuli in relation to this visual stimulus* to which he responded (namely, Pepperberg's questions). If he had not known what the questions meant—that is, that they were addressing different characteristics of things—he would have only answered the same thing three times or guessed the correct answer merely by chance: obviously, he didn't do that, given his high rate of success. In so doing, he convincingly demonstrated the ability to modify his response *depending on the different, additional, auditory stimulus, while the visual stimulus remained the same.* This speaks to the required relative stimulus independence (C3). Such a modification of the behavior with regard to a new stimulus in addition to the visual trigger stimulus did not exist in the experiments with the pigeons. In that instance, there was always only a visual stimulus. Even if the stimuli have become increasingly complex and unnatural, and even if it is indeed verifiable whether the animals can distinguish "fish situations" from "non-fish situations," it is, nevertheless, not verifiable whether they have the concept FISH at their disposal and somehow know that, for example, fish belong in the category WATER ANIMALS.

For Newen and Bartels (2007) and Bartels (2005), the form of modification of the answer given by Alex, depending on the question posed, is furthermore sufficient for the *normative dimension* of concepts. This, according to Donald Davidson, implies that a speaker, in order to truly hold a belief, must necessarily have *beliefs about these beliefs*—or at least *the belief that his belief can be wrong*. In Davidson's account, there are no beliefs without such meta-beliefs (Davidson 1982, 324). Davidson explains:

> Much of the point of the concept of belief is that it is the concept of a state of an organism which can be true or false, correct or incorrect. To have the concept of belief is therefore to have the concept of objective truth. If I believe there is a coin in my pocket, I may be right or wrong; I'm right only if there is a coin in my pocket. If I am surprised to find there is no coin in my pocket, I come to believe that my former belief did not correspond with the state of my finances. I have the idea of an objective reality which is independent of my belief. (326)

Newen and Bartels (2007) deliberately weaken this criterion. Even if one agrees that the belief of seeing a cat requires that one knows what a cat is (i.e., requires one to have the concept CAT), it would be far from clear whether a person must therefore know, in a general sense, what a belief is. It would be particularly contentious if he or she had to have the certain belief that beliefs can be wrong. Davidson could offer no convincing argument for his radical thesis why a cognitive system must have the concept BELIEF at its disposal in order to have beliefs. Newen and Bartels therefore agree that having conceptual representations implies having misrepresentations: "Davidson is right in insisting that conceptual representations, like all cognition-based representations, presuppose the possibility of misrepresentation." Yet they contradict Davidson's further implication: "But he's wrong in thinking that misrepresentations can only occur in cognitive systems in which the concept HAVING A FALSE BELIEF may occur." The reason for their caution is this: "If all cases of misrepresentation were characterized as cases in which the concept of HAVING A FALSE BELIEF occurs, then representation would be restricted to cognitive systems with second-order conceptual capacities from the beginning" (289). It seems that the assumption that conceptual representations are limited to cognitive systems with meta-beliefs is wrong. Empirical findings showed that children and toddlers, who (still) had no theory of mind and above all (still) had no understanding of false belief, were nevertheless able to communicate verbally and had concepts. Thus, according to Newen and Bartels (2007), it is wrong to make a theory of mind, including an understanding of false beliefs, a precondition for having concepts. Instead, in terms of the normative dimension of concepts, the authors draw a connection to the simpler ability to recognize misrepresentations and the ability to modify behaviors that build upon them—and Alex was certainly capable of modifying his responses in reaction to additional information gleaned from Pepperberg's questions (Bartels 2005, 178).

Donald Davidson also requires—and Newen and Bartels also recognize this in principle—that beliefs always occur in the context of other beliefs. The question, for example, as to whether a dog could believe an object, such as a cat, to be in the tree, must be answered as follows: he can only hold that belief if he also has other, general beliefs about trees (e.g., that they need soil and water, have leaves or needles, grow, etc.). According to Davidson, beliefs are thus localized in dense networks that contain other, related or connected beliefs (1982, 320–21). Regarding this requirement, which is present in Newen and Bartels's criterion 4, it can be said that Alex's concepts are obviously organized in semantic networks. These include upper categories such as COLOR and subcategories assigned to them such as YELLOW, RED, and GREEN (Bartels 2005, 184). This demonstrable conceptual net positions Alex ahead of the pigeons. Even Rico, the border collie, had criteria 1–3, but not (demonstrably) criterion 4 (Newen and Bartels 2007, 297–98).

It is not so clear, however, whether Alex is indeed much cleverer than other animals tested so far or if he just had better training and testing conditions. Pepperberg's metholdology might be exceptionally suitable for parrots. She had her parrot classify the objects in terms of abstract classes, but did not resort to synthetic concepts that were unnatural for Alex and that were irrelevant in his environment. Alex was an animal that had long been in captivity and been socialized there. He interacted with all these objects. Their color and material properties were recognizable to him, and he had experience in dealing with the objects. The reward for his cooperation in the experiments was being allowed to play with the objects, to hold them, to take them in his beak and gnaw on them. He knew that cork feels different than metal. This also makes his discrimination performance meaningful and runs somewhat counter to the criticism of Monen et al. (1998), who were still highly critical with regard to the concepts of pigeons. Their fundamental objection as to why there was indeed discrimination in the pigeons, but not demonstrable conceptualization, was already based on a fundamentally incorrect basic prerequisite of the experiments, and was elaborated as follows:

> An obvious requisite for studying categorization is … that the organism has access to the information required to make the correct distinction. Imagine training human subjects to distinguish between photographs of girls called Jenny and ones of other girls. If all Jennys in the training set had dark hair and all other girls were blond, subjects would quickly learn to discriminate between the photographs. However, the categorization would have nothing to do with the girls' names. One experiment with a blond Jenny would suffice to demonstrate this. The subjects could not learn to categorize the girls according to their names, because they had no access to the critical information. Learning to categorize the girls by the color of their hair obviously does not demonstrate an understanding of the concept of "girls called Jenny."
>
> MONEN et al. 1998, 224–25

Monen et al. recommended, therefore, that in animal experiments stimulus classes such as YELLOW or FOOD be used, classes that are subjectively recognizable for the animals. Their belonging to a concept is, on the one hand, *given physically* (in contrast to a distinction attributed only art-historically, such as that between impressionism and cubism or the naming of "Jenny"). In addition, their belonging to a class is also dependent on *characteristics of the tested animal*, that is, on the animal's vision or manner of eating. The classification of a banana in the category FRUIT or the classification of fruits in the category FOOD only makes sense for an animal that eats bananas. This sense, we could argue, arises especially for this animal with its individual living environment, its biology, and its habits, through the semantic positioning of this object in a network of other objects relevant to the animal.

According to Newen and Bartels (2007) and Bartels (2005), and in line with the presented considerations, Alex also fulfills criteria 3 and 4, which means that he actually holds and operates with perception-based concepts or conceptual representations. How can his concepts be described more precisely? Bartels takes a look at their *function* (2005, 11, 185–87) and argues convincingly that Alex's concepts are functionally describable.[24] According to Bartels, concrete conceptual representations exist in the case of a living organism in the

24 Bartels (2005) chooses this path because most other theories on concept possession are linguistically based and thus hardly applicable to animals. It would go beyond the scope of this book to present and discuss different theories of concept possession and different

form of specific neuronal states or processes, and thus are ultimately realized via neurobiology. He argues that concepts in animals first become apparent on the basis of manifest abilities (for example, in a discrimination capacity). Then the determined abilities can be analyzed and abstracted by means of functional roles. To characterize the meaning of a conceptual representation thus means to distinguish functional neural properties and relations, which define a neuronal structure suitable for fulfilling the functional roles manifested in behavior (Bartels 2005, 183ff.).

What this suitable neural structure looks like is, for sure a neurophysiological question. However, the functional roles of Alex's perceptual concepts, which become apparent in his behavior, can be determined philosophically in terms of their meaning or content. Alex's conceptual perception of YELLOW obviously has three functional roles: an *indicator function*, a *classification function*, and a *comparison function* (see fig. 7) (Bartels 2005, 84). The adequate integration and separation of concepts within the semantic network is performed by the classification function and the comparison function. Alex can represent equality and diversity of perceived objects with respect to upper categories. Of particular importance, because it sets Alex apart from the other animal examples discussed, is the comparison function of his concepts. It allows one to be sure that Alex represents red as a color and not as a material or anything else. Newen and Bartels (2007, 297) required that in order "to have a concept of red, a cognitive system must have a representation of the extensional

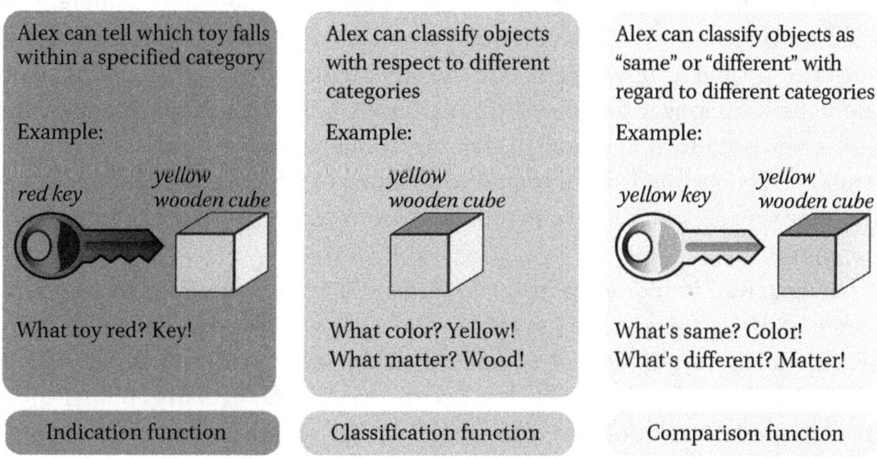

FIGURE 7 Conceptualization of Alex the parrot, according to the interpretation of Bartels (2005)

---

theories of representation. More details can be found in Bartels (2005, 181ff.) and Newen and Bartels (2007) as well as in the literature to which both refer.

as well as the intensional features of the concept of redness." Accordingly, the intensional feature of redness is precisely being a part of a minimal semantic network, which determines redness as an occurrence of the color dimension. Alex seems to have concepts that are both extensional (defined by the class of objects they apply to) and intensional (defined by their position in a conceptual network).

5.3 Functional and Intentional Aspects of Natural Animal Communication

Conceptual ability is certainly a necessary prerequisite for linguistic ability, and the referential use of sounds in the sense of terms with a semantic content is certainly a major building block of language. The last section considered the question of whether animals have concepts in the philosophical sense. The focus was on how Alex the grey parrot responded when one tested his concepts and their position in a conceptual network of upper and lower categories. Unfortunately, such targeted querying of the networked mental organization of concepts is difficult in animal experiments. Alex is therefore an outstanding example wherein philosophical considerations can be explored.

By contrast, the *function* of sounds is easier to read from behavioral observations. Therefore, the present section addresses the question of how animals use sounds for communication. On the one hand, the following considerations then move from the individual, with her mental representations, toward the social group and its interactions. On the other hand, other aspects of linguistic ability will no longer be viewed from a purely philosophical perspective. Instead, we again turn to ethological observations as well as linguistic theories. The central question is what semantic meaning or which functional reference do gestures and sounds have in animals, and whether these are intentionally communicated.

In 1934, Karl Bühler introduced an influential characterization of language into linguistics, the so-called organon model. He wanted to provide a descriptive model of the full-fledged, concrete speaking event understood in the light of the living conditions in which it occurs reasonably regularly (Bühler 1934, 24). His model is relational and assumes that, in a speech situation, *one* is communicating *to another* something *about something*. At the center of this concept is the expressed, sensually perceptible, usually acoustic phenomenon. It is anchored in three relational foundations between, on the one hand, two psychophysical systems—i.e., that of the sender and that of the recipient—and, on the other hand, that of the objects or facts about which communication is being generated (cf. Bühler 1934, 25, 28). The semantic function of the (complex) speech symbol, says Bühler, results from three variable moments arising from this anchoring. These elevate the sound, as it were, to the rank of a sign:

> It is *symbol* by virtue of its association with objects and facts, *symptom* (indicium) by virtue of its dependence on the sender whose innermostness it expresses, and *signal* by virtue of its appeal to the listener whose external or internal behavior controls it like other traffic signs. (28, transl. MK)

These are not functions of the sign itself, as we have come to know in the case of the functions of Alex's perception-based concepts, but (at least in the case of expressive and conative functions) Bühler describes functions of the sign *in its use*. All three semantic functions of the language character described here—that is, its *representation function* as a symbol, its *expression function* as a symptom, and its *imperative* or *appeal function* as a signal—are due to the pragmatic function of language. Language is both knowledge and interest simultaneously; it always intends something to achieve something. In the sense of Bühler's model, what is achieved is then a representation, an expression, or an imperative/appeal, with each linguistic sign basically including all three aspects, one of which may become dominant depending on the speech situation. Thus, the organon model encompasses the semantic functions of complex speech signs from the perspectives of both the sender and receiver, and from both perspectives not only speech signs of humans but also sounds and gestures of animals can be considered; and these can be interpreted in their semantic-linguistic function. In so doing, I will initially concentrate on the sender's perspective, especially on the *expression of intentions* in the gestures of orangutans. In a second step, I will explore the receiver's perspective, especially the *appeal function of alarm calls* in monkeys.

The fact that animals issue sounds as *expressions of their own internal states*, for example when they feel fear, joy or pain, is rarely disputed. However, these could be involuntary, similar to when a human burns her hand on the stove and reflexively cries out in pain. Although this sound would not be intended as an appeal, those close by would be warned about the hot stove, provided they could adequately interpret the outcry. On the other hand, a mother's appeal to a child in the form of the shout "Careful, traffic light!" is no inarticulate cry of pain, but she resorts to a specific sequence of sounds. Her cry is intended as a warning call. A sound sequence that is consciously chosen which only fits for this warning and which carries a clear meaning, is flexibly adapted to a prevailing situation (for example, it is only expressed if the child runs too fast at a red traffic light). It is aimed specifically at a recipient. If the reaction of the recipient, in this case the child, does not correspond to the sender's request, in this case the request of the mother, because the child, for example, simply keeps running at the traffic light, the mother will react accordingly: she will if necessary repeat her appeal, intensify it, and finally take action to stop the child. The challenge, in the case of animals, is to find certain vocalizations or gestures with clear meanings in their communication and determining which function these signs fulfill from the recipient's perspective. It's even more difficult to determine whether a sender actually wishes to express specific intentions. Both aspects are needed to better understand what functions the sounds and gestures of animals fulfill.

Cartmill and Byrne (2010) attempted to identify exactly these aspects in the gestures of twenty-eight orangutans (*Pongo pygmaeus* and *Pongo abelii*) during three months of observation in three European zoos. The focus was on the intentional meanings of the animals' gestures (and in Bühler's words: on their appeal function). It turned out that out of the sixty-four clear communicative gestures,[25] forty were regularly used for one particular effect. The animals seemed to use these forty gestures to achieve certain goals. Six kinds of goals were observed, including (1) initiating an interaction or contact (for example, in the context of grooming or play), (2) asking for an object, (3) suggesting that the partner share an object, (4) suggesting that two or more orangutans move together, (5) suggesting that a partner should retreat, or (6) suggesting that she should stop a certain action. Twenty-nine of the observed gestures were even used with only one meaning. The authors thus conclude that, "despite their contextual flexibility, orangutan gestures are made with the expectation of specific behavioral responses and thus have intentional meanings as well as functional consequences" (Cartmill and Byrne 2010, 793).

25 The authors classified different movements of the animals and excluded some in principle. Only those movements that were executed in a social context and did not fulfill a recognizable other (primary) function for the animal were considered as *potential* gestures (for example an animal that scratches executes a motor-effective movement, not a communicative movement). Potential gestures are motor-ineffective movements, i.e., movements "that did not directly perform a practical act such as scratching or picking up an object" (Cartmill and Byrne 2010, 796). Furthermore, social interactions involving continuous active contact between the animals (such as physical play, cuddling, suckling, or reproductive behavior) were excluded (796). Only those potential gestures that were firstly recipient-oriented and secondly goal-oriented were then evaluated as actual communicative gestures (795–96).

The question of intentionality with regard to communicative gestures and sounds is central, as the authors rightly suggest, since, as already mentioned above, language can be understood as an intentionally structured transfer of information. On the sender's side, information is actively selected and coded, whereas the receiver actively receives, processes, and interprets it (Cartmill and Byrne 2010, 793–94). If no information is available about the sender's intentions, then only the functional meaning of a gesture or a sound can be determined, by observing what causes it or what reaction it triggers in the receiver. Then the focus is on the recipient's perspective and the competence of the recipient as the one who decodes signs. But if there are indications that the sender has produced a gesture or a sound with the aim of triggering a certain behavior in the receiver, then it is possible to analyze which intentional meaning the gesture or the sound has (794). The intentional meaning is provided to the sign by the sender, whereby the sender's voluntary and controlled competence to address a recipient can be focused on. In an ethological approach, in terms of behavior, the intentions must be as clear as possible even if they are in themselves mental phenomena that we cannot directly see or prove. This certainly ensures residual uncertainties. However, the criteria for intentionality that Cartmill and Byrne created are widely accepted in behavioral biology. They require the sender to focus on one or more receivers, orient on a specific target, and show flexible behavior in contrast to simple stimulus-response automatisms (797). In their observational material, the authors looked for indications that a sender is expecting a particular response from a receiver, and not only sending signals following an automatic response to environmental stimuli. Such an expectation could be seen, for example, if the sender continued to orient herself toward the recipient after completing the gesture without performing another action herself, when she looked back and forth between the receiver and the object or location, or if she remained where she was and behaved in accordance with the recipient's state of attention (i.e., using inaudible, quiet gestures only when she could see her). Also, the intensification of a gesture and the combination of several gestures while waiting for a response were understood as indications that the sender was expecting a certain reaction, pursued a specific goal, and had specific intentions. In addition, gestures had to be used at least once in a clearly intentional way in order to count as an established gesture that is part of the communicative repertoire of an individual (Cartmill and Byrne 2010, 797).

Different sender goals as well as reactions could be clearly determined from the senders' observed gestures, the reactions of the receivers, and the counter-reactions of the senders. In this way, a series of gestures have been identified in orangutans that are predictably linked to appropriate responses.[26] On the one hand, these are flexibly adapted to the context and to the attentional status of the recipients, and, on the other hand, they can be described as intentional, since the sender evidently expected, respected, and/or urged the appropriate response if it did not take place immediately. The strength of the intentionality varied and resulted from how fixed or variable the goal of the sender was in connection to a certain reaction:

> Gestures that were used with essentially a single goal-outcome match, i.e. more than 70% of the time and three times as often as the next most common goal-outcome match, were defined as having "tight meanings." Gestures used frequently with a single goal-outcome match, i.e. 50–70% of the time and twice as often as the next most common goal-outcome match, were defined as having "loose meanings." We defined gestures that had a single goal-outcome match less than 50% of the time as having "ambiguous meanings."
>
> CARTMILL and BYRNE 2010, 798

Of the sixty-four communicative gestures identified, twenty-nine had narrow meanings, seven were less clearly identifiable in their intentional significance, four were ambiguous, and twenty-four were used too seldom to be analyzed (Cartmill and Byrne 2010, 800).

Six intentional meanings of the gestures could be determined, including appeals to play requests, appeals to stop an action, to look at an object or take it, and to share food. Neither the manner of expression nor the meaning of these gestures seemed to vary systematically among the orangutan populations: "Only one gesture, *fake*, was limited to use in one group, and it was idiosyncratic

26 In order to determine the goals, the authors continue, "We gave each gesture one of eight presumed goals: Affiliate/Play, Stop action, Sexual contact, Look towards signaller, Look at/Take object, Share food/object, Co-locomote, or Move away. Inter-observer reliability of two experimenters' attributions of gesturers' presumed goals was performed on 6 % of the total dataset (90 gesture tokens) and achieved a high level of agreement (k = .83). Presumed goal was deemed therefore to be a replicable measure and a useful tool in further analysis of gesture intentional meaning." To determine the reactions, the authors write: "For every social interaction involving one or more gestures, we recorded the immediate reactions of the recipient to each gesture. Reactions were categorized as: moving towards or away, sharing or taking food or objects, following or picking up the other, looking towards the other, playing, affiliating (including nursing and grooming), mating, ceasing an action, performing a particular gesture, or not responding at all. We defined the final reaction of the recipient as the interaction outcome, as it caused the signaller to stop gesturing" (Cartmill and Byrne 2010, 798).

to a single adult female and observed only during a particular teasing exchange with her juvenile daughter. It is possible that this gesture was unique to the female or it could be that the context was so specific it was never observed in another group" (800). However, individual and group-specific variations in the sounds of great apes are known in the wild and in captivity, and are to be expected when language is culturally transmitted. For example, orangutans in the wild use sounds within a certain behavioral context that may even be acoustically completely different in different populations. This applies to sound sequences that accompany nesting behavior, for instance. In addition to social and anatomical factors, this could be due to vocal learning processes that ensure that new sound sequences are acquired and culturally transmitted. It has also been proven that orangutans in captivity are capable of sound imitation and can even incorporate completely new human whistling sounds into their own phonetic repertoire (Wich et al. 2009; van Schaik et al. 2003, 2006; Hardus et al. 2009; Hopkins et al. 2007). Of course, those gestures identified by Cartmill and Byrne (2010) only represent the beginning of a gesture lexicon of orangutans in captivity. The authors believe that the six semantic meanings found reflect a semantic system with remarkably few differences compared to human speech. The gestures are used flexibly, despite their defined importance, in view of the attentiveness of the other individual, and seem to form a limited set of appeals that are invitations to help or appeals for diadic interaction. While this does not bring them the strongly referential meaning of human words with regard to the communication of information, the results add to a growing body of observations. Together, these findings suggest that the gestures of orangutans are forms of diadic interaction between two or more individuals (Cartmill and Byrne 2010, 800ff.). The experiments conducted by Cartmill and Byrne do not necessarily verify the representation function of semantic linguistic signs in animals. What they do show is their functional reference and the intentionality of the sender with a view to achieving these functions. The results show that the expression function (in the sense of the expression of intentions) and the appeal function (in the sense of the expectation of a specific reaction) can also be found in the gestures of orangutans. Both are therefore semantic aspects that characterize the gestures of these animals. Their function is to seek the initialization and maintenance of social interaction between two individuals. Thereby the six intentional meanings found "can be grouped into positively and negatively valenced categories of requests" (802).

When assessing these findings, however, one should not forget that social interaction is not just about conveying facts, though this is often emphasized. Classic language concepts, which assume that the sender knows something about which he deliberately informs the recipient, have long been dominant—perhaps wrongly (Tyack 2000, 270). Bühler also builds on the idea of such a reduction of ignorance through the linguistic transfer of knowledge when he emphasizes the dominance of the representation function of signs or symbols in humans. However, such concepts of language emphasize only one semantic aspect out of three, making it very one-sided and neglecting the appeal and expression function, which is particularly important in everyday language. This deals with the communication of multiple goals and intentions, not only the symbolic representation of facts and objects but their meaning for the social group. Tyack (2000) therefore generally proposes a more modern concept of language that puts the multifunctionality of language in focus. He convincingly identifies six different functions of communication, which he describes using the example of communication in whales and dolphins. At least one of the functions, namely the sending of signals for the confusion or active deception of others, seems clearly to presuppose intentionality and volitional control and to have the sender's perspective in view.

Here, I emphasized initial research results on the intentional importance of the gestures of animals. I have done so because I believe that we need a better understanding of the expression function of signs in animals. This chapter now turns to the functional significance of sounds in animals and thus focuses on the understanding of appeals or imperatives by the recipient. It is true that the expression of intentions has an appeal and is somewhat functional in that it is linked to certain expectations of the recipient. However, it is no surprise that in Bühler's case all three semantic functions of the language character were dominating phenomena in the sense that they hardly exist isolated from each other. Depending on the speech situation, sometimes one or the other function becomes dominant. This semantic multidimensionality—at least a two-dimensionality with regard to expression and appeal functions—is also present in the gestures of orangutans. The emphasis here can therefore be placed on something that had not been considered to be on equal footing for a long time: both the sender and recipient perspectives require competencies. While the achievement of an intention and a goal and the persistent and adequate enforcement of the same is up to the sender, the power of decryption, on the other hand, lies with the recipient. With regard to the semantics of calls, this decryption competence has long been specifically investigated using the example of alarm calls in animals, and it is less controversial than the intentionality of the sender. It can be read more clearly in the reactions of the animals.

Vervet monkeys (*Cercopithecus aethiops*) use different sounds for different predators, which have been discussed as acting as words with a semantic content. Obviously, these calls are functionally referential: if an animal in the group launches the specific alarm call for a leopard, then the group members respond with a specific reaction. They retreat to nearby trees. If the alarm call sounds like a snake call, then they line up on their hind legs and inspect the ground around them. If it sounds like the alarm call for eagles, they search the sky or run into dense bushes.

Both Struhsaker (1967), who observed these calls for the first time in the late 1960s, and Seyfarth et al. (1980), who began to systematically investigate them in the 1980s, suggested that the importance of the calls was clear to the animals. In playback experiments that played recorded calls to the monkeys, the researchers found that the sounds were actually differentiated by their different meanings. The sounds first evoked the corresponding response without a real sender being visible in his real alarm condition and without a corresponding predator approaching. All the animals heard in the experiments were the alarm calls via a loudspeaker. It is thus ensured that the reactions of the receivers were based solely on the hearing of the alarm call by a sender and not on information that they themselves could detect, for example, from the presence and reaction of the sender or from the behavior of the predator. Moreover, a variation of the state of excitement in the recorded and played sounds, which could be simulated by the change of the amplitude, did not cause any other sorts of reactions (Seyfarth et al. 1980). The decisive factor, therefore, is not the state of excitement in the caller but the content of the call.

Finally, researchers checked whether the vocal sounds function as words in the sense that they are actually perceived primarily in their different meanings rather than in their acoustic properties. The fact that humans do this is particularly evident in the fact that sounds that are phonetically similar do not always lead to the same reactions in the listeners—because they can each have different meanings. For example, "offense" and "defense" are immediately assigned different meanings despite their similar sound. Conversely, very different types of sounds can lead to similar reactions—because they can mean the same thing despite their different sound. For words like "offense" and "crime," we tend not to pay attention to the different sounds but immediately consider them to be synonyms. According to Seyfarth and Cheney (1993, 90), we tend to value "the relation between the words and the things designated … [rather] than the acoustic characteristics." According to Seyfarth and Cheney, this is a significant transfer achievement that means that a real call "is essentially no longer attributed to the characteristics of the sound, but to those of the designated object" (90).[27]

---

27 Of course, this does not mean that every sound does not have an (albeit sometimes slightly different) sound profile. The meaning of a sound can always only be conveyed via the sound itself; a meaning completely without sound cannot be conveyed in spoken words.

How strongly we pay attention to the meaning of a word is revealed in experiments on the habituation effect. Here, one becomes accustomed to listening to a word through repetition of the sound. However, upon hearing a new, sonically similar and meaningfully different word, one recognizes the new meaning (that might have been barely audible) and orients oneself toward it. We encounter this phenomenon even in everyday life. Let us imagine, in the context of the above-mentioned example words, a lawyer who hears how the public prosecutor's office is describing the crime of his client, and let us also imagine that he already knows enough about the crime. He is much more interested in what the public prosecutor thinks about the arguments he himself has put forward in defense of the client. All the prosecutor's pronouncements concerning the offense start to bore the lawyer after some time. When the prosecutor speaks of a "crime" instead of an "offense," he still finds it hard to pay attention. He knows that, in spite of the phonetically different synonym, the whole discussion still revolves around the same fact. The fact that they sound different is obviously not the decisive moment that would make him listen carefully. Only when the word "defense" is spoken instead of "offense" does his attention return. This is astonishing considering the small phonetic difference between the words, and can only be explained by the fact that the lawyer distinguishes the words on the basis of their different meanings.

The transfer of the habituation effect takes place in words that are phonetically similar and similar in meaning. It also occurs in words that are sonically different but nevertheless alike in meaning. The habituation effect, however, is not transferred when it comes to sonically similar sounds with a different meaning, nor when it comes to phonetically different sounds with a different meaning. In addition, the identity of the speaker plays a role for the listener. It is also possible to check whether the habituation effect is transmitted to the calls of other individuals. When it comes to a new meaning, it makes sense to consider new calls, regardless of who issues the call (the individual whose first sound one has become accustomed to, or another individual). If, however, the second call has a similar meaning, then it might make sense to react to a new originator with new attention, but no longer to the same issuer of the call. It is in these contexts and their possible combinations that conceivable tests are devisable with regard to Vervet monkey calls (see fig. 8).

What researchers needed for the experiments were suitable sounds. Both a rumble and a chattering of the animals are *sonically different but, when it comes to their meaning, very similar*. Both sounds are emitted in an encounter with a neighboring group: The rumble, a loud, long-rolling *wrr*, is used as soon as the other group is spotted. The sound evidently makes the members of each group aware of the mere presence of the other. A rough-hued barking or chattering

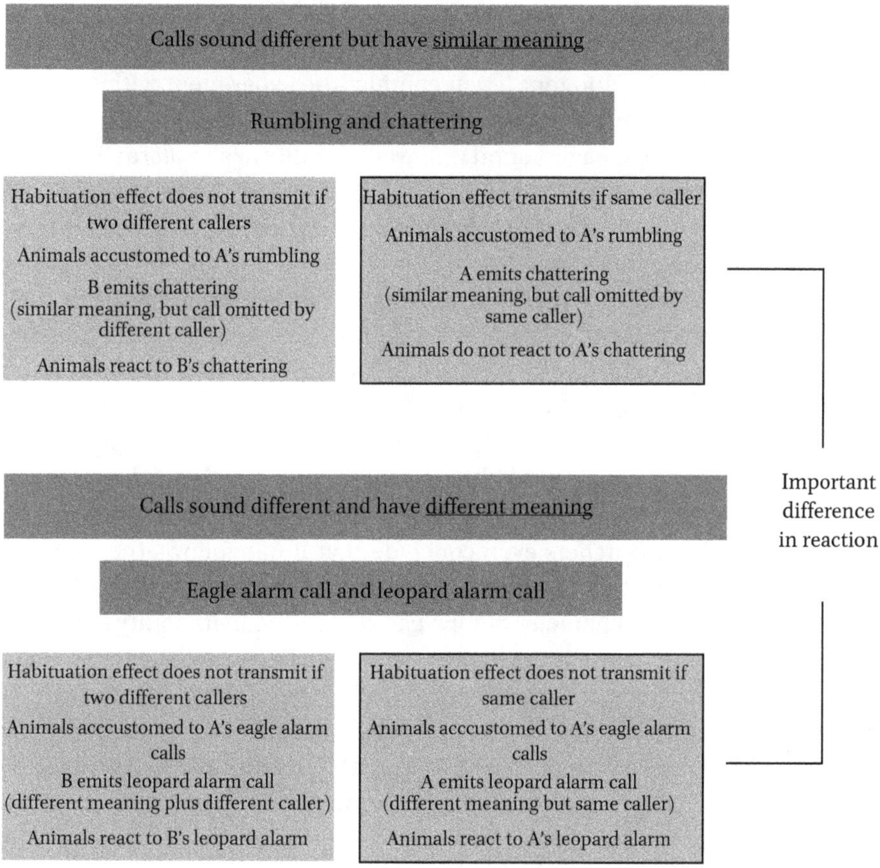

FIGURE 8 Vervet monkeys habituate differently, depending on whether the situation concerns sounds that are the same in meaning or different (cf. Seyfarth and Cheney 1993)

sound only accompanies territorial clashes, as the clash escalates into threats, chases, or fights. The two calls are very similar in meaning, at least insofar as they inform about another group; in the quality of the sound, on the other hand, they are very different. In a habituation test, the monkeys would have to keep them separate if they were solely based on the acoustic impression. Otherwise, if they had no significant difference in meaning, they should not differ much from one habituation to the other (Seyfarth and Cheney 1993, 91). It turned out that if the animals got used to the rumbling, and if one then played them the chattering that is similar in meaning, then the habituation effect was indeed transmitted—if the rumble and the chatter came from the same issuer. If the chatter sound came from an individual other than the senselessly repeated rumble had, then the habituation effect did not transfer (Seyfarth and Cheney 1993, 92).

The picture is quite different if you conduct the experiment with two *calls that are different in sound and meaning*. For this purpose, the animals were accustomed to a repeated eagle sound that was in reality a false alarm. However, if they were presented with the meaningfully different leopard alarm, they still reacted attentively, regardless of whether the new call came from the same individual or from a new originator. It was obviously important that it meant something new. The experiments succeeded in causing the monkeys to compare two vocalizations, to judge their similarity, and to recognize their distinguishing criteria. The response of the animals suggests that these primates have an idea of the importance of the calls of their peers. If a second scream sounds, then the animals do not just compare the acoustic quality (otherwise, similar sounding calls would lead to the same reactions), but, according to Seyfarth and Cheney (1993), they paid attention to the meaning of both sounds. The researchers even conclude that if one follows the idea that a Vervet monkey sound can feature as a word as soon as it is no longer primarily attributed to its sound character but instead recognized in its ability to characterize an object or a situation, then one must already interpret the rumbling and the chattering noises as words. In this sense, the monkeys seem to have an elemental semantic system in which some calls—such as alarms for leopards and snakes—signify quite different circumstances. Others—the rumbling and the barking or chattering, for example—seem to refer to a similar situation for the monkeys; these could be interpreted as representing graded content within a higher-order semantic class or category (92–93). The fact that snake and leopard, obviously, count as different instances of the category of natural predator certainly does not tell us much about the animals' understanding of such higher categories. In addition, it remains unclear here what the animals exactly know about the information status of their conspecifics.

However, what already became apparent in the early experiments by Seyfarth and Cheney is that young animals practice the calls. They issue imprecise but not completely inappropriate alarm calls, for example, a leopard alarm as other potentially dangerous land mammals approach or an eagle call even with the approach of pigeons and other birds. This indicates that the competence of appropriate alerting of young animals is practiced and refined (Seyfarth and Cheney 1993).

The experiments surrounding monkeys' alarm calls are by now in an advanced state. Ouattara et al. (2009) presented a very extensive long-term study. They analyzed the sound repertoire of Campbell monkeys (*Cercopithecus campbelli*) and demonstrated a previously unknown vocal complexity in these primates:

> Adult males produced six different loud call types, which they combined into various sequences in highly context-specific ways. We found stereotyped sequences that were strongly associated with cohesion and travel, falling trees, neighboring groups, nonpredatory animals, unspecific predatory threat, and specific predator classes. Within the responses to predators, we found that crowned eagles triggered four and leopards three different sequences, depending on how the caller learned about their presence. Callers followed a number of principles when concatenating sequences, such as nonrandom transition probabilities of call types, addition of specific calls into an existing sequence to form a different one, or recombination of two sequences to form a third one. (22026)

This goes beyond specific sounds for different predators as well as indicating their mere presence. In light of their findings, the researchers even speak of a "proto-syntax" in monkeys: "We conclude that these primates have overcome some of the constraints of limited vocal control by combinatorial organization. As the different sequences were so tightly linked to specific external events, the Campbell's monkey call system may be the most complex example of 'proto-syntax' in animal communication known to date" (Ouattara et al. 2009, 22026). Other primates, such as Diana monkeys (*Cercopithecus diana*), have similarly complex alarm calls (Zuberbühler et al. 1997, 1999). In suricates (*Suricata suricata*), in addition to the nature of the danger, its immediate imminence and thus the degree of urgency of the appeal to flee are graduated accordingly. The recipients react more or less vehemently in a corresponding way (Manser 2001; Manser et al. 2001, 2002). Also, the alarm calls of Siberian jays (*Perisoreus fistus*) are known to not only indicate the type of predator but also its current behavior. Their alarm calls for resting or attacking birds of prey are different and will be answered accordingly with different responses (Griesser 2008).

Since alarm calls from animals are complex systems, their interpretation must be correspondingly complex. It is clear that the Vervet monkeys can distinguish different types of events, provide adequate alarm calls, and react appropriately to them without seeing the designated event itself. Accordingly, Newen and Bartels (2007, 304) reason that, without the need for the visual stimulus, the animals can distinguish between different types of dangerous situations that require different reactions. In addition: "The monkeys are not only able to understand a certain alarm call (e.g., eagle-alarm) as a warning with respect to a special predator, but they also modify their reaction if some other stimulus appears in their proximity which indicates that the alarm call was incorrect (e.g., if there is an indication of leopards, but no further sign of leopards in the immediate environment)." False alarms could be recognized as false positives: "The monkey who calls the eagle alarm can modify it into a leopard alarm in the case of new evidence" (290). In this sense, the animals can recognize when they have made mistakes, and they seem to learn over time to better distinguish among predators. Discrimination ability and a stable mental representation are given. But there is also a certain stimulus-independence:

> The response behavior of the animals has been observed to depend on further "neutral" stimuli like the caller's reliability; for instance, there is significantly more attendance to the calls of older, experienced animals.
>
> SEYFARTH et al. 1980

There is even evidence that alarm signals can be used for deception in rare cases: When two groups of vervets were in a territorial fight and one was about to lose, a member of the inferior group gave a leopard-alarm, with the consequence that the fight was immediately terminated and the territorial hierarchy of the groups remained unchanged. These cases nicely demonstrate the stimulus-independent use of alarm calls and the best way to describe their behavior is to presuppose that they systematically classify events according to the properties involved. Since their alarm system constitutes a minimal semantic net we have some evidence that they have the perception-based concepts EAGLE-ATTACK, LEOPARD-ATTACK, PYTHON-ATTACK, and BABOON-ATTACK. An important generalization here is that the first two conditions of concept possession may not always involve an object-property structure but sometimes an event-property structure. (Newen and Bartels 2007, 304)

In summary, it can be said that the expression function of linguistic signs is also present in animals and is generally beyond doubt. On the contrary, it is often deliberately attempted to emphasize the appeal and representation functions of animal sounds in addition to an expression function, or it is deliberately attempted to establish that a variation of an alarm call is not solely an expression of emotional excitement (see Seyfarth and Cheney 1993, 88). In the meantime, however, research has gone beyond this. That animal sounds have a convincing appeal function has been presented; however, the clear, comprehensible competence here seems to lie in the intelligent listener who decodes the appeal. Whether and to what extent deliberate arbitrary control and targeted information transfer exists on the part of the sender is still controversial. The fact that the sender pursues intentions with his signs seems, after all, to be deduced from the context of situations and the counterreactions of the sender to unwanted receiver reactions.

According to Bühler, expression, appeal, and representation functions are dominance phenomena and exist as such in humans as well, for whom one of the three basic roles of the speech sounds is alternately in the foreground (1934, 32). Bühler still develops a difference between humans and animals with a view to the dominant function here. In humans, this is the representation function of language, which he believes to be absent in animals (8, 30). Strong representation functions can be found in theoretical statements. According to Bühler, this is "most pronounced" in scientific language and "reaches its peak" in systems of logic (32). Now, it is of course clear that very few people express themselves in everyday life using logical-mathematical formulas. Even if Bühler were right that referential information in the sense of the (pure) representation function only occurs in humans, important questions remain. First, the function still might exist in animals in a weak(er) form. Secondly, it could be questioned whether such a pure or dominant occurrence of the representation function should indeed be a crucial condition for basic semantic understanding per se. Mathematical-logical, semantic signs reduced to their pure representation function might not be present in animals. Also, a pure narrative for the sake of representation, not as an expression of the speaker and not as an appeal or intention conveyed to the listener, has so far not been observed in animals in a way we sometimes find it in humans. But is a dominant representation function needed to speak generally of language ability, or is it only necessary that this function should emanate from a specific scientific-descriptive language? And how pronounced and how isolated from expression and appeal function is human representation for the sake of sole representation, and human narrative for the sake of sole narrative? Are not expression and appeal function more dominant in everyday human language than such a pure representation function? And should we not, in terms of language in animals, orientate ourselves to the functions of everyday human language rather than to the functions of special human languages?

5.4 Propositional Representations and Basic Grammar: First Notes from Language-Teaching Experiments

So far, vocalizations have been understood as utterances of individual "words" and, with the exception of the question of intentions behind the gestures of orangutans, it was primarily discussed from the recipient's perspective whether and to what extent individual sounds in animals are like concepts, with a semantic content serving specific functions in animal communication. The discussion has not yet turned to the *combination of concepts* as a specific, meaning-generating achievement of a speaker, who not only has individual terms at her disposal but can express herself in sentences with their help. Language ability in the sense of the creation of sentences through the combination of concepts—that is to say, language *production* beyond language *comprehension*—may be especially controversial in animals. Nevertheless, some important results of research, which refer to the fact that some animals have so-called propositional or sentence-like representations, are now to be presented. In the process, we are looking for sentence-like structures (up to a basic understanding of grammar) in animals. With regard to content, I will look for indications that some animals can also communicate about complex events that are not actually present, but only thought. We turn now to the language-teaching experiments with great apes, especially the experiments with Kanzi the bonobo.

In the 1970s, there was a scientific breakthrough in the field of language teaching experiments with animals; for the first time ever, researchers trained apes in American Sign Language (ASL). Since then, many other language-teaching attempts have followed, in which apes learned to understand or speak ASL, spoken language, or lexigram languages (consisting of symbols that are selected by the animals using tablets or computer touchpads, see fig. 9). They delivered results that were as interesting as they were controversial. There have been reports of word creations, generalized uses and even spontaneous sign communication between individuals.[28]

FIGURE 9 Lexigrams from the language teaching experiments with Kanzi: "No," "Jump," and "Log Cabin" (a cabin in the woods)
ADAPTED FROM SAVAGE-RUMBAUGH ET AL. (1998, 16)

28 On the experiments with great apes, see Savage-Rumbaugh et al. 1998; on language in animals in general, see Gould and Grant Gould 1997, 99–221, and Hauser 2001, 223–65.

The most interesting results came from the language studies with a bonobo (*Pan paniscus*) called Kanzi. Kanzi was initially not trained in a lexigram language. He was only present in his mother's training sessions, saw how she learned symbols for individual words, and then, by his own volition, acquired some of the signs as an observer. Kanzi is now more than thirty years old, has been using lexigrams since he was about three years old, and is expanding his repertoire, in constant communication with people. Accordingly, he has amazing skills. He now understands the spoken English language of his trainers on the level of two- to three-year-old children and knows more than 384 lexigrams, with which he can report, ask, and answer himself (Savage-Rumbaugh et al. 1998). Only in a few cases do the symbols he uses resemble the objects they describe, in most cases they are of an abstract nature or based on letters.

Kanzi proves an enormous understanding of language through his correct responses to very different requests and questions, which are also completely new to him (67ff.). The following examples are just a few of many that could be documented in the course of the experiments:

> When [Kanzi] heard "Would you put some grapes in the swimming pool?" he got out of the swimming pool, walked over to where a number of foods were placed on a towel, and picked up the grapes and tossed them in the water.... When asked on a walk in the woods "Would you please give Panbanisha [Kanzi's adoptive sister] an onion?" Kanzi looked around for a patch of onions and, when he found one, pulled out a bunch and took it back to Panbanisha. When he had a new ball that he wanted to take into the colony room with him when he went to visit Matata [Kanzi's adoptive mother], but he was told "Your ball stays here"; he placed it by the door while visiting Matata but of course retrieved it immediately after he left. One day when Kanzi was visiting Austin [another bonobo], he wanted some cereal that had been prepared specifically for Austin. He was told, "You can have some cereal if you give Austin your monster mask to play with." Kanzi immediately found his monster mask and handed it to Austin, then pointed to Austin's cereal.... We noted that when Kanzi heard a sentence like "Can you throw a potato to the turtle?" he never made a mistake such as throwing both items or throwing the turtle towards the potato. Typically, he either responded correctly, or in some way that suggested he had not clearly heard a specific word but had nonetheless grasped the general gist of what was said. For example, he might throw a tomato to the turtle, rather than a potato, if he were not listening carefully.
>
> SAVAGE-RUMBAUGH et al. 1998, 67, bracketed glosses by JBS

Kanzi began independently to combine the lexigrams: "He recognized that two symbols could be combined to form meanings that neither symbol in isolation could ever convey" (63). He chose combinations like "chase ball" or "chase you" to express exactly what he wanted to play, and with whom. He combined "backpack lighter" to ask for his backpack, from which he needed the lighter to turn on a fire, and "childside food surprise" to say that he wanted to go to a certain terrain area to find hidden food there (62–63). Kanzi comments on what will happen next in the video he knows, even before the event takes place (45), and he expresses what he wants to do himself and what others should do. In games where the roles are interchangeable, he determines who should catch or tickle him, who is to be caught or tickled by him, or who is to catch or tickle someone else while he is watching. This requires specifying the respective roles and the respective game by tapping corresponding lexigrams in a specific syntactic order (49). According to Savage-Rumbaugh, Kanzi is thus able to understand the spoken language of people communicating with him to an amazing degree, and he can also use the learned symbols in new situations and combine them into new meanings—an activity that he was not trained to perform. In addition, he uses the lexigrams to refer to objects or facts that are not actually present in the speech situation.

This is different from the experiments with Alex (who could correctly name items that lay in front of him and compare them with one another) and is also different from the alarm calls of the monkeys (if they reacted to real, approaching predators). Kanzi also describes earlier facts or personal desires and reports on what he did earlier or would like to do later. Savage-Rumbaugh states that Kanzi clearly knows that words can be used to communicate about things he wants or intends to do, even if the actions signed are not happening at the precise time of the communication (63). By thus combining symbols independently, he can describe events that took place when the listener was not present and express preferences that he later realizes, for example, walking on a certain route through the forest (36ff.).

The combined results of Savage-Rumbaugh's studies go beyond the correct use of individual terms by an animal. According to Newen and Bartels (2007), the results of these language-training experiments indicate that Kanzi the bonobo possesses propositional representations and that his symbolic system represents a basic natural language (2007, 300). Representations are propositional if they represent a systematic combination of at least two different concepts, thus enabling new combinations in a productive way, and if they are strongly stimulus-independent: "The combination of the concepts RED and BALL constitutes the representation that the ball is red" (Newen and Bartels

2007, 298).[29] Although there is no consensus as to what a natural language is, the authors propose a minimal definition following their consideration of propositional representations, which is ultimately also applicable to Kanzi: a symbolic system forms a natural language when, first, the symbols are conceptual representations and, second, they are linked together in a way that satisfies the principles of compositionality, systematicity, and productivity described for propositional representations, and if they are stimulus-independent. These criteria may not be sufficient, but they certainly are the key, necessary criteria for natural languages (Newen and Bartels 2007, 298–99). Newen and Bartels come to the conclusion that, even if this is merely the minimal definition of language ability, Kanzi does at least satisfy it.

The authors, however, are aware that the research findings of Savage-Rumbaugh are controversial and in some (though not all) cases give the impression of being based more on individual observations of the researchers than on clearly documented, systematic investigations. Now, anecdotal observations are nevertheless observations, and as such they often herald controlled experiments that could confirm them. Newen and Bartels recognize that the observations about Kanzi's behavior must therefore be acknowledged with caution. Provided the description of his abilities is correct in principle, Kanzi would be a paradigmatic case of an animal possessing propositional representations. Because he possesses the capability to produce such sentence-like representations, one can speak of a natural language in a simple sense. Note that this does not mean that conceptual as well as nonconceptual representations cannot already exist without a natural language (Newen and Bartels 2007, 301).

---

29 "Propositional representations are defined as a combination of at least two concepts (not related as dimension and instance to each other) while the composition of the concepts is systematic (according to the generality constraint), productive (it can be used to produce new combinations) and independent of any characteristic stimuli (strong stimulus-independence)." For the stimulus-independence of propositional representations required here, the authors write: "An additional feature must also be satisfied to make it a paradigmatic case of a propositional representation: the strong stimulus independence of the representation. This feature is closely connected with the so-called 'Generality Constraint' …: If a cognitive system has two concepts F and G and can represent two objects o1 and o2 then it is able to classify (a) object o1 as having the property F, (b) object o1 as having property G, (c) o2 as having property F, and (d) o2 as having property G." So if a cognitive system is able to represent, based on its perception, the fact that the ball is red, and if it also has a concept of other colors (say, the concept BLUE and the concept YELLOW), then it must basically be able to represent that the ball is blue or yellow "even if it is neither the case that the ball is yellow nor that the ball is blue. If the other systematically possible representations can only be formed when the relevant perception is available, then the representation is not propositional" (Newen and Bartels 2007, 298).

Newen and Bartels clearly differentiate between Alex's and Kanzi's abilities when they delve into the limitations of their theory: according to their analysis, Alex has the perception-based concept of RED and can represent red as a color. But while he does use the term COLOR, it cannot be proven in the experiments how far he understands this upper category. With regard to his terms for the upper categories, he simply was not asked by the researchers: "He definitely has neither the concept SAMENESS nor the concept of DIFFERENCE nor the concept PROPERTY"—even though he was able to assess individual colors and materials in terms of their affiliation, their equality, and their diversity with respect to these concepts (Newen and Bartels 2007, 300). In order to have such concepts, a cognitive system must have propositional knowledge: "Such complex notions can only be learned in a holistic way by acquiring the relevant propositional knowledge" (300).

Perhaps comparable to Kanzi's abilities are the abilities of other language-trained great apes (such as those of Washoe the chimpanzee or of Kanzi's adopted sister, Panbanisha), who mastered either ASL or lexigram languages. There is a growing research interest in the question of whether animals can structure longer sequences using grammatical composition rules. Trained apes seem to be able to understand simple rules of composition and, to a lesser extent, their application. Savage-Rumbaugh et al. (1998) report that Kanzi has few problems decoding embedded phrases and also observes word-order rules in his own use of lexigrams (63–73, esp. 64, 72).

Let us move from great apes to another species in search of clearer and perhaps less controversial data. Dolphins, too, understand signals generated in artificial languages and correctly execute complexly formulated instructions. The artificial languages used here consist of acoustic signals, computer-generated whistles, and hand gestures.[30] The dolphins seem to understand these arbitrary symbols and also the rules by which they are combined. The sentences they understand can be nonrelational (naming only an object and an action to be performed on that object). But they can also be more complex and relational if, for example, two objects and one action (that relates in a specific way to these two objects) are named. This would be the case, for example, for a request such as "Bring the tire on the right side of the basin to the basket on the left of the basin." Such a request would typically be given to a dolphin in the experiments of Louis Herman and his colleagues at the Kewalo Basin Marine Mammal Laboratory in Honolulu, Hawaii, while the dolphins were surrounded

30 Acoustic language was used with the dolphin Phoenix, and gestures were used in the experiments with Akeakamai (Herman 2009, 8). Akeakamai died in 2003.

by many different objects on the left and right. In order to know what she should do, the dolphin needs to understand what she has been told.

In the experiments with the dolphin Akeakamai, an inverse grammar was used. Therefore, a signed sentence like "Person—Ball—Fetch" would mean "Get the ball and bring it to the person" (the person could either be in the pool or on the edge of the pool.) The correct decryption direction for the objects is backwards (or inverse); the transport object ball is located behind the person for whom the object is meant, and the verb comes only at the very end. Inverse grammar, therefore, does not allow you to perform the required answer step by step after every single gesture without waiting for the other gestures. So the dolphins cannot follow a word-for-word approach, but have to decrypt the complete sentence to know what relation there is between object 1 and object 2. In the sentence "Person—Ball—Fetch," the relational action between object 1 (transport-object-ball) and object 2 (destination-object-person) is only given by the verb (fetch) at the end of the sentence. The sequence of the action to be carried out is thus different from the arrangement of the gestures due to the inverse grammar: swimming to the ball, picking it up, and bringing it to the person must be executed. An animal that already swims toward the ball at the first gesture "person" has already failed in the task. Akeakamai, however, understood up to five-word sequences in the experiments (in which not only two objects and their relation to each other were modified but also the location of the objects—by the gestures for "left" or "right"), and she already understood such sequences upon first hearing them (Herman 2009, 9).

Crucial in Herman's experiments is that semantic contrasts can be formed through the formulations, since semantically reversible sentences are possible as prompts to the dolphin. Sentences in which the direct and the indirect object exchange their position or in which the location information for the objects changes suddenly take on different meanings. The response of the animals to this challenge shows that for humans and dolphins the phrase "Swim to the tire at the water's surface and take it to the basket at the bottom of the pool" means something different when the objects are transposed in their grammatical position ("Swim to the basket at the bottom of the pool and bring it to the tire on the water's surface") (White 2007, 96, with reference to Herman et al. 1984; see also Herman 2002, 2009, 9ff.).

In addition, the language-trained dolphins react rationally to so-called semantic and grammatical anomalies (Herman 2009; Herman et al. 1993; Holder et al. 1993; Herman and Uyeyama 1999). A *semantic anomaly* violates semantic rules and semantically possible relationships between objects. For instance, the phrase "Hoop—Phoenix—Fetch" prompted Akeakamai to take the dolphin Phoenix to the tire. This is impossible: Phoenix cannot be transported.

The verb and the second object do not fit together semantically. An animal that has understood that something should be brought to the tire and that the object to-be-transported represents a problem can, however, react in a comprehensible way while preserving the word order. Possible rational responses to the anomaly would either reject the invitation from the outset or transpose it by bringing another, transportable item to the tire (i.e., performing a so-called substitute-action). In 72 percent of cases, Akeakamai declined to act;[31] in 22 percent of cases, she chose another (transportable) object and brought this to the tire. Yet she never turned the meaning around and tried to bring the tire to Phoenix rather than Phoenix to the tire. This shows her strict control of the word order for the interpretation of a meaning (Herman 2009, 10).

While dolphins' reactions to semantic anomalies are already impressive, there are also interesting findings on their reaction to *syntactic anomalies*. Syntactic anomalies break syntactic rules or relationships. For example, in the grammar known to the dolphins, there is no syntactic rule for three object nouns in succession. Sentences like "Speaker[32]—Surfboard—Basket—Fetch" are not syntactically allowed. In figure 10, for this sentence all semantically and grammatically possible as well as impossible decodings are listed. On the one hand, a rational reaction to this syntactically anomalous sentence is to reject the request. Alternatively, however, according to the inverse grammar, *grammatically possible* structures that are embedded in the inadmissible sequence could be filtered out. To do so, the animals would have to ignore one of the given object nouns (indicated by x in the figure) and recognize the possible structures for the other two, plus the verb. The dolphin was capable of doing so. Since the speaker was attached to the edge of the pool and only the surfboard and basket were transportable items, the surfboard could be brought to the loudspeaker (as if the sentence had been: "Speaker—Surfboard—Fetch") according to the valid inverse grammar rules. The basket could also be brought to the surfboard ("Surfboard—Basket—Fetch") or the basket to the speaker ("Speaker—Basket—Fetch"). Decoding the sentence as if it had been "Basket—Surfboard—Fetch," on the other hand (meaning "Bring the surfboard to the basket") is indeed *semantically possible*. But it is not a *grammatically permissible structure* if an inverse grammar applies to the original command with its three object nouns. Interestingly, this decryption of the sentence, listed in the third position in figure 10, was not performed by the dolphin.

31 She first turned to object 2, but then broke off this movement, turned back to the trainer, and undertook no further action (Herman 2009, 10).

32 "Speaker" here means a loudspeaker attached to the edge of the pool.

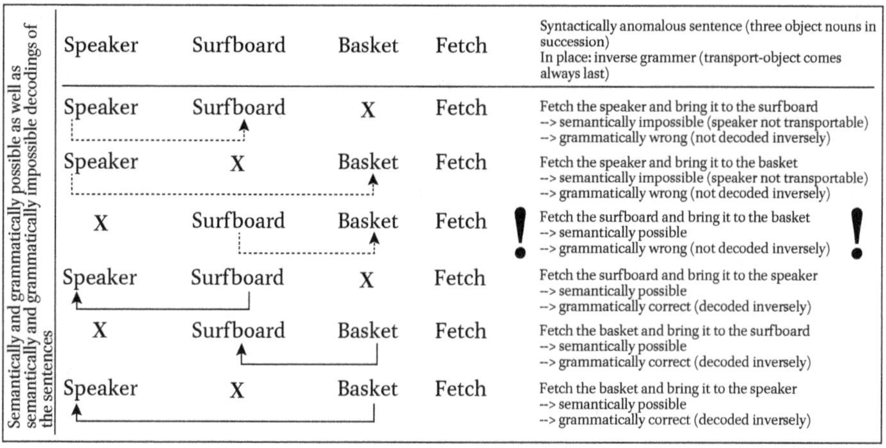

FIGURE 10 Dolphins are able to filter grammatical structures in language-training experiments. Therefore, they do *not* bring the surfboard to the basket, even if it was semantically possible.

The animal seems to know that the sentence must be not only semantically but also grammatically possible to be a valid command. In twenty-four out of thirty-seven anomalies, Akeakamai extracted grammatically possible subsets and executed them. In six cases she refused the request from the start and in seven cases she carried out a substitute-action. Here too, however, she never simply reversed the word sequences in ignorance of her familiar grammar (Herman 2009, 10). Thus, in language training experiments there are the first promising results for the understanding of simple word-order grammars in dolphins as well.

Additionally, the ability to connect individual sounds in a rule-guided way is being sought in the natural communication of animals. The combination and recombination of sound sequences in the sense of a "proto-syntax" in monkeys has already been mentioned. Vogelsang also refers to a simple syntax (see Berwick et al. 2010). Evidence of a complex, rule-based embedding of sound sequences and their variation can furthermore be found in the songs of gibbons and whales (Clarke et al. 2006; Suzuki et al. 2006; Helweg et al. 1992).

In order to assess the previously sparse findings on grammatical abilities in animals, the central question seems to be whether grammatical competence as a form of hierarchical thinking and of this hierarchical organization of modules is completely different from other modes of hierarchical thinking. Although evidence of grammar in animals has been sparse so far, there is much evidence of hierarchical, structured, and even logical thinking in animals in domains other than language. Moreover, grammar is just one of several features of language and, finally, it is unclear how complex a grammar must be to be considered a basic grammar—and a sufficient criterion for language. In 1956, Noam Chomsky created a hierarchy of formal grammar classes that in turn generate formal languages. Regular grammars (finite-state grammars) are at level 3 because the permissible production rules are rigidly restricted, and these grammatical forms have little power of generation with regard to possible creatable sets. Their rules are more local and do not include hierarchies over the long distances of long sentences and beyond their embeddings. Type 0 grammars, such as the phrase-structure grammar found in natural human languages, are more unrestricted and lead to a more creative language. The Chomsky hierarchy of grammars in this sense correlates with a Chomsky hierarchy of languages (Chomsky 1956, 1959).[33]

The language of animals, following Chomsky's theory, is therefore something other than the language of humans, because it does not include an infinite embedding of phrases. But it is, nonetheless, not entirely grammar-free, and that could become more and more important in the light of recent studies. It follows more rigid rules of composition, and there is—independent of the findings of the language-training studies with apes and dolphins—evidence that regular grammars exist in cotton-top tamarins (*Saguinus oedipus*) and rhesus macaques (*Macaca mulatta*) (Hauser et al. 2002a, 1573, table 1, 1577ff.; with reference to Hauser et al. 2001, Hauser et al. 2002b, Terrace et al. 2003, and Fitch and Hauser 2004). Even skeptics like Hauser, Chomsky, and Fitch acknowledge this (Hauser et al. 2002a) and differentiate between "the faculty of language in the broad sense" and "the faculty of language in the narrow sense." Language in the broadest sense includes a sensory-motor system, a conceptual-intentional system, and the compositional rules for recursion, which provide the ability to embed phrases in phrases and thus generate an infinite number of statements from a finite set of elements. Language in the narrow sense, however, is—and this is confined to human beings alone—the latter element of recursion. This

33 In more recent publications comparing human and animal speech, Chomsky also emphasizes that animals do have regular grammars, but not phrase-structure grammar: "At the lowest level of the hierarchy are rule systems that are limited to local dependencies, a subcategory of so-called 'finite-state grammars.' Despite their attractive simplicity, such rule systems are inadequate to capture any human language. Natural languages go beyond purely local structure by including a capacity for recursive embedding of phrases within phrases, which can lead to statistical regularities that are separated by an arbitrary number of words or phrases. Such long-distance, hierarchical relationships are found in all natural languages for which, at a minimum, a 'phrase-structure grammar' is necessary. It is a foundational observation of modern generative linguistics that, to capture a natural language, a grammar must include such capabilities" (Hauser et al. 2002a, 1577).

element, so the authors believe, did not develop in relation to language, but was borrowed from other cognitive domains in which hierarchy plays a role (such as complex social interaction, number quantification, or navigation) (Hauser et al. 2002a, 1569).

This leaves us with the need, first of all, to differentiate between questions about language as a communicative system and questions about the rules of composition or recursion (Hauser et al. 2002a, 1569). With respect to the former, there are promising (functional) responses about animal abilities; regarding the latter, there may never be answers to animal competencies. Second, the studies on regular grammars in animals need to be more closely scrutinized and followed up to see if the abilities evident there could not indicate that these animals are capable of understanding more complex and distant hierarchies and dependencies in composition rules. So far, it has at least not been ruled out that some animals may in fact be able to learn the basics of a phrase-structure grammar.[34] Third, the ability for recursion in cognitive areas other than language needs to be sought in animals:

> Comparative work has generally focused on animal communication or the capacity to acquire a human-created language. If, however, one entertains the hypothesis that recursion evolved to solve other computational problems such as navigation, number quantification, or social relationships, then it is possible that other animals have such abilities, but our research efforts have been targeted at an overly narrow search space.… If we find evidence for recursion in animals, but in a non-communicative domain, then we are more likely to pinpoint the mechanisms underlying this ability and the selective pressures that led to it.
>
> HAUSER et al. 2002a, 1578

Of course, this would raise new questions. Why have human beings expanded this recursive thinking ability to the development of a recursive language system while other animals have not? If recursive thinking occurs in animals but is restricted to a particular function in behavior while it transcended in humans

34 Hauser et al. (2002a) cite one of their attempts at the grammatical understanding of cotton-top tamarins: "Tamarins have a limited capacity to learn the type of long-distance hierarchical dependencies necessary to achieve the class of phrase-structure grammars. If true, this limitation would place severe restrictions on their capacity to learn any natural human language. It is currently unclear whether this limitation generalizes to other animals, and whether it is similarly imposed on humans at different stages of development" (1578).

and becomes permeable to other intellectual domains, then one wonders what evolutionary mechanisms might have effected this. Either the language of humans was a byproduct of other developments, or there was a unique selective pressure that was responsible for it (Hauser et al. 2002a, 1578). But a third alternative might also be conceivable: at least some of the supposedly nongradual limitations in the language ability of animals, which seem to push them into the area of "mere" communication, may be due to the still-sketchy state of research.

5.5 Can We Speak Meaningfully of "Language" in Animals?

All in all, while many might think that language is a uniquely human ability the issue seems much more complicated than that. In some respects, differences between humans and nonhuman animals are apparent, but in many important aspects of language (especially language *understanding*), humans do share a commonality with nonhuman animals. Finding a simple and broad definition of language that encompasses all of its features, and is in principle applicable to animals, seems more difficult than in the case of culture, if it is not altogether impossible. Language as a total phenomenon can only be theoretically grasped and defined through many individual aspects and subdisciplines of linguistics and philosophy of language. Therefore, different aspects of language and their combination should also be used in comparison to animals. As far as semantics and pragmatics are concerned, the criteria of Newen and Bartels (2007), the idea of the functional roles of animal concepts (Bartels 2005), as well as Bühler's organon model can be used as an interpretive framework. As far as syntax is concerned, we can approach animal's capacities by applying the differentiations provided by Hauser et al. (2002a) or, in conscious opposition, argue that a particular elaborate version of grammar is not necessary for the possession of language. Different degrees or single aspects of semantic, pragmatic, and syntactic understanding seem to also exist in animals according to the results and interpretations presented in this chapter. However, major research gaps become more evident here than in the case of culture. With regard to the following points, one can speak of a limited semantic, pragmatic, and syntactic understanding in animals as well as similarities in the acquisition of communicative competences in humans and animals: the way of acquiring names for things, in animals—as in the case of fast mapping—can be similar to the acquisition of words by human children. In their natural communication, a wide range of animals such as pygmy marmosets (*Cebuella pygmaea*), greater sac-winged bats (*Saccopteryx bilineata*), and some birds go through phases of sound acquisition that resemble the babble phase in children (Elowson et al. 1998; Knörnschild et al. 2006; Marler and Slabbekoorn 2004). It may not be a coincidence that the most succesful language-teaching attempts were those in which animals such as Alex, Kanzi, and Panbanisha acquired communicative function and meaning very early in their development and (in a way similar to human toddlers) by listening to or watching individuals use words, gestures, or lexigrams. In addition, when it comes to acquiring the semantic meaning of different alarm calls, young monkeys seem to learn step by step, by listening and watching. At first they can only roughly differentiate the categories EAGLE, SNAKE, and LEOPARD; however, they achieve specificity in their use of alarm calls over time.

Animals have pragmatic competence as senders and receivers of signals. Alarm calls seem to have a functional reference that needs to be successfully conveyed and decoded. For Peter Bieri, these are also semantic representations, since the animals realize that something is the case (a predator is approaching) and they classify the events (into different types of predators). In this sense, nonlinguistic animals also measure their phenomenal information using a repertoire of templates, and these templates are their *concepts* (Bieri 1997, 24 and Chap. 3). How exactly their classification performance takes place and whether they also think in terms of entire networks of concepts was not very clear in the alarm call examples. At least the language *understanding* of the receiver in the sense of a successful semantic decoding of the characters is now regarded as well documented.

Language *production* on the part of senders—i.e., the question of whether animals as senders send out targeted information and thus use signs primarily as symbols in the representation and processing of information—remains a matter of controversy. The more complex the combinations of sound sequences to be found, and the more adapted they are to changing and complex contexts, the more likely it is that information will be transmitted intentionally and controlled arbitrarily. Animals that vary sounds in many ways—for example, depending on the danger of a situation—and combine them routinely are probably doing more than expressing an involuntary state of excitement. From the sender's point of view, there are initial indications of intentionality in orangutans' gestures. Even skeptics like Hauser et al. (2002a) today acknowledge that humans and animals hardly differ with regard to sensory-motor as well as conceptual-intentional aspects of language (and if they do differ, they do so only minutely) (1573). Recent studies indicate that monkeys and apes, at least, communicate multimodally, taking into account the attentional state of their counterpart (Leavens et al. 2010), and that they use complex combinations of sound sequences with regard to several different aspects of the context (Arnold and Zuberbühler 2006). They even design such complex calls by embedding sequences according to combination rules into other sequences (Ouattara et al. 2009).

From the ethological side, functional definitions of language are emphasized, and it is worth highlighting that language alone cannot be grasped as linear and schematized information-processing between a sender and one or more recipients. Instead, there is an increasing emphasis on language (and language acquisition) as a creative, co-regulated process that operates in constant reciprocal social interaction and in which the communication partners constantly adapt their behavior to one another. In such models, the differences between the complex communication of some animals to the language of humans may seem far less significant than in classical models of communication based on theories of information technology (see King and Shanker 2003 and, in a similar vein, Tyack 2000).

While ethologists have long studied the communication of animals in captivity and in the field, philosophy of language and philosophy of mind have only lately turned to the recent findings of communication and language in animals. They show, however, that we can account for semantic language competence in animals in a strict philosophical sense (Newen and Bartels 2007; Bartels 2005). Concept possession could be demonstrated in Alex the grey parrot, who could refer to different colors and materials of objects with the help of perception-based concepts and classify them as similar or different in terms of their affiliation with the respective upper category. His perception-based concept YELLOW was describable through the functional roles it fulfilled. It had an indicator function, a classification function, and a comparison function (Bartels 2005).

Finally, animals have nonconceptual as well as conceptual mental representations. According to Newen and Bartels (2007, 294), we should not forget that "possession of concepts is not an either-or question, but something that develops gradually. So, in practice we will have many "borderline" cases where we cannot decide whether the conditions of concept possession are already satisfied." What concepts are is mainly a philosophical and linguistic question. Whether animals operate such concepts or not is a purely empirical question. To answer this question in the affirmative would in some cases most accurately reflect what we find in animal behavior, according to Newen and Bartels. Furthermore, provided that the perceptual faculty of the animal is similar to that of human beings, it is not necessary to apply stricter criteria to animals for the attribution of mental states and processes than we do to human beings. Basic residual uncertainties always remain if one is not concerning oneself with one's own psyche but rather those of other living beings: "Given our ability-based approach of concept possession, it seems to be very unlikely that even different human beings share exactly the same concept RED. Thus, to speak of 'our concept RED' can be understood only with a grain of salt" (Newen and Bartels 2007, 304–5).

Regardless of the existence of unclear cases, we have to distinguish between perception-based concepts (which animals like Alex possess) and more complex concepts and combinations of concepts. There are certainly concepts that animals can hardly be said to possess. These are based on sentence-like mental representations and propositional knowledge and are found, for example, in logical concepts such as AND and OR. You can only have such concepts if you can deal with sentences that contain them and if you can draw the right conclusions from these sentences. In addition, according to Newen and Bartels, in the course of their development human beings acquire concepts based on scientific theories. This is true for concepts such as ATOM and ELECTRON, for instance. These concepts can only be obtained if you have at least a minimal (true or half-true) understanding of the underlying scientific theory. However, the acquisition of such complex terms is preceded by simpler terms, resulting in a process of concept acquisition: "Acquiring concepts from an ontogenetic as well as from a phylogenetic perspective is a process of first acquiring perception-based concepts, then sentence-based and finally theory-based concepts" (302).

Bühler had also referred to the dominance of the representation function of human language and, as an example of this, to the possibility of a theoretical-scientific language. We can now answer that even perception-based concepts such as those found in Alex, and simple sentence-based concepts, as they may have appeared in Kanzi, represent forms of comprehension and even simple forms of language. Whether animals (especially animals other than great apes) have complex and sentence-like, let alone theory-based, representations, is an open question. Linked to this is the largely unresolved research question as to whether animals have a simple grammar. Research seems to indicate that complex phrase-structure patterns are not found in animals. However, simpler, more localized, and more rigid compositional rules, as well as simple rule-based embedding, are understood and applied by animals. Such embeddings, and with them other protosyntactic capabilities, must be further investigated in the use of learned artificial languages, as in the natural communication of animals, and compared with the beginning of grammatical competence in infants and young children. The ability of recursion should also be further explored in animals in nonlinguistic areas.

Differences among individual species as well as between humans and animals are certainly also based in the more or less pronounced multimodality of communication. Apes, for example, communicate through more vocal and nonvocal channels than do suricates.

Finally, differences could also lie in the dominance of the representation and information functions of language in humans—if one wishes to continue to emphasize these as strongly as classical language-models do. These two functions could correlate with an understanding of the mental states of other creatures. Communication that seeks to convey information and thereby reduce insecurity in others is at least to some extent based on the ability to assess mental states (such as the knowledge or nonknowledge) of others. If a living creature knows that a conspecific does not know something, it may make sense to specifically inform it about that thing. Research gaps on the question of language in animals (in the sense of targeted linguistic information for others) therefore point to research gaps on the issue of understanding mental states (theory of mind) in animals. Nonetheless, as the following chapter will show, we are facing initial indications that a theory of mind—like culture and language—is made up of many individual abilities, some of which are already found in animals.

CHAPTER 6

Theory of Mind in Animals?

6.1 Theory of Mind as Everyday Psychology

The simplest definition for theory of mind might reads as follows: a child (or an animal) has a theory of mind when it distinguishes between the real world on the one hand and, on the other, her own mental states and processes as well as the mental states and processes of others, thereby assuming that such states and processes motivate (re)actions and interactions. The term "theory of mind" was coined by David Premack and Guy Woodruff. Their 1978 article "Does the Chimpanzee Have a Theory of Mind?" linked an interest in theory of mind with a comparative research program. The definition given there was the following:

> In saying that an individual has a theory of mind, we mean that the individual imputes mental states to himself and to others (either to conspecifics or to other species as well). A system of inferences of this kind is properly viewed as a theory, first, because such states are not directly observable, and second, because the system can be used to make predictions, specifically about the behavior of other organisms.
>
> PREMACK and WOODRUFF 1978, 505

What is described as theory of mind can be deduced from the way in which we normally understand ourselves and our fellow human beings, as we have already seen in Premack and Woodruff's definition. In this sense, the question of theory of mind is a question of human's self-understanding, and more specifically, an epistemological question about specific metacognitive competencies.

Psychologists understand theory of mind as our everyday, common-sense, or folk psychology (Bartsch and Wellmann 1995, 4; Hülsken 2001). Our ordinary understanding of ourselves and other people relies on our understanding of beings that have an inner life. We assume that behind the visible physical world and beyond the observable external actions in each individual there exists a mental world of thoughts, convictions, hopes, ideas, intentions, and emotions. So we not only have knowledge about things and facts, but also knowledge

 | DOI:10.1163/9789004415072_007

about our own mental world and the mental world of others.[1] In Chapter 5, the words of Donald Davidson were cited regarding such metabeliefs. Daniel Dennett also stresses that such knowledge is second-order knowledge. It is not just simple beliefs, desires, and other mental states, but beliefs, desires, and other mental states *about* and *regarding* beliefs, desires, and other mental states. For Dennett, second-order states are a necessary criterion for personhood (Dennett 1981a, 308–09; see also Dennett 1981b). So, our metacognitive knowledge comes together in our theory of mind. Mental processes and states are believed to guide human action, and an understanding of them is thus used to explain and predict actions and behavior. It is the latter function of a theory of mind that leads some scientists and philosophers to believe that a theory of mind is a theory per se. Bartsch and Wellmann say:

> The phrase 'theory of mind' emphasizes the coherence of our everyday mentalistic construal of people and action by linking it to a theory (1995, 5).

It is important to see that our theory of mind truly shapes our daily lives and social interactions. If you want to make someone happy, then you give her something that she would like to have. If you would like to keep something that someone else wants, deceive her by pretending you no longer have it. If you know something someone else doesn't know, then you tell her—and so on (cf. Bartsch and Wellmann 1995, 5). At the center of the theory of mind are actions, beliefs, and desires, and it is a sensible and frequently chosen approach in the literature to conceive of our folk psychology as a "belief-desire understanding of mind and action" (Bartsch and Wellmann 1995, 5; Bischof-Köhler 2000, 9).

Beliefs and desires have in common that they are intentional, in the philosophical sense, since they are directed at something. One is convinced of *something*, assumes *something*, or believes *something*. Likewise, a desire is always directed *at* something. "Belief and desire," according to Bartsch and Wellmann, "are hybrid constructs, spanning mind and world in a particular way by

1 The definition of theory of mind is not limited to an understanding of the mental states *of others*; it also entails a knowledge about one's *own* mental states. One could argue that a certain level of knowledge of one's own mental states constitutes a central part of self-consciousness. In that sense, the phenomena of self-consciousness and theory of mind overlap. It is furthermore possible that a certain understanding of one's own mental world (i.e., a form of self-awareness) is a necessary precondition for having an understanding of the mental world of others. This chapter asks questions in both regards: what animals know about their own perceptual states as well as the perceptual states of others.

describing an internal mental state about, for, or towards the world" (31). It has to be more precisely clarified what is to be understood by a belief or a desire:

> Beliefs are meant to refer to a general category of thoughts encompassing knowledge, opinions, guesses, convictions, and hunches, that is, all mental states that attempt to reflect something true about the world. More broadly, thoughts include not only serious beliefs but fanciful ideas, states of imagination and dreams—mental states that represent fictional worlds. (5)

Thus, a speaker who makes an assumption does not always have to invoke the observable reality. She can also appeal to what she herself only suspects, which is merely her personal opinion or even her purely fictional imagining. In addition, her high degree of persuasion does not mean that what she claims or knows is true. This fact allows for deception and misbelief, which we are aware of thanks to our theory of mind.

Desires are determined as follows:

> Desires are also to be understood as a general category including wants, urges and states of caring about something; that is, a whole range of 'pro-attitudes' toward or about something.
>
> BARTSCH and WELLMANN 1995, 5

If one understands our folk psychology as a schema of beliefs and desires such as this, then the following, for example, would be conceivable: based on our perceptions, certain beliefs arise, which lead to actions, which in turn are followed by reactions such as happiness or anger. On the other hand, our sensations and physiological circumstances (such as love, hate, hunger, and pain) could lead to certain desires, which also trigger actions, which are again followed by reactions. This is illustrated in figure 11. However, a schematization of psychological processes always risks not coping with complex mental phenomena and processes. Of course, many things could be criticized in this scheme: Why, for example, should our perceptions only lead us to believe something, but not to want something? Or vice versa: Why do sensations only result in desires, but not in beliefs? Perceptions and sensations (basic emotions) stand side by side, unconnected, which seems like an artificial separation. Also, the possibility that beliefs may be constitutive of desires is not expressed here. It is also problematic that the schema acts as if perceptions and sensations precede the mental event, merely triggering it, and are not themselves mental states or processes. This is not so. What we see is processed

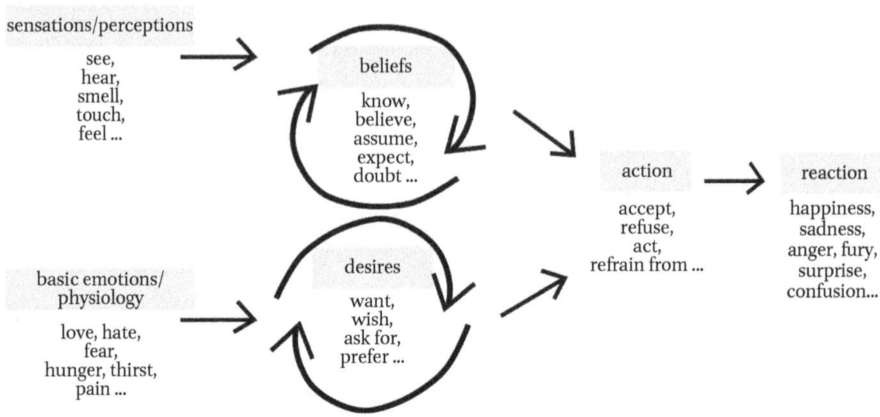

FIGURE 11 Scheme of human everyday psychology based on beliefs and desires
Adapted from Bartsch and Wellmann (1995, 7)

mentally in complex *states of perception*. Knowing what others see or do not see, for example, is also knowledge of their mental act of *seeing*. In addition, emotions can also be triggered by mere thoughts, such as memories, without there being specific (external) stimuli. Thus, although theories of folk psychology focus on beliefs and desires as mental states, one should not forget the role of feelings and perceptions in their complex interconnection with the imagination within the theory of mind.

No matter how one attempts to schematize psychological processes, one will at least have to assume something like the following: human beings are exposed to stimuli that trigger sensations, and this leads to an extremely complex perceptual activity. This already entails cognitive processing and is not to be understood as a linear process. Through constant feedback, what is perceived is recognized and evaluated during the reflection process. Our perceptions are compared with the known, systematized, weighed, and much more. Desires and beliefs arise in varying degrees of intensity (and establish even more complex emotions, such as pity) that we express through our actions—including our verbal actions. As social beings, thanks to our theory of mind, we are able to understand these activities in ourselves and others.

As in the case of other complex cognitive competences, the abilities described here are largely acquired in the course of ontogenesis and refined in social interaction. Section 6.2. will go into theory of mind research in children. My aim is to introduce some crucial research questions and to introduce some methodological shifts that are also of relevance when it comes to theory of mind research in animals.

6.2 Theory of Mind Research in Children

6.2.1 *False-Belief Tests and Their Critics*

Children know quite a bit about their social environment from an early age on. They recognize fellow humans by paying particular attention to their faces (Workman and Reader 2004, 125). Babies spend significantly more time looking at facial-like stimuli than at other stimuli, and at as early as four months of age they show a greater interest in realistic faces than in schematically depicted ones (Workman and Reader 2004, 121ff.; with reference to Johnson and Morton 1991). However, complex social interaction requires more than knowing what conspecifics and group members look like. It is also about how children learn that other people have a complex mental inner life consisting of beliefs and desires, perceptions and sensations. In developmental psychology, it is assumed that comprehensive changes take place in the consciousness of children during the fourth year of life. Some scientists even speak of a "qualitative change" (Kasten 2005, 135; with reference to Maehle 1999, 59). According to this scheme, three-year-olds do not know that there is a difference between what they or others perceive of reality and reality itself. Four-year-olds, however, begin to discriminate between their own perception of reality and that of others. They understand that their ideas have content that differs from the content of other people's ideas, especially when those other people have not had access to a certain source of knowledge (Bischof-Köhler 2000, 11; Kasten 2005, 135).

In developmental psychology, an *understanding of false beliefs* stands as a milestone in social competence. This is because the presence of an understanding that beliefs may be wrong clearly marks a shift to a representational theory of mind (Onishi and Baillargeon 2005, 255). When someone understands that the action of their counterpart is sometimes based on false beliefs, she has understood two very important things: First, the individuality of mental states. Someone else can belief something else than myself (which includes the possibility that one of us beliefs something wrong while the other knows the truth). Secondly, she has understood that mental states are in a special relationship to reality: they *represent* reality which doesn't mean that they coincide with reality. This implies that beliefs can sometimes represent something false, since they are not necessarily accurate in relation to reality.

The so-called Maxi story and a number of similar tests[2] show how three-to-five-year-olds differ in their understanding of false beliefs. The children are told a story, mostly with the help of dolls and a dollhouse, as seen in figure 12.

2 The Maxi-Test is also known as the "Location-Change Task." It was developed by Wimmer and Perner (1983) and modified by Baron-Cohen et al. (1985) to become the Sally-Ann Test.

FIGURE 12 Dollhouse scene typical for tests like the Maxi test per Frith and Frith (2005, 644)

In this story, Maxi puts a bar of chocolate in the green kitchen cupboard and leaves the room to play outside. Meanwhile, his mother takes a piece of the chocolate and puts the rest in a second, blue cabinet. The kids are now asked where Maxi will look when he comes in to eat chocolate. The three-year-olds usually answer "in the blue cupboard." Nearly half of the four-year-olds and over 80 percent of the five-year-olds say that Maxi will look into the green cupboard.

Three-year-olds are, according to this classic experiment by Wimmer and Perner (1983), as yet unable to distinguish between the knowledge *they* possess and the knowledge *Maxi* has. They do not yet understand that knowledge depends on a person's level of information, and that different people can consequently know different or more or less correct things. That someone can have a false belief because she had no opportunity to correct her knowledge does not seem to come to their minds. They do not seem to realize that knowledge is acquired or corrected through new information. Even if one explicitly points out to them that Maxi has not seen the mother put the chocolate from the green to the blue cupboard, they answer, until they have reached the age of four to five years old, that Maxi will look in the blue cupboard (Kasten 2005, 136). The conclusion of the false-belief tests seems to be that children only understand the nature of false beliefs when they are four to five years old.[3] But

Further false-belief tests are the Deceptive-Box Test, by Perner et al. (1987), as well as the Appearance-Reality Test by Flavell et al. (1983) (Call and Tomasello 1999, 381).

3 For an overview of the previous theory of mind research and a good compilation of the relevant work, see Onishi and Baillargeon (2005, 255) as well as Frith and Frith (2005) and, more recently, Andrews (2012).

does this also mean that it's not until they are four years old that they acquire a theory of mind? Or are these false-belief tests simply incapable of revealing the theory of mind in children younger than this age, who are less skilled at language? If this were the case, then we may be dealing with methodological issues that may also be evident in experiments on theory of mind with other beings not endowed with language, such as nonhuman animals. Such methodological issues might reflect in a problematic way on the interpretation of theory of mind tests in children and in animals. If an understanding of false beliefs is overemphasized as a criterion for the existence of a theory of mind, then it may also be taken too seriously as a missing theory of mind criterion with respect to animals. So how could the focus on false beliefs obscure a living being's general theory of mind?

Tests based on the false-belief paradigm often require a number of skills that are not necessarily theory-of-mind-related, such as linguistic or grammatical skills, which children under the age of four possess to a lesser degree. The supposed failure of children in these tests might then be due to the lack of these other skills and not necessarily due to the lack of a theory of mind.

The described studies were also all-or-nothing tests. Either they prove a capacity to exist, or they do not. Intermediate stages do not become visible. This is problematic because such a design corresponds to a monolithic view of theory of mind. Rather than being something that you do or do not possess at all, a theory of mind may be better viewed as a capacity that gradually evolves through numerous stages and that exists in many individual building blocks. Many test designs do not capture such an understanding. In this case, there is a problem that also addresses the question of other socio-cognitive abilities: if these abilities are understood as large, closed complexes of phenomena, then there is a danger that preliminary stages and individual building blocks are being underestimated or neglected.

Research slowly became aware of the methodological problems of the false-belief test and its variants: "Since the test is all-or-none ... it has the effect of making what might be a gradual developmental profile look like a sudden stage-like shift" (Workman and Reader 2004, 128). In addition, researchers started to criticize the rigid insistence on giving children tasks which they needed to master in the lab: "There is an unfortunate syndrome loose in developmental psychology: Call it 'neurotic task fixation'"(Gopnik et al. 1994, 157).

Two new research directions, which will be discussed in more detail below, justify these criticisms. For one thing, in non-language-based tests, children who are already only fifteen months old demonstrate a basic understanding of false belief. On the other hand, even children who are much younger than four years old use mental expressions in their everyday language. They refer to their own mental states and processes and the mental states and processes of other people. They also differentiate their own mental states from the mental states of other people and earlier (false) beliefs from current (true) beliefs. The fact that children of this age refer to mental states and processes in their everyday language shows that they may possess some important building blocks of a theory of mind, even before they pass classical (false-belief) tasks.

6.2.2 *What Do Nonverbal False-Belief Tests Reveal?*

In this section, a study by Onishi and Baillargeon (2005) will be presented, which provides evidence for a very early understanding of false beliefs in children with the help of a nonlinguistic test design.

The researchers used the so-called violation-of-expectation method: instead of researchers telling the children a story or playing with dolls, as in previous false-belief tests, and then asking them questions, the children watch a silent experimenter. During the test, this person first hides a toy in a green or a yellow box. Then a change takes place. For example, the item is taken from the box it was placed in and is then hidden in the other box. In some experimental conditions the experimenter can observe this change; in others she cannot. Depending on what she was able to see, she now has a false or true belief regarding the location of the toy. She believes, for example, that it is still in the yellow box, while it has actually been moved to the green box without her knowledge.

The key question for children's understanding is: Do the children expect the experimenter to find the hidden toy according to *what she believes* (based on both true and false beliefs)? Do they expect, for example, that when she returns, she will look for the toy in the yellow box, even though the children themselves know it is no longer there? The duration of the children's observations is measured. In the case of a *violation of their expectation* (hence the term "violation-of-expectation") their observations, measured by the amount of time spent looking, last longer (Onishi and Baillargeon 2005, 255). The connection between a violated expectation and a longer looking time is well known in psychology and is often used for tests.

If the children expect the subject to search in the yellow box because they suspect the toy is there, but the subject instead reaches into the green box, it will be likely that the children are amazed and spend significantly more time looking. The *violation* of an expectation measuring the longer period of observation is thus connected to the *existence* of such an expectation in the child. In the first trial runs, the experimenter only plays with the toy and then puts it in the green box. The window through which the children can watch her is then closed. After a short delay, the children can watch as the experimenter returns and grabs the green box (255–56). Upon returning, the experimenter acts as we would all expect. She looks for the toy where she left it behind.

In the following stage, the children each observe one of four possible variants of the event after the experimenter has placed the toy in the green box. This step of the experiment is called the "belief-induction trial" (see fig. 13). Two of the variants of the belief-induction trial ensure that the experimenter still returns with the correct conviction as to where the toy is (see fig. 13, variants A and B). The other two variants of the belief-induction trial, however, ensure that her conviction is no longer correct, because something has changed in the meantime without her noticing. The toy is no longer where she thinks she has hidden it (see fig. 13, variants C and D) (Onishi and Baillargeon 2005, 256).

In the actual test run, the experimenter now accesses either the green or the yellow box after her return. The duration of the child's gaze observing the experimenter's action is measured at this stage. The study could show that in all four conditions the children actually expect the experimenter to look where she *rightly or wrongly* thinks she will find the toy because she placed it there before. If she does not do this, the children's gaze is held significantly longer (257). The children themselves have seen that the toy is no longer in the place where the experimenter suspects it to be. Nevertheless, they assume that she will follow her false conviction and seek to find it where she believes it should be. Thus, the children do not simply transfer their own conviction, corrected by their observation 2 ("The toy has changed its location"), to the experimenter. Instead, they expect her to still be guided by her false conviction ("The toy is still where I hid it") if she has had no opportunity to correct that false conviction. However, when she has had the opportunity to observe, just like the children, that the toy was moved, the children in turn expect that she has also corrected her opinion. These results undermine the theory that children cannot distinguish between their own (correct) knowledge and the (false) knowledge of others, depending on what they and what others could each see. It does not seem to be the case that an understanding of false beliefs only arises in four or five years old, and that there is no prior knowledge at all of false beliefs. The researchers themselves conclude:

> These results suggest that 15-month-old infants already possess (at least in a rudimentary and implicit form) a representational theory of mind: They realize that others act on the basis of their beliefs and that these beliefs are representations that may or may not mirror reality.
>
> ONISHI and BAILLARGEON 2005, 257

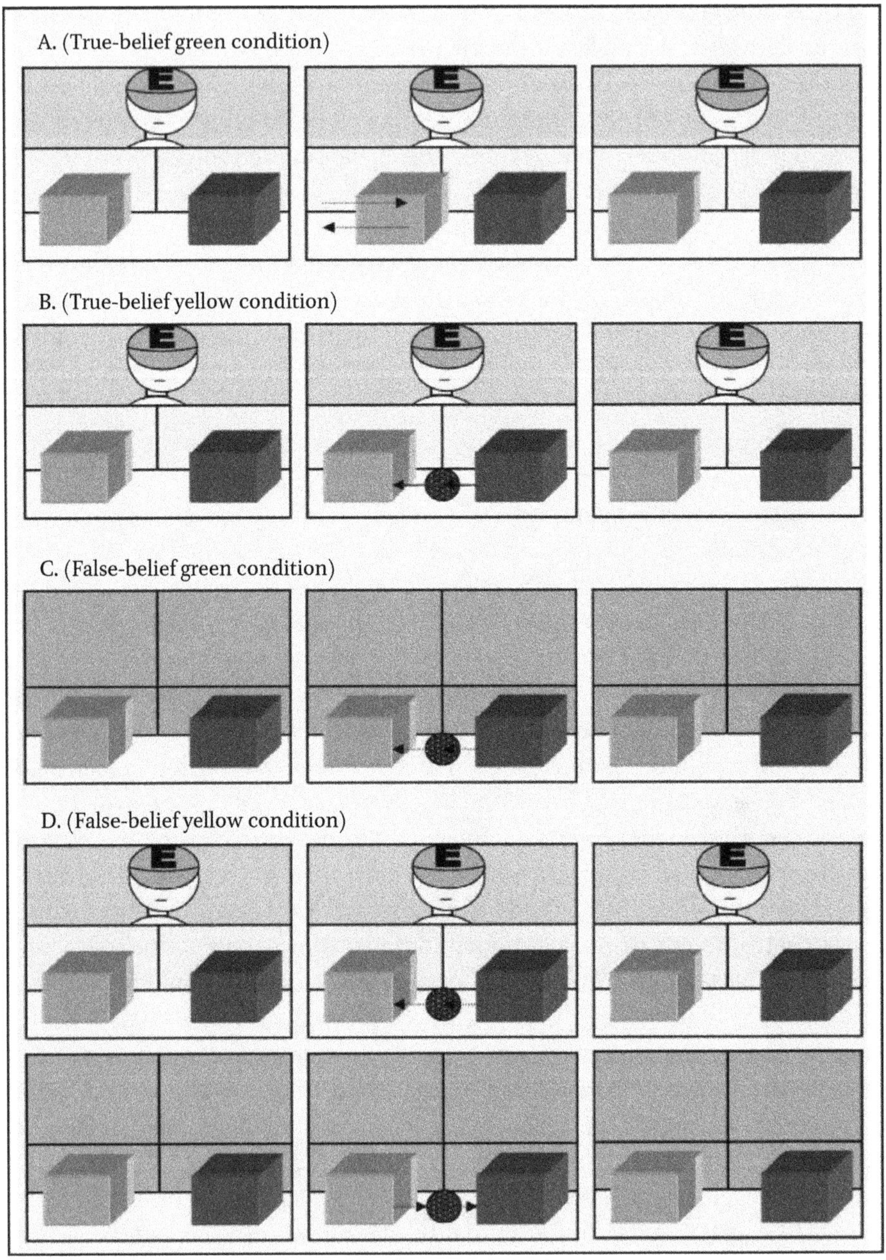

FIGURE 13 Belief-induction trial after Onishi and Baillargeon (2005). A. (True-belief green condition): the experimenter watches; the toy location does not change; only apparent movement is carried out. B. (True-belief yellow condition): the experimenter watches; the toy location changes (green → yellow). C. (False-belief green condition): the location of the toy changes (green → yellow), but the experimenter does not watch. D. (False-belief yellow condition): the experimenter watches; the toy location changes (green → yellow); the viewing window is completely closed; the toy is moved back again (yellow → green), but the experimenter does not watch (Onishi and Baillargeon 2005, 256)

Many questions about the acquisition of a theory of mind in children are still open, and the development of suitable test designs seems difficult. In the meantime, however, Buttelmann et al. (2009b) for example have published first positive results obtained in a paradigm for active assistance in young children that indicate a simple understanding of false beliefs. Further nonlinguistic false-belief tests like these must be developed and discussed in their relevance to the onset theory of mind in infants. While methods, like anticipatory looking measures, are still challenging with regard to validity and replicability in human children and adults (Kulke et al. 2018), researchers have recently used them for testing great apes (Krupenye et al. 2016). The tests show that the animals anticipate an actor to act upon his beliefs, even if they are false. It could also be shown that animals distinguish true from false beliefs in another, nonverbal, implicit test that is mastered by human 1-year-old infants (Buttelmann et al. 2017). We will come back to these tests with animals but stay a bit longer with human children first. In the following section, it will become clear that research results that do not originate from experimental settings but rather from the recording and analysis of the everyday language of children, also question the widespread belief that a theory of mind does not arise in children until the fourth or fifth year of life. This gives us another reason in addition to the growing criticism on classical false-belief tests to suspect that we might easily underestimate the capacities of human children. Only if we understand this in the case of toddlers, we will be careful enough with the paradigms we confront other animals with.

6.2.3 *What does the Everyday Language of Two- to Four-Year-Olds Reveal about Theory of Mind in Children?*

The content of a theory of mind is also expressed in language. Formulations with mental verbs such as "I know/I believe/I think/that x" indicate that an understanding of mental states exists in the speaker. We express certainty or uncertainty in what we know, believe, think, etc. by choosing which of these verbs to use: if I'm sure that what I think is true, then I say, "I know that x," while in states of greater uncertainty I prefer the phrase "I think that x." The speaker can also make her degree of certainty clear by means of so-called epistemic adverbs, such as "surely" and "perhaps." The preference expression "x rather than y" allows us to describe two elements in a comparative way and we can use it, for example, to express a preference for our desires. Using phrases like these, we voice our beliefs and desires and thus also our mental states and processes in their gradations. So-called *contrastive* have special meaningfulness for language acquisition research. Contrastives are contrasting statements in which a speaker makes clear within a single expression or within a very restricted context that she is able to distinguish between different forms of mental states or processes in herself and others, in the use of mental verbs. Such statements (not single words) usually refer to mental verbs and may read: "I believe x, but you believe y"; "I just believe it, but I don't really know"; or "I didn't know before, but now I know."

From what age do children use such mental expressions, and when do they begin to employ the most meaningful contrasting expressions? Bartsch and Wellmann point to linguistic expressions that clearly relate to beliefs in American children from an average age of two years and nine months (1995, 43). Additionally, the studies of Shatz et al. (1983) suggest that "mental references" appear for the first time in the second half of the third year of life (301, 317). Bartsch and Wellmann found contrasting expressions at an average age of three (the children were between two years and eight months and three years and eight months old when they used contrastives for the first time). Perhaps such expressions even occurred earlier, when no recording device was present (Bartsch and Wellmann 1995, 47).

A CHILDES[4] database analysis of the everyday language of two German girls named Caroline and Simone confirmed these findings for the German language (Benz 2004). Contrastives regarding mental states and processes were found in Caroline's data as early as two years and one month. In Simone's data, the first contrastive was recorded at two years and seven months.[5] Thus, at the end of the third year of life, at the latest, German children distinguish between mental states—this is well before they pass classical false-belief tests. They differentiate between different mental states and processes, both their own and those of other people. For example, they express a difference between what they did not know before and what they know now, and they distinguish what they themselves believe from what another person believes (Benz 2004).

The findings on the occurrence of mental expressions in the language of German children and the results of American studies could be an indication that children, even before they are three years old, have an idea of their own thinking and the thinking of others in terms of a simple theory of mind. They may even have a particular understanding before they have the necessary words or grammatical rules that allow them to communicate that understanding to the outside world. Nonlinguistic theory of mind tests seem to indicate

---

4 CHILDES stands for Child Language Data Exchange System. CHILDES is a database for the analysis of transcribed corpora of children's speech (see https://childes.talkbank.org, 2019/05/17; MacWhinney (2000)). A crucial element of these conversations is that they were recorded in everyday situations over several years.

5 For the different types of contrastives, see Benz 2004, 50ff., 75ff., 106ff.

this. Here again, it should be emphasized that until ten to twenty years ago, most psychologists assumed that a theory of mind only exists in children when they are around four to five years old. Some might still hold on to this position:

> Children before the age of three and a half years are naive realists. They view the way the world appears to them as unquestionably true and public. Thus, they do not yet understand that their beliefs about facts are only assumptions that can or may not correspond to a real fact. With the same self-assurance, they assume that other people also live in the same reality, that is, they perceive and know the same as themselves. They do not yet understand that others can have a different opinion about the same facts.... With the onset of a theory of mind in the fourth year, that changes.
>
> BISCHOF-KÖHLER 2000, 11

With the advent of transcribed corpora of children's speech about the mind, a new, startling problem came to the fore. There is obviously a discrepancy between what children reveal about their theory of mind in their actions in the domain of psychological experimentation and what they say about mental phenomena in their everyday language. The reason for this could be that the tests require an artificial situation and apply a problematic methodology. The methodology aims for a core capacity (false-belief understanding) that, from a developmental perspective, might come last, not first, in theory of mind acquisition. Also, this core capacity might not be the only relevant capacity that is indicative of theory of mind.

6.2.4 *Implications for Theory of Mind Research in Animals*

Increasingly, it should be taken into account that theory of mind is not necessarily a monolithic capacity. It is probably not something that individuals possess either wholly or not at all, and it is not something completely absent one moment and then suddenly "there" in the next. If theory of mind develops gradually in ontogenesis, then there must be phases before the fourth birthday in which children have already acquired building blocks of it, and it would be extremely simplistic to say that this was suddenly present at three and a half or four years, and before that absent in every way. This thesis is supported by the research presented here (Onishi and Baillargeon 2005; Buttelmann et al. 2009b; Bartsch and Wellmann 1995, 47; Benz 2004, 106 ff.). When discussing the issue of theory of mind in animals, these considerations should not be forgotten. They play a similarly important role in that domain.

If theory of mind is not a monolithic ability, there are significant implications for future research. If one takes theory of mind as a kind of "toolkit of social cognition" that contains many different components, then the possibility of exploring exactly these individual components unfolds before us in place of the examination of theory of mind as a complete phenomenon or one reduced to an understanding of false belief. Rather than asking only if one person knows what others (wrongly) believe, one can increasingly research other theory-of-mind-related skills, such as knowing what others see or have seen. In this instance, it is also possible to investigate whether some animals possess such social knowledge—even if they may not understand false belief.

For a long time, no convincing evidence for an understanding of false belief in animals could be presented (Call and Tomasello 1999). Here, too, the reasons for negative findings were perhaps at least partly due to test design (Call and Tomasello 1999, 391ff.). Consequently, further experiments and the development of nonlinguistic, animal-appropriate false-belief tests were needed. But before some researchers came up with new paradigms which I will come back to, scientists turned to other types of mental states than false belief and discovered perceptual states as interesting candidates. It seems that some animals are capable of understanding those.

6.3 Theory of Mind in Animals: What Do Animals Know about Perceptual States and False Beliefs?

6.3.1 *Understanding One's Own Perceptual States*

Knowing that (or what) you know might be the basis for dealing with others' knowledge and ignorance. The kind of knowledge I mean here comprises for example the ability to *discriminate between knowing and not knowing* and to *know when you remember*. This section will first address the question of what understanding animals have of their own perceptual states, especially that of their own sight and the resulting knowledge.

Being able to use one's own knowledge—for example, the distinction between one's own certainty and uncertainty—makes flexible behavior in different situations possible. Personal advantages may arise from such knowledge and its comparison to the knowledge of others. For example, some animals hide food from conspecifics. They should know which food is hidden where and be sure about their own memories. However, the act of hiding itself, along with visiting hiding places, can be risky, as observers may later become thieves of the hidden food. It may therefore be helpful to understand what others see and cannot see in situations like these. After all, the comparison of what you have seen with what others have seen is often the basis for deceptive behavior. This takes advantage of the ignorance of others, or leaves others intentionally in a state of ignorance, while you yourself benefit from what you know. Do animals know what they themselves know? Do they understand that they can get information by looking, and distinguish between their own certainty and uncertainty?

The question regarding *knowledge about one's own vision* in animals has been addressed in Hampton et al. (2004), among other sources. In the authors' experimental setup, rhesus monkeys (*Macaca mulatta*) sit on a board, onto which tiltable metal tubes are screwed. Food is placed on the end of one of the tubes. A mechanism allows the monkeys to tilt the selected tube and thus fetch the food. If one tube is selected, the other tubes are locked; i.e., the animal cannot select a tube for a second time (Hampton et al. 2004, 240 ff.). In addition, the height of the board is adjusted, ensuring that the monkeys have to actively look through the tube when they want to check to see which tube the food is hidden in. From their sitting position, they cannot see the food at the end of the metal tube. So if they want to check which tube the food is in, they have to actively bend down and look through the tubes. In some experimental conditions, the monkeys can watch the placement of the food; in others they cannot. If they are aware that they lack the knowledge to choose the right tube in those tests where they cannot observe the food being hidden, then they would be expected to check these experimental conditions before making their choice. Otherwise, they would lose the chance to get the food, since they get nothing if they make a mistake.

The results of the study showed that most monkeys, if they could see where the food was hidden, made their choice without checking. On the other hand, in those experiments where they had not seen which tube the food was hidden in, they tended to look in the tube and make sure the food was in it before trying (Hampton et al. 2004, 244). They seem to know *whether or not they have seen where the food was hidden*. Thus, they have a simple form of metacognitive access to the content of their own short-term memory (239).

In addition, however, rhesus monkeys also seem to *know if they've forgotten something*. Hampton (2001) was able to verify this with a visual and delayed match-to-sample experiment. The animals were first shown a randomly selected image (such as a rooster) on a computer screen. Then the picture disappeared from the screen. After a time delay, in which the animals often forgot what image they had been shown, they now had the opportunity to undergo a test. They could choose to test or not to test by tapping a button—this happened, it should be emphasized, even before they saw the test screen. The test would require them to recognize the previously viewed image among four images in total and select it. The question at hand was to determine under which conditions the animals decided to quit the test. If they decided to subject themselves to the test, they would receive a delicious piece of food if they correctly recognized the rooster. If they chose a wrong object, they would get nothing, so any of the monkeys that could not correctly remember the picture shown would go away empty-handed if they still decided to risk making a (wrong) choice instead of simply opting out. However, if they chose not to participate, they would still get a small reward (Hampton 2001, 5359–360). This situation created the pressure of balancing the risk of being left empty-handed with the degree of certainty of one's own memory of the image.

The results of the experiment showed that the rhesus monkeys made their selection relying on their own knowledge, and only risked taking the test if they had sufficient knowledge to make the correct choice. In so doing, their choice indicated that they knew when they remembered and when they did not. They were thus able to assess and differentiate their level of knowledge. Compared to the control condition, in which an exit option was not offered (because the test demanded that a choice be made), the number of passes in which the object was correctly recognized was significantly higher when given the option to make a choice (Hampton 2001, 5359–360). It is controversial on what considerations and what knowledge the animals base their decisions in these and similar experiments, preferring either to opt out of the test or to gain additional information. And when it comes to great apes (gorillas, chimpanzees, bonobos, and orangutans) Call (2010) has demonstrated that these animals actually know when they might be wrong under more stringent and more extensive control conditions.

The experiments by Hampton and his colleagues are just two examples of controlled attempts to assess the animals' knowledge about their own mental states and processes. Many species have been shown to act according to their own memories. Corvids like the Florida scrub jay (*Aphelocoma coerulescens*) and Clark's nutcracker (*Nucifraga columbiana*), for example, are considered to be particularly adept storers of food, and they are able to find food they have hidden several months ago. They remember hundreds of places where food is hidden, but also the nature or type of food: something that is perishable is dug up and consumed much sooner than longer-lasting foods (Balda and Kamil 1992; Clayton and Dickinson 1998). In order to feed successfully in this way, an outstanding memory is necessary. However, it is also particularly noteworthy in such strategies that, as indicated at the outset, the concealment of food can be observed by conspecifics. Is it clear to the birds that these observers now also know where the food is hidden because they have seen it? Again, this would be based on knowledge of the connection between vision and recalled information and would fall within the scope of theory of mind in animals.

6.3.2 *Understanding the Perceptual States of Others*

6.3.2.1 Knowing What Others Have Seen: Corvids

Studies on the question of whether or not ravens use their knowledge about what others have seen to determine their own behavior build on a long tradition of research (Lorenz 1935; Heinrich 1999). It was proven early on that ravens (*Corvus corax*) visit remote and carefully selected locations both in the wild and in captivity to hide food from conspecifics. Experiments and observations from recent years have shown that ravens also hide their food behind objects in order to block the view of their conspecifics (Bugnyar and Kotrschal 2002). Furthermore, in the wake of a suspected theft, the animals carry already hidden food to a new location and thus conceal it from potential foraging thieves. A purely random misdirection of competitors for food can be ruled out, and the birds do seem capable of tactical deception (Heinrich 1999; Bugnyar and Kotrschal 2002, 2004). Bugnyar and Kotrschal (2002) argue, for example, that ravens, both in the role of food storers and in the role of forage thieves, are capable of withholding their intentions from others. They either do this to prevent them from learning by observation, or simply to avoid conveying their intentions: "Such interactions may qualify as 'tactical' deception and may have created a considerable pressure selecting for social cognition in ravens" (185).

In the meantime, intensive research is being conducted into whether ravens actually take into account the memories and perceptions of others in the competition for food. Test designs exploit the antagonistic interactions between individuals who hide food ("storers") and those who rob the food shelters ("pilferers"). In one of these experiments, the animals were confronted with the following situation: A raven hides food and perceives visually and audibly that she is being observed from an adjacent cage by a higher-ranking competitor (a so-called observer). On the other hand, she can only audibly perceive a second competitor, the nonobserver. She cannot see her, and the nonobserver's view of the area where the raven is hiding her food is also blocked (Bugnyar and Heinrich 2005, 1642). Thus the food-hiding raven can detect the presence of two competitors. These two differ in their perceptual states, or in what they see and cannot see. In a second step, how the ravens behave is observed under different experimental conditions. In some tests, the animals are left alone or without any competitors on the terrain containing the hidden food: either only the raven who hid the food is left in the room (without observer and nonobserver) or the observer with the nonobserver are left in the room (without the first, food-hiding raven). In other test runs, however, a concrete competitive condition is created whereby the food-hiding raven is given access to the area together with the observer or together with the nonobserver. The central question is: Does the raven who hid the food beforehand take into account, whether or not she could have been observed by the competitors? Is she aware that the food in her hiding places is not safe when the observer is there, too?

The experimenters found that the raven who had hidden the food would display a great deal of behavioral variance and behave differently depending on which conspecific she was competing with—as if she knew that the hiding places were no longer safe when the raven who had observed her was in her vicinity. In the competitive situation with the observer, she leaves a smaller percentage of the food in the vulnerable hiding places; i.e., she retrieves more of her hidden food from her hiding places than she would if she were alone or in competition with a nonobserver in the hiding area (Bugnyar and Heinrich 2005, 1643). In addition, she acts extremely flexibly when competitors are present. Only when the potential thief comes near a hiding place does she become active, but she remains inactive when a nonobserver approaches the hiding place. Thus, she does not simply act according to a rule that goes something like: "Go to the food when another raven comes within two meters of it." The raven allows nonobservers (who were unable to see where the food was hidden and therefore cannot know that they are approaching the hiding place) to approach the hiding places without acting to remove the food from those places (1643). Given this behavior, the authors conclude that the ravens have a flexible understanding of the knowledge of others: they not only exploit their own knowledge, but also consider what others could or could not see. They seem to know that their observable access to the test area will influence their later actions:

> Ravens show fairly sophisticated skills when competing with conspecifics over hidden food. As storers, they selectively retrieved caches with individuals that were visible at the time of caching, and as pilferers, they hurried to the caches only with those competitors that had also had visual access to the caching room and thus were likely to pilfer the caches themselves. These results suggest that ravens not only recall whom they had seen during caching … but also know that the others' view can be obstructed by opaque structures … and that this affects their chances of finding cached food.
>
> BUGNYAR and HEINRICH 2005, 1645

Ravens thus demonstrate, according to the researchers, a level of understanding with regard to perceptual states of others that could be comparable to that of chimpanzees. This most likely exceeds simple behavioral rules (1645); the animals do not simply orient themselves to whether another animal was present or absent at the time the food was hidden, but rather likely represent whether the individual present saw the hiding process or not.

Chimpanzees, with whom Bugnyar and his colleagues compare the ravens, have already been the subjects of more numerous and detailed theory of mind experiments. Fifty years have passed since Premack and Woodruff made their inquiry into the theory of mind in these animals. These years have yielded a large number of impressive research findings, which increasingly suggest that chimpanzees can understand some mental states and processes, including the visual perception of others (see for instance Tomasello et al. 2003a; Call and Tomasello 2008).

6.3.2.2 Knowing What Others Have Seen: Chimpanzees

In 2000 and 2001, a group of researchers led by Michael Tomasello of the Max Planck Institute for Evolutionary Anthropology, in Leipzig, published results of experiments that were, according to the authors, for the first time indicative of an understanding of mental states in chimpanzees (Hare et al. 2000, 2001; for a detailed review of the studies, see Tomasello et al. 2003a, 2003b; Call and Tomasello 2008). The paradigm also expands on the food rivalry between a dominant animal (D) and a subordinate (S) (see fig. 14).

An initial series of experiments (Hare et al. 2000) has already shown that a lower-ranking chimpanzee (S) very flexibly makes use of her advantage in knowledge if she can see one of two pieces of food hidden from the higher-ranking chimpanzee (D) by a physical barrier. If both animals are left in the room with the food, she avoids the food that D can see as well, since she would be inferior to D in a direct fight over the food.

Instead, she approaches the food that D cannot see because it is hidden by the opaque barrier (Hare et al. 2000; Tomasello et al. 2003a, 154). To ensure that S actually acts on the basis of what D sees, various control conditions have been introduced (for the control conditions, see below and cf. Hare et al. 2000, 779ff., as well as Tomasello et al. 2003a, 154).

In a second series of experiments (Hare et al. 2001), S observed an experimenter hid the food on his side of the barrier. At the same time, S observed whether D also saw the hiding because her door to the test room was either open or closed (see fig. 15). If both animals were subsequently allowed into the room, S seemed to know whether D had followed the hiding process and behaved in an accordingly reserved way or not (Hare et al. 2001; Tomasello et al. 2003a, 154).

The experiments could be interpreted as showing that S acts on the basis of what she knows about the perceptual state of the other animal. Then S does not just seem to know what D sees or does not see. The lower-ranking

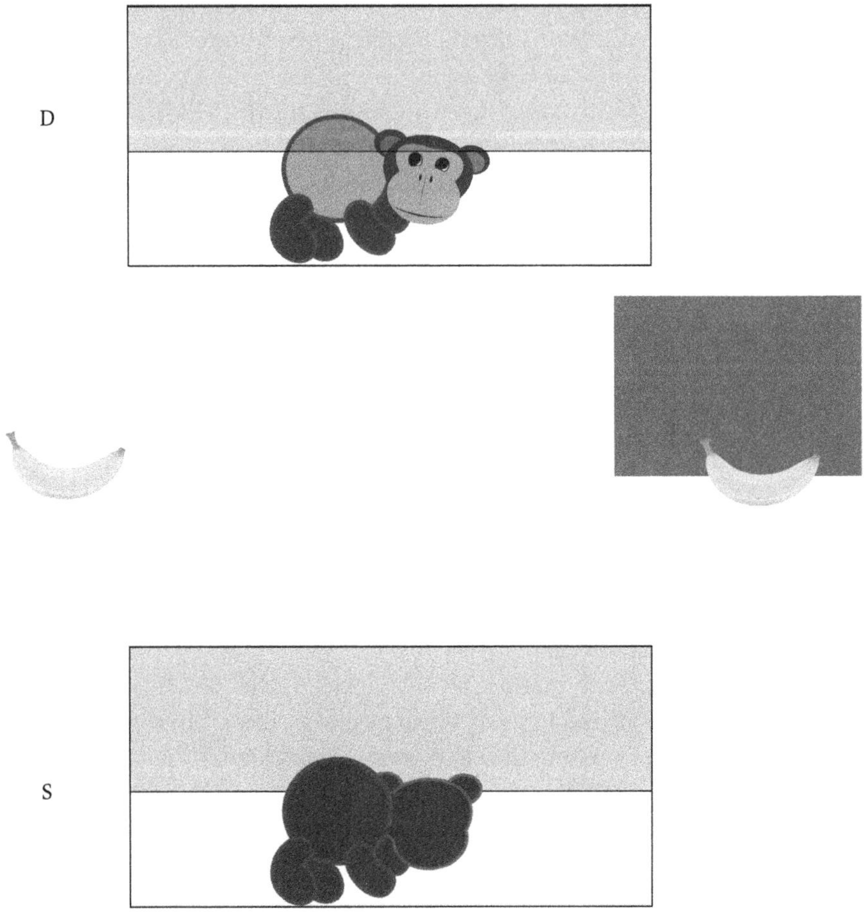

FIGURE 14 Food rivalry in the experiments of Hare et al. (2000). Adapted from Tomasello et al. (2003a, 154)

chimpanzee also knows that what the dominant chimpanzee has seen some time ago will affect her later behavior. In other words, she knows that what she saw at a time t1 (when the food is hidden) is available to her at a time t2 (when the animals are let into the test room and can fetch the food), and this knowledge will direct her behavior.

The researchers themselves come to the conclusion that at least some nonhuman primates (namely, chimpanzees) can attribute at least some mental states (namely, the perceptual state of seeing) to others and that the formation of theories should include these empirical findings. They go on to remark that "this does not mean that chimpanzee social cognition is the same as human social cognition, but it does mean that telling them apart just got harder"

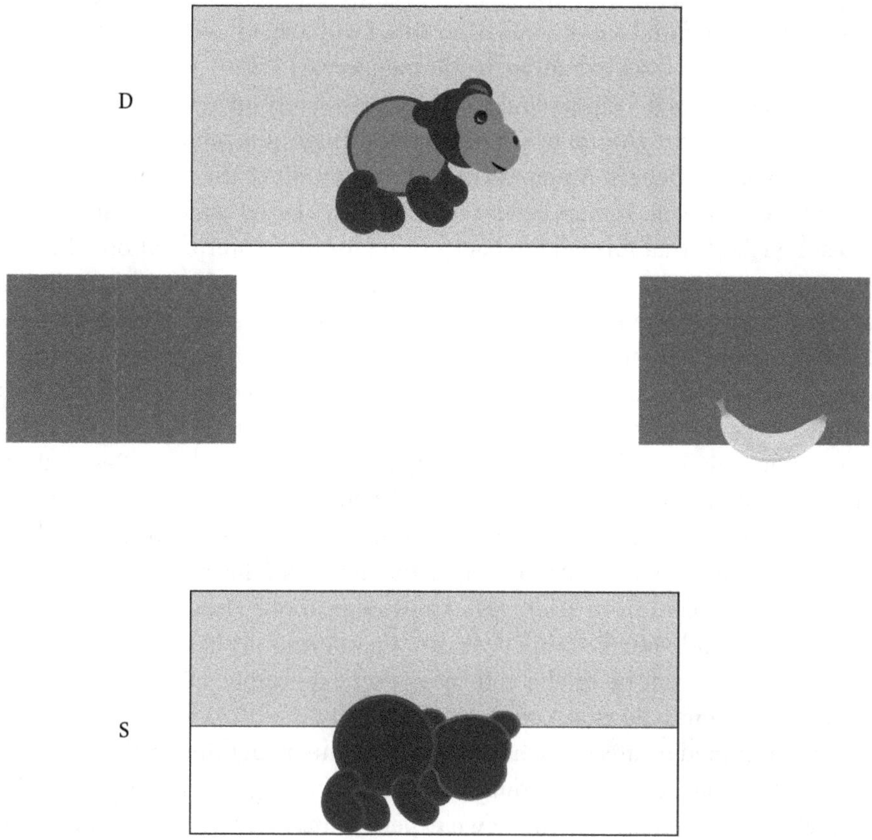

FIGURE 15 Crucial test condition of Hare et al. (2001), in which S can see if D also observes the concealment process. Adapted from Tomasello et al. (2003a, 155)

(Tomasello et al. 2003a, 153). The interpretation suggested here refers to a competence that falls within the area of theory of mind. The animal can assess the perceptual state (seeing) of another animal and impute to her conspecific that she will act on the basis of this perceptual state. S predicts that D's behavior will be based on what she could or could not see and adjusts her own behavior accordingly.

6.3.2.3 On the Interpretation of the Chimpanzee Studies: Tomasello vs. Povinelli

The interpretation of these studies is controversial. It has lead to a vivid debate on meta-cognition versus behavior reading in chimpanzees, started by a dispute between Michael Tomasello and Daniel Povinelli. I will summarize some important arguments from this debate because they are symptomatic for debates on animal minds in general. After this, I will turn to two very recent studies on false-belief understanding in chimpanzees.

The researchers in Leipzig could already in 2003 rely on a broad body of data that suggested that chimpanzees understand intentions and goals as well as perceptions and beliefs. A mentalistic interpretation of the test results therefore fit the data well. Tomasello and his colleagues could refer to a series of interesting results that formed the background for their interpretation. They argued that chimpanzees know what others see in terms of visual perception, and that they pay attention to the gaze direction of others and follow their gaze behind barriers. They do this even if they have to change their position in order to see *precisely* what the other one sees. They realize that opaque barriers block visual access at a certain angle, whereas transparent barriers do not. They also know if an individual has seen something in the immediate past. Researchers also understand that they know something about intentions in actions: chimpanzees can see effort, frustration, and contentment as a sign of what others are doing or will do next.[6] According to Tomasello and his colleagues, the animals are in a position to relate this knowledge about visual perception and behavior: "An individual chimpanzee can use information about what an other sees to predict what he or she will try to do next—especially in competitive situations" (Tomasello et al. 2003a, 156).

The combined results of such studies on the theory of mind in chimpanzees provided the background for Tomasello and his colleagues in interpreting the behavior of chimpanzees in their own experiments. And they were prepared to go a step further than a few years ago: they now asserted that chimpanzees—and perhaps other animal species—have a social-cognitive schema in the sense of a theory of mind, but this does not mean that it can be described as "full-blown" and "human-like" (Tomasello et al. 2003a, 156). However, not everyone agreed.

Daniel J. Povinelli and Jennifer Vonk responded to Tomasello et al. (2003a) with a critique centering on the question of what the chimpanzees in the experiments of Hare et al. (2000, 2001) are actually responding to—to what others have seen or can see, or simply to their observable behavior? In the opinion

6 As far as the animals' understanding of the intentions of others is concerned, Tomasello and his colleagues could today additionally refer to the studies of, inter alia, Call et al. (2004) and Buttelmann et al. (2009a). The fact that chimpanzees can read emotional expressions on the face of experimenters in order to draw conclusions about their desires is of great importance when it comes to the question of empathy in apes.

of Povinelli and Vonk, the latter was sufficient to explain the results of the Tomasello group, while the former represented an exclusive skill of humans. They pointed out that in order to attribute the mental state of seeing to another individual, behavior must first be observed. Instead of going for the explanation that posits a "detour" in terms of reasoning about mental states, the prediction of their conspecifics' behavior could also rest directly on behavioral observations and learned statistical regularities, without the chimpanzees knowing anything about mental states. It is precisely this "behavioral abstraction" hypothesis that would be the appropriate explanation for animal behavior. According to this hypothesis, chimpanzees can construct abstract categories of behavior, making predictions about future behavior based on previous observations and adapting their own behavior to all of this (Povinelli and Vonk 2003, 157ff; see also Povinelli et al. 2000; Povinelli and Vonk 2004; Povinelli and Barth 2005).

According to the arguments of Povinelli and his colleagues, both humans and chimpanzees are able to abstract behavior, but only humans *reinterpret* behavior *in terms of mental states*. Povinelli and his colleagues therefore proposed a "reinterpretation hypothesis" with regard to humans and their unique abilities (Povinelli and Vonk 2003, 158ff.). Experiments that want to prove that chimpanzees can also mentally interpret behavior and thus have a theory of mind would have to exclude the logic of Povinelli and his colleagues as a less demanding explanation (Povinelli and Vonk 2003, 159–60).

It is very difficult to escape some of Povinelli and Vonk's criticism. Tomasello and his colleagues pointed to their control conditions but from a philosophical point of view it is questionable whether this was a sufficient reply. Let's explain this by ways of an example. In the experiments by Hare et al. (2000), so goes Povinelli's and Vonk's criticism, S may turn to the food behind the barrier, and not in the open space, simply because she pays attention to D's behavior. D orients herself in the direction of the food she can see and runs toward it to fetch it. These are all *ways of behaving* that S observes—in addition to an observation of D's perceptual state of *seeing*. Tomasello and his colleagues countered, however, that in one of the control conditions of the experiment, S was given a head start over D when her door to the test room was opened a little earlier. This lead time forced her to choose her own behavior even *before* she could see that D was moving toward the food in the open space. Also when given this head start, S turned to the food hidden behind the barrier. Tomasello and his colleagues argued that this cannot be explained by D's behavioral response, but only by what kind of behavior S has predicted for D due to her perceptual state. But the behavioural abstraction hypothesis can easily deal with this too, since it's not a behaviouristic account and so doesn't require that S have perceptual access to D's behaviour to make a prediction about it. As a behaviour reader S can make predictions based on behavioural regularities alone.

Thus, it was not clear whether the control conditions, Tomasello and his colleagues referred to, were actually sufficient (or could ever be sufficient) methodologically. In principle, Povinelli and Vonk proposed that all experiments be modified in the future in such a way that any possibility of orientation toward the behavior of others can be ruled out entirely (Povinelli and Vonk 2003, 159–60). Although this may be desirable in terms of content, with the modification of the experiments in this direction, another, equally fundamental methodological error could be made. The requirement for an exclusion of mere behavioral observation must be compatible with the need for test structures that are as natural as possible (Tomasello et al. 2003b, 239). This seems to be one of the biggest methodological problems of current research. Tomasello et al. (2003b) pointed out that their experiments confront the animals with new situations, and thus that the possibility that animals in previous situations have learned behavioral rules that they now simply transfer to a similar context, at least, can be ruled out. The use of transparent barriers, for example, was new to the animals. But although the animals had never experienced the transparent barriers before, they did not get confused about their effect on the vision of others: "They knew that it did not obstruct the dominant's visual access to the food. Perhaps this situation is not novel enough for Povinelli and Vonk, but in our opinion the novelty ought to stay within the bounds of principles of ecological validity" (Tomasello et al. 2003b, 239).

Much of the discussion about theory of mind experiments with chimpanzees focused on the arguments presented here. But even regardless of whether the control conditions implemented by Tomasello and his group can invalidate the behavioral abstraction hypothesis, there are problems associated with the approach of Povinelli and his colleagues.

For example, it is not entirely clear to what extent the ability to interpret behavior mentally is dependent on language ability from the outset. After all, terms such as "reinterpret" and "encoding" are used and the supposition is also directly expressed that theory of mind could have evolved in the course of evolution in conjunction with language ability (Povinelli and Barth 2005, 713; Povinelli and Vonk 2003, 158). Does the interpretation of Povinelli and colleagues ultimately tie the ability to attribute mental states to linguistic capacity? If so, then that interpretation would be following a problematic definition of theory of mind, due to the fact that such a definition is a very specific one and in need of justification.

Moreover, in Povinelli's interpretation it is not very clear what is meant by a (nonmentalistic) behavioral abstraction in general and how it relates to theory of mind abilities. Although the behavior of animals in some of the previous theory of mind experiments may theoretically be explained by mere behavioral observations and by rules derived from them, it is profoundly implausible that chimpanzees deduce statistical relationships between such a large number of possible behaviors and potentially learn and remember the resultant actions of others. To avoid this dilemma, Povinelli and Vonk attribute to chimpanzees the ability to *abstract* behavior. The question now arises, however, as to what exactly the difference is between the "forming of a behavioral abstraction of seeing" and the "understanding of a mental state of seeing." Can the former be gained without the latter? What is the functional difference between the two? For example, an individual could learn the behavioral rule: "Avoid food when a particular individual has previously looked for food in this very hiding place." But how are such rules acquired, and how can such rules be understood by the animals? And is it really more parsimonious to assume that a being acts on the basis of such behavioral abstractions?

Finally, it must be asked why it should be so that humans—in contrast to chimpanzees and other animals—also mentally interpret behavior. What is the benefit of a double interpretation that is behavior-oriented and mentalistic in comparison to a purely behavior-oriented interpretation? The authors write that the ability of behavioral abstraction exists in humans and chimpanzees, and probably in their last common ancestor. But humans "added" a mentalistic system of interpretation to this system of interpretation. Human beings have a new theory of mind system "woven" into their inherited cognitive architecture, so that now both systems exist harmoniously, side by side. How precisely this addition and interweaving is to be understood is unclear. The authors also seem to think that human's theory of mind system has emerged from the behavioral abstraction system: "Theory of mind evolved by *exploiting* the existing systems for behavioral abstraction" (Povinelli and Vonk 2003, 158; italics added). It may be true that chimpanzees have simply not developed such a system. But even if they have not, this requires further explanation.

Taken together, Povinelli's critique left us with many challenging questions. The philosophical debate got rather complicated after the debate between the Povinelli and the Tomasello group and many new empirical results which we have gained since then are in need of interpretation (see for example the discussions presented in Andrews 2005; Buckner 2014; Clatterbuck 2015; Halina 2015; Andrews 2017). To keep things simple, I will bracket this further debate now and turn back to recent attempts to test for false-belief understanding in great apes.

6.3.3 *False-belief Understanding in Great Apes*

For some time, research seemed to have focused on perceptual states rather than false belief in animals which was a methodologically clever and important strategy that also yielded interesting discussions as we have seen. But recently, scientists touched upon the false-belief issue again and published two very promising studies on chimpanzees.

Krupenye et al. 2016 used an anticipatory looking test (which was originally developed for human infants) to show that three species of great apes reliably look in anticipation of an agent acting on a location where she *falsely* believes an object to be, even though the apes themselves know that the object is no longer there. They argue that their results "suggest that great apes also operate, at least on an implicit level, with an understanding of false beliefs" (Krupenye et al. 2016, 110). In the two experiments they conducted, the animals were confronted with new situations on videos which they had never encountered before.

In the first experiment an antagonistic scenario was shown to them in which a human actor was attacking with a stick another actor dessed up in a King Kong costume. King Kong was hiding in one of two haystacks and the human actor was hitting on those haystacks. Infrared eye-tracking showed that the majority of observing apes expected the human actor to hit the haystack where the human actor himself rightly or falsely believed King Kong to hide. They do so, even if King Kong had changed his hiding place while the actor went outside to get his stick, a move they themselves had observed. Thus, the animals knew where King Kong was hiding last, but the human actor couldn't know.

In the second experiment King Kong stole a stone from the human actor and hid it in one of two boxes. A belief-induction trial caused the actor to rightly or falsely believe where the stone was, based on King Kong moving the stone from one box to another while the actor couldn't see it. When the human actor attempted to retrieve the stone, anticipatory looking of the apes watching the scenario again showed that they expected him to search for the stone based on where he rightly or falsely believed it to be.

In all conditions of the experiments, the desired target (King Kong or the stone) was removed from or had left the scene after the belief induction, so they were not actually present any more. And all videos ended when the human actor came back to the scene and positioned himself for a central approach or central reach for the target, but before he actually grabbed or hit the target. Thus, no directional cues from his side revealing his beliefs or intentions were given to the apes. The authors also controlled for several behavioral explanations (for example the apes could not solve the task by just searching in the place where the object was first or last located or by searching in the place where either the actor or King Kong had last attended to). Furthermore, Krupenye et al. emphasize the new, antagonistic (and thus competitive) scenario that they have used to encourage the subjects' engagement:

> Apes were never shown the actor's search behavior when he held a false belief, precluding reliance on external behavioral cues learned during the task. By requiring subjects to make predictions in situations that involved a constellation of novel features (e.g., a human attacking an apelike character hiding in a haystack), we also minimized the possibility that subjects could apply behavior rules acquired through extensive learning during past experiences.
>
> KRUPENYE et al. 2016, 113

While it might be true that all change-of-location false-belief tasks are in principle explainable by assuming that the animals apply an abstract behavioral rule that says "agents will search for things where they last saw them," the authors believe that such an explanation does not suffice to deal with the diverse results on animals' understanding of mental states in general (Call and Tomasello 2008). Also, if they always applied such a rule we couldn't for example explain why apes (but also children) appear to infer in other experiments "whether others can see through objects that look opaque, based on their own experience with the occlusive properties (i.e., see through or opaque) of those objects" (Krupenye 2016, 113). What they assume about the others' informational state based on their perceptual information and experience and based on their mental states like beliefs and desires seems much more complicated. While they have not passed false-belief paradigms that ask them to make *explicit behavioral choices* (a failure that might also be due to task demands or contexts), we should now at least assume that they have an implicit understanding of belief (Krupenye 2016, 113).

Krupenye et al. are not the only ones who conducted interesting false-belief tests with great apes recently. Buttelmann et al. 2017 demonstrated that in a nonverbal (implicit) false-belief test (which is also passed by human 1-year-old infants), chimpanzees, bonobos, and orangutans distinguish between true and false beliefs in their helping behavior. The researchers were driven by two line of thoughts arising from the Krupenye et al. study.

First, they conceded that the apes in this study could have mastered the test simply by predicting that the actor would go to the last place where he saw the object. This means, that we need converging evidence from other tests to support an interpretation of the results based on the animals' false-belief understanding.

Furthermore, they argued that is is important to show that the animals "can go beyond predicting others' behavior and also use their understanding of beliefs appropriately in real social interactions" (Buttelmann et al. 2017, 2). Thus, their test design used *active behavioral measures*.

The apes were confronted with a scenario where they had to help a struggling experimenter to open a lid of a box containing an object. In the crucial false-belief condition, the experimenter was unsuccessfully trying to open a box where he thought the object to be, but the object had been moved without his knowledge. Like human infants did in a similar task (Buttelmann et al. 2009b), in the false-belief condition, the apes help by unlocking *the other* box, the one, the experimenter was not tugging at. They did so significantly more often than in the true-belief condition. The authors write: "Given that at test everything except what the experimenter believed about the location of the toy was identical in both conditions, apes had to go beyond the observable facts of the situation to figure out the experimenter's goal" (Buttelmann et al. 2017, 6).

If we assume that the apes connected the experimenter's false belief about the location of the object with his actual goal, it makes sense to open the box for him where the object is located and not the one where he falsely believes it to be. In a second test the researchers could show that the apes behave differently, when the experimenter doesn't hold a true or false belief about the location of the food but is completely ignorant about the location (the animals then simply open one of the boxes by chance).

The results of the study show that the recognition of a false belief in the experimenter makes the difference in the animals' helping behavior. Thus, the study provides "another indication that great apes might have some level of false-belief understanding." It is furthermore the first study "to demonstrate that apes are able to use this understanding in their social interactions" like when actively helping someone else (Buttelmann et al. 2017, 10).

Like Krupenye et al. Buttelman and his colleagues think that their results point to the fact that great apes possess false belief as a "key marker of advanced human social cognition" and that we can therefore no longer entertain the view that this capacity is "the exclusive province of the human species" (Buttelmann 2017, 1).

6.4 Can We Speak Meaningfully of "Theory of Mind" in Animals?

Tomasello and his colleagues rightly emphasize that a theory of mind encompasses many building blocks. If chimpanzees had only some of them, it would not be surprising. Children also acquire an understanding of mental states gradually. They fail to understand the false beliefs of other in the beginning, which does not mean that they therefore have no understanding of others' mental states and processes such as seeing, wanting, and intending—and thus already possess a simple theory of mind. This means, that even if we didn't have first evidence of false-belief understanding in animals, or even if this evidence remains scarce, we do not have nothing pointing towards theory of mind in some animals. The early understanding of other mental states and processes should be recognized as theory of mind ability in children as much as in animals. It is likely that progress will not be made in answering questions of evolution, nor those concerning the ontogenesis of social cognition, as long as a monolithic view of theory of mind continues to be advocated. A theory of mind is not something that individuals either have wholly or not at all, and it does not begin when people acquire an understanding of false beliefs (cf. Tomasello et al. 2003b, 240).

This chapter showed how difficult it is to research theory of mind in children and how an overemphasis on one core capacity (in our case false-belief understanding) can easily cloud our perception of theory of mind that is in fact, conceptually speaking, not limited to this core capacity. Likewise, the question of whether animals have a theory of mind is thus not only solely a matter of empirical research on the cognition of animals, but rather a question of which understanding of theory of mind one uses in the future.

When the larger concept of theory of mind is to be broken down into its individual components, we can find interesting research results on theory of mind in animals as well. This becomes obvious as soon as we turn to some animals' understanding of perceptual states. It seems likely that these tests continue to be refined, in an attempt to master the balancing act between excluding mere behavioral observation and preserving natural test designs. Melis et al. (2006), for example, following their colleagues' food competition studies, ask whether chimpanzees can actively manipulate what others see.[7] This would be strong evidence that they have an understanding of what others see or do not see, and would tell us what effect this has on their actions. To understand the experiment of Melis and her colleagues we must have a quick look at preceding experiments by Hare and colleagues.

Hare et al. (2006) already demonstrated that chimpanzees prefer pieces of food that a competing experimenter cannot see or those that it can approach

7 The intentional deception considered here is a subform of tactical deception in which the agent has the goal of manipulating perceptual states or other psychological states in others (Melis et al. 2006, 154).

without being seen. Barriers between the experimenter and the animal have been used for these studies. Among other things, the barriers ensured that the chimpanzee and experimenter could no longer see *each other* in some experimental conditions (due to a continuous barrier between the two) or that, though the experimenter's face was no longer visible for the chimpanzees, *for the experimenter* the body of the chimpanzees could still be seen (as a result of a shared barrier between the two). To the chimpanzees in these experiments, it seemed clear that, even if they positioned themselves so that they could not see the experimenter's face, the experimenter could still see how they crept up to the food (Hare et al. 2006, 510). Nevertheless, in most conditions in this series of experiments, there was still the possibility, on the one hand, that the chimpanzees could observe the experimenter and read his behavior, or, on the other, that they acted on a simple egocentric rule: "If I cannot see the experimenter, I am safe" (Melis et al. 2006, 157). Following the Povinellian requirement, Melis et al. (2006) wanted to bypass such a simple, nonmentalistic orientation to the behavior of the experimenter. In their follow-up study, therefore, in the starting position of the experiment, the experimenter was only visible to the animal through a gap of just twelve centimeters in the otherwise completely shielded front pane of the test room (Melis et al. 2006, 157–58).

The animals now had the choice of reaching into the test room from left or right through different tunnels in order to steal the food lying there. The tunnels were either transparent, opaque, or they made a loud noise. Thus, approaching hands could either be seen or not seen or heard or not heard by the experimenter (the animals knew that the experimenter was there but they were unable to observe her behavior) (see fig. 16). Once the chimpanzees had left their starting position, they could no longer observe the experimenter and her behavior and were thus forced to *imagine* whether or not she could see or hear their hands approaching through the tunnel (Melis et al. 2006, 157).

The chimpanzees actually preferred the opaque tunnel in the visual conditions in which the experimenter was present, and in the auditory conditions in which the tunnel did not make a noise. On the one hand, in control conditions without such present but not-visible test subjects, Melis et al. could show that the animals did not display these preferences (157). On the other hand, Melis and her colleagues showed that chimpanzees seem to know not only what others see, but also what others hear. In addition, they tried to refute Povinelli's criticism and rule out direct behavioral observation as much as possible. Nevertheless, following Tomasello's requirement, they remained faithful to the competitive test design of the food competition paradigm, because this has already proofed to be a suitable test design for chimpanzees who are in fact rather competitive when it comes to food.

FIGURE 16 Experimental condition according to Melis et al. (2006, 156)

It was rightly pointed out in the debate that one should not resort to unnatural test designs for the sake of excluding behavioral observation. Unnatural test designs might be able to rule out behavioral observation in a more rigid way but the price might be that they cannot show the full spectrum of what the animals can do. Attempts that have rigidly established the behavioral abstraction hypothesis of Povinelli and his colleagues, have come under criticism for their unnaturalness. In these experiments (Povinelli and Eddy 1996), chimpanzees had to distinguish between a subject who could see them and one who could not, with the nonseeing subject wearing a bucket on her head and the seeing one carrying a bucket on her shoulder. The tests did not lead to a positive result. The animals performed "begging gestures" directed at both test persons—regardless of their different perceptual state. The person with the bucket over her head clearly could not see these gestures.

This paradigm may involve a number of difficulties that, at first glance, are not so obvious and go beyond the understanding of mental states of others. First, sharing food and begging for food in chimpanzees that are otherwise in intense competition for food is rather unnatural. The food competition paradigm therefore corresponds more closely to chimpanzee behavior, which is determined by competition and hierarchy, than cooperative test designs involving begging gestures (Tomasello et al. 2003b, 239). In fact, Hare and Tomasello (2004) have shown that chimpanzees perform better in competitive situations than in cooperative ones. Second, in similar experiments, the animals demonstrated that they were quite sensitive to the experimenter's attentiveness (Kaminski et al. 2004). The difference here was that they only had to interact with *one* experimenter (who had, for example, either turned the body and the head or just the body, not the head to the animals) instead of having to choose to interact with *two* experimenters who displayed one of the two behaviors. So perhaps it is particularly difficult to have to choose between two individuals, each with a different attentional state (Tomasello and Call 2006, 376–77). Finally, the bucket-head experiments were even problematic for children: they were unable to complete the task before the age of five, even though they already passed classic false-belief tests earlier and certainly had earlier knowledge of the perceptual states of others (Kamawar and Olson 1998; Tomasello et al. 2003b, 239). Again, this is an indication that the design of the experiment may presuppose more than intended. All in all, while it seems desirable to rule out behavioral observation this attempt should not override other important considerations, like the necessity to offer the animals as natural test designs as possible.

In terms of method, the possibility of that the chimpanzees may be merely reading behavior can maybe never ruled out completely. This is because reading behavior is an *intrinsic component* of a mentalistic interpretation. Having an understanding of perceptual states means to understand the *context* of what an individual *does* and what an individual *has seen*. Having a theory of mind means understanding the *interplay of both behavior and mental states*. The aspect of behavior and behavioral observation is thus inscribed in the theory of mind competence. The only thing that can be attempted to ameliorate the problems in experiments is to exclude a concrete act of observation of the other precisely when and where the animal makes her decision—this has been done in the studies of Melis et al. (2006) and in some control conditions of the studies of Hare et al. (2000 and 2001).

Moreover, Andrews (2005, 534) rightly emphasizes that careful consideration should be given to applying the same scientific rigor to studies of animals as to human studies: "We must not ask more of the chimpanzee than we ask of the child." In this sense it will be helpful to have a close look at the experiments with human children and ask ourselves whether we would be willing to apply the same interpretation to the same results if we were dealing with non-human animals. The fact that it is hard to figure out what is going on in the minds of animals is a challenge but it is not necessarily a unique challenge that we only face when dealing with animals. The philosophical problem of other minds, which basically states that we are granted no insight into the mind of others, no matter whether they are humans or animals, can be met with a rather relaxed and pragmatic attitude:

> The new problem of other minds is not whether the other has a mind, but whether the other knows that we have a mind. Fortunately, there may be a way out of the problem, and it is the same as the way out of the traditional problem of other minds: given the current evidence and evidence from future research, we must be willing to make an inference to the best explanation on the basis of the entire body of evidence. Though a definitive proof is lacking, that shouldn't concern us, since most of us are utterly unconcerned about the lack of a knock-down argument for other human minds.
>
> ANDREWS 2005, 534

If we simply reject the results of the Leipzig researchers as insufficient evidence for a mentalistic understanding in animals, we also lose sight of what the results can tell us about interesting differences and similarities regarding theory of mind in different species, according to Andrews. There may be such differences not only between chimpanzees and humans, but also between chimpanzees and other species: "Chimpanzees behave as if they have the concept of seeing, and so do most children at the age of four. Domesticated cats, presumably, do not behave in this way. Where the behaviors are different, the interpretations of those behaviors will be different. And where they are the same, we ought to see some similarity between them" (534). To date, the theory of mind of domestic cats and most other species (except for the chimpanzee) remains indeed largely mysterious to us. It should be understood, however, that future research could confirm the path that Tomasello and his colleagues have laid out. And this most likely has consequences beyond understanding the socio-cognitive abilities of our closest relatives.

Chimpanzees and other nonhuman animals have an understanding of the goals and intentions, perceptions and knowledge of others, and connect these areas of understanding. We even have some evidence of an understanding of false beliefs in them. This most likely is just the beginning of theory of mind research in animals. The pronounced belief-desire psychology of the six-year-old child, let alone the adult human, may not exist in nonhuman animals (Kaminski et al. 2008; Call and Tomasello 2008). However, this is not absolutely necessary when speaking about the central building blocks of a theory of mind and, in general, of a simple, de-intellectualized theory of mind in animals.

CHAPTER 7

Summary and Transition

Up to this point, this book has dealt with the question of whether animals have cognitive abilities such as culture, language, and theory of mind that have long been attributed to humans alone. My interdisciplinary analysis of the research results on this question revealed that we can speak of the presence, in at least some animals, of a simple culture, central aspects of linguistic ability, and certain areas of a theory of mind. Different forms of consciousness (from phenomenal to semantic to propositional cognition) have been shown to play a role in animal thinking, which apparently follows inner templates or consists of mental representations of concepts, and even occurs as recursive patterns, metacognition, and propositional knowledge. Equally, time and again I emphasized the flexibility and enormous complexity of animal behavior, as well as the importance of socially learned and socially practiced interactions. Basic prerequisites for conscious thinking, (for example according to the distinctions of McFarland, Griffin, and Bieri), and the criteria put forward in the theories of different scholars (for example, Beck, Haidle, Bartels, Newen, and Tomasello) have been discussed and applied to the interpretation of a wide range of empirical studies. The ground upon which competences, such as culture, language, and theory of mind, are cited as long-held criteria of differentiation between humans and animals, have become unsteady. Such capacities are no longer clear indicators of a fundamental difference between humans and animals. They are no longer firmly rooted as such from the perspective of empirical research and philosophical conceptualization.

The overall impression is that monolithic ideas of culture, language, and theory of mind are of little further help. This applies both to the acquisition of these cognitive abilities in childhood and during human evolution, as well as to the interpretation of these abilities and their occurrence in animals. The reason for the inadequacy of monolithic definitions cannot therefore be found solely in the fact that we are dealing with animals. It must exist in a common element or problem that is effective both in the development and occurrence of cognitive abilities in humans as well as in animals. This common element, which renders the description of cognition in animals so difficult, is the gradual nature of complex socio-cognitive abilities. Culture, language, and theory of mind are not adequately understood when described as overall complexes that one either possesses or does not. Human beings, too, acquire such capacities through a gradual process with intermediate stages. Generally speaking, then,

 | DOI:10.1163/9789004415072_008

it is therefore not a question of proving so-far "uniquely human" capacities in a form by which animals' competencies can be rated as "fully legitimate" and "humanlike." It is neither probable nor necessary that animals' cultures, languages, and everyday psychologies are exactly as ours, nor that they include all the building blocks of these complex phenomena known to humans. When the abilities of human beings are used for comparison, it does not make sense to compare what animals can do with the elaborate skills that adult human beings possess. It is more interesting to use a child's competences in those stages of development in which the sought-after competences begin to develop or to concentrate on the central components of complex abilities. At the empirical level, we need even better experimental designs and more comprehensive studies on as many species as possible. At the analytic-theoretical level, the task is to continue to search for appropriate interpretations and determinations of abilities and characteristics that describe similarities and differences between different species.

The results of the book so far can be summarized as follows:

The question of culture in animals was introduced by way of many different animal examples. Central terms and interpretive backgrounds were worked out on the basis of these case studies. For example, the notion of learning and the concept of goal-directed flexibility were introduced, as well as the importance of life in the social fabric. As a detailed example, culture was discussed in chimpanzees, among whom a great variety and complexity of traditions and social learning can be observed. In addition, tool fabrication and tool use in orangutans, gorillas, dolphins, and New Caledonian crows was presented, and in the latter case, meta-tool use was presented as well. In describing and explaining these behaviors, a simple and simultaneously broad (biological) definition of culture as socially transmitted behavior has proved useful. Based on the cognigrams detailed by Haidle (2012), it became clear that the complexity in the tool behavior of different animals compared to early hominids can be clearly dissected, for example, by observing that in humans, observation focal points often unfold simultaneously, or the production of a tool takes place over a longer period of time. However, the central elements, such as future planning or opening up of a number of focal points of attention, is generally already evident in some animals. Differences seem to be more gradual than essential.

Within the chapter on language in animals, the phenomenon of concept possession was examined in a deeper perspective. Theories of interpretation from philosophy of language and philosophy of mind, such as those outlined by Bartels (2005) and Newen and Bartels (2007), could aid in interpreting existing research results. Accordingly, Alex the grey parrot did display the ability for conceptual mental representations. His perceptual concepts were organized in small networks with subcategories and upper categories. He had the capacity for correct discrimination based on the ability to classify and proved this in a stimulus-independent way. His concepts corresponded to functional roles in the sense of an indicator, classification, and comparison function. An impressive *understanding* of language, even to the level of sound combinations with a basal grammar, were also evident in other animals, for example in the complex alarm calls of some monkeys or in the reaction of dolphins to commands in sign language. These animals also correctly follow semantic-reversible commands and react rationally to semantic and syntactic anomalies. An understanding of sentence-like representations and even individual production of simple sentences was discussed with regard to language teaching studies with chimpanzees.

Finally, in the chapter on theory of mind, a comparative approach between the study of this phenomenon in animals and the research with human toddlers was presented. Children and animals have in common that they are not fully linguistically gifted and not fully developed when compared with a healthy adult human being. In both cases, degrees of graduality can be observed, due to different evolutionary development (in the case of different species) or in terms of ontogenetic development (in the children). The theory of mind abilities of both groups were not directly compared with each other as the same tests are usually not run with both groups in one and the same experiment. But I developed an argument through a methodological critique: methodological weaknesses in child research refer to fundamental *content-related* weaknesses in the definition of theory of mind itself, which is problematic because this definition, with its weaknesses, also underlies animal research. The methods in classical theory of mind research in children (primarily false-belief tests) are problematic because they correspond to a very demanding and one-sided understanding of theory of mind. "Real" theory of mind understanding in children has long been reduced to an understanding of false beliefs and was often spoken about as if it was something that you either totally possess or do not, (this conception includes the false-belief building block). Research with animals had to break free, so to speak, from this understanding and did so by focusing on (chimpanzees') attribution of perceptual states, mostly visual. There has been comparatively little research on the attribution of other modes of perception, or the attribution of other entirely different types of mental states, such as emotions. Another challenge researchers on animals' and childrens' theory of mind have to deal with is the fact that theory of mind is often interrogated in language-based tests that are equally unsuitable for infants and animals. This, too, demands from the researchers to adapt existing test designs to their subjects' individual characteristics and, if necessary, to come up with completely new test designs. Because of these methodological challenges, however, we run the risk that our tests for both groups are unsuitable. This links with the risk that we misrepresent or (in case they can't master a task) underestimate the abilities of children and animals. Non-verbal false-belief testing, or a focus on other mental states than false beliefs, as well as the study of everyday language in children can lead to new results. It has been shown that, at least with respect to chimpanzees and ravens, their understanding of perceptual states, visual perceptual states but also other perceptual states, might be a fruitful candidate for a more de-intellectualized theory of mind capacity.

A proper interpretation of what animals know and can do only seems possible when one looks back at the core areas and the core capacities of complex abilities. In this book, in contrast to monolithic definitions, the following areas have been proposed as possible core areas: in the case of culture, the existence of social learning and targeted flexibility within the transmission of traditions; in the case of language, concept possession, a comprehensive (semantic) *understanding* of language as well as a basal regular grammar; and, in the case of theory of mind, the understanding of perceptual states. To be sure, additional building blocks and core areas might be of interest for the question of complex socio-cognitive abilities in animals.

Certainly there are consequences for the conceptualization of the human-animal relationship if animals share with us capacities and characteristics that we have always believed to be uniquely human. "If you look at the history of (especially, but not exclusively philosophical) treatments of the human-animal theme," says Forster (2007), "it is remarkable how often a particular structure repeats itself: the theory identifies a single central property that supposedly distinguishes all (normal) humans from all (other) animals and which is highly valorized, either explicitly or implicitly, by the theorist" (Forster 2007, 761, transl. MK). Forster calls this structure the *gap and superiority structure*. There are several reasons for approaching this structure with suspicion, he argues. The first two reasons were made clear earlier in this book. "Prima facie, at least, some species of animals seem to possess the characteristics mentioned," says Forster. The fact that these theories stick with their claims and just alter from time to time the characteristics they cite, suggests, furthermore, that they "may ultimately be more about the structure than the characteristics." And a reason for that can easily be found: "after all, as theoreticians in this area, we humans are hardly impartial" (Forster 2007, 761–62, transl. MK). After discarding a hitherto valid criterion of differentiation we tend to simply come up with the next, monolithically defined, supposedly unique characteristic that humans possess. This has indeed become, after several repetitions of such a method, a questionable procedure. Forster proposes an alternative thinking structure. Instead of looking for a single, privileged criterion of differentiation, one can invoke a variety of qualities and work with cluster concepts. These may in some cases distinguish humans from all other animals, and in other cases humans will be differentiated from some species but not from others—either merely gradually or in essence. However, the qualities invoked might be of very different biological "utility" or moral value. Some might be positive, some negative, some ambiguous, and some neutral. Sometimes humans may be superior to other animals in individual characteristics, sometimes they may be inferior (Forster 2007, 763).

It may be difficult to follow Forster's suggestion to engage in consistently gradual thinking, especially because the criteria of differentiation emphasized in philosophical theories are indeed traditionally attached with value, or "valorized" as Foster says. It is not normatively neutral to argue that humans have complex abilities like culture, language, theory of mind, self-consciousness, autonomy, morality, and many more, while animals don't have all of this. Such capacities are often considered crucial for the moral status of a being, for the rights she holds and the treatment she gets. The assumption that human beings (and so far only human beings) have certain abilities is thus value laden. Now, if animals have such abilities too, then, in Schweitzer's words, they enter the animal-free "good room" of moral philosophy. Moral philosophers may then have to bid adieu to honoring certain abilities as "values." But does this in fact make sense if the sole reason for doing so is that other animals than humans *also* have these characteristics and skills? What if there is, in Forster's sense, a set of abilities that, despite their possibly gradual nature, actually represent value in the interaction between living things? Now we are faced with the question of whether, when it comes to the human-animal relationship, we can no longer put emphasis on complex socio-cognitive abilities, or whether there are opportunities to integrate these as gradual phenomena in ethical theories and considerations of species protection. If some animals have vital building blocks of such capabilities and, at the same time, we do want to assign an important role in our ethical behavior to those skills, then we owe the animals a number of further considerations, and we should be committed to their real-world implementation.

From the outset, this book has followed two key questions. The first question, whether animals have cognitive abilities that have long been attributed to humans alone, is now to be connected with the second question of the book: What does it mean for our relationship to animals and our dealings with them if we are learning of increasingly complex socio-cognitive skills they too possess? In other words, to what extent is the previously discussed question about the *presence* of socio-cognitive abilities in animals of interest at all? Does answering this question only provide results that are interesting for various disciplines such as psychology, ethology, or philosophy of mind and the theories they work with, or do these results present us with ever greater *ethical* or *socio-political* challenges?

I begin the next portion of the book by explaining the basic positions and central concepts of animal ethics and demonstrating how animal rights are justified. The argument from cognitive similarity between humans and animals is presented as one of the central arguments in animal ethics and scientifically-based animal welfare considerations. The central question of animal ethics is: Why should we protect animals? Three possible and widely cited answers are as follows: we should protect animals (1) because they are capable of suffering, (2) because many of them are already critically endangered, and (3) because they are similar to us in many ways, especially in their social and cognitive abilities. I will pay particular attention to the last answer, and will demonstrate how cognitive kinship leads to strong claims for the protection of animals. Two approaches will be presented in this regard: the approach of the Great Ape Project founded by Paola Cavalieri and Peter Singer (Cavalieri and Singer 1993) and the argument by the American philosopher Thomas I. White (White 2007), which transfers the claims of the Great Ape Project to dolphins.

Two aspects will be highlighted here: First, how well justified arguments for animal welfare and animal protection are from an ethical standpoint when referring to animal sentience, to the harm done in captivity and the wild, and to cognitive similarity. Secondly, to what limited extent the animal welfare requirements derived from these arguments are actually put into practice. The example of great apes and dolphins is particularly illuminating because these species are clearly sentient and socially and cognitively complex. The strong ethical demands arising with regard to these two species are faced with the worrying fact that these animals, both in captivity and in the wild, are massively impaired in their well-being due to human influence. In order to contrast the theoretical demands from the field of animal ethics with this lack of implementation, the first chapter of the following part of the book, “Kinship and Responsibility: The Moral Status of Animals,” is followed by a second chapter entitled “Kinship and Responsibility: The Discrepancy between Ethical Claims and the Status Quo.” It is only against the background of this discrepancy that the importance of the demand for stronger protective rights for animals becomes understandable. And it is only against the background of this discrepancy that the relevance of cognitive abilities for animal ethics—their scope and their interaction with ethical arguments—can be discussed.

PART 3

The Relevance of Socio-Cognitive Abilities in Animals for Animal Ethics and Animal Welfare

CHAPTER 8

Kinship and Responsibility: the Moral Status of Animals

8.1 Introduction to Basic Concepts in Animal Ethics

Animal ethics is classified as a subdiscipline of (applied) bioethics.[1] It deals with issues of ethical appropriateness and the impact of human behavior on nonhuman animals (see Birnbacher 2002). In animal ethics we ask whether and to what extent non-human animals can be potential victims of the effects of our behavior and whether and to what extent we should consider them morally. This question can be reformulated as "Do animals belong to our *moral community*?" The members of this community are to be considered not only because of their monetary value or their instrumental usefulness, but also because they have *intrinsic value*[2] to which a moral status and moral rights can be attached. In moral terms, they are considered *directly, for their own sake*, whereby *direct moral consideration of them* also means that we have *direct obligations to them* (see Regan 2004; Blumer 1997, 38ff.; Krebs 1993, 996–97; Gorke 2000, 87).[3]

1 A classification within the fields of environmental ethics or ecological ethics would be possible. Animals can be regarded as a subgroup of the nature surrounding us as much as they are part of our environment or ecology in general. A good overview on the basic concepts and questions of ethics of nature is presented in Krebs (1999).

2 In environmental ethics and animal ethics, the terms "intrinsic value" and "inherent value" are sometimes used as synonyms and sometimes as different concepts, and their meaning must therefore be clearly indicated. Following Regan (2004), Blumer (1997), and Krebs (1993), I refrain from defining these terms synonymously. An intrinsic value is first attributed to all living beings who, for their own sake, are to be directly and morally considered and therefore belong to our moral community. An "inherent value," on the other hand, means a distinctive, particularly serious intrinsic value, according to which a living being is an end in itself in the Kantian sense and therefore has rights as a bearer of interests and needs (see below).

3 Even if only one group falls within the range of *direct* responsibility (in the case of anthropocentrism, for example, we only have direct duties to humans), this does not mean, of course, that in these accounts of ethics there are not also other groups that can be indirectly protected. They are not to be considered morally for their own sake, then, but rather, for example, because of their value to humans. Even in this way, protection obligations and protective rights can be justified.

 DOI:10.1163/9789004415072_009

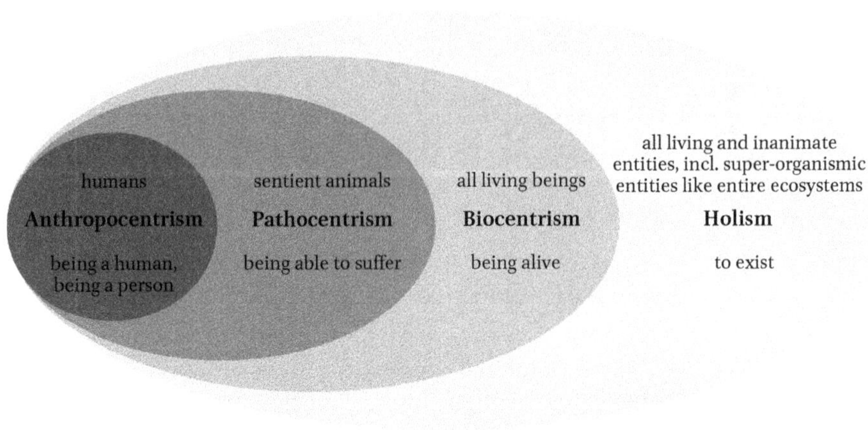

FIGURE 17 Basic types of environmental ethics, their areas of direct human responsibility, and their criteria for moral consideration according to Gorke (2000, 88)

Documented, positive laws can then be derived in legislation from the moral status of a living being and its moral rights.[4]

The question of who belongs to the moral community and who does not is based on criteria. The majority of normative approaches in environmental or animal ethics, including that of Birnbacher (2002), derive their standards from a certain theory of value. Depending on which entities are assigned an independent value and/or a moral status with regard to which qualifying conditions, one distinguishes between *anthropocentric, pathocentric, biocentric,* and *holistic* approaches (Birnbacher 2002).[5]

The four basic types of environmental ethics differ not only in the entities to which they grant intrinsic value and therefore direct moral consideration, but also *on the basis of which properties or qualities* they do so. Gorke (2000), with recourse to Frankena (1979), presents the four basic types of environmental ethics in a scheme of concentric circles, as shown in figure 17. The circles developed around the moral agent as the center of consideration symbolize moral

4 The following discussions will first deal with *moral* rights of animals in ethical arguments. Their role in *animal welfare or conservation (law)* is addressed in some of the following sections and chapters. The discrepancy between ethical demands and the status quo of animal conservation and welfare is clearly seen in the fact that moral rights are far from being followed with respect to legal rights and their implementation (see Chapter 9). Animal rights requirements (as they are formulated in the philosophical debate) usually go far beyond established law.

5 Instead of "holism," the term "physiocentrism" is often used. Biocentric and physiocentric approaches are often combined as "ecocentric" (Birnbacher 2002).

communities of different range (Gorke 2000, 87).[6] Due to the unique and specially valued characteristics of human persons, only human beings are included in direct, anthropocentric moral responsibility.[7] Anthropocentric approaches assume direct obligations exclusively vis-à-vis other people, whereas nonhuman nature only has a value to the extent that it is either a means of achieving human purposes or the subject of a contemplative relationship (be it scientific, religious, or aesthetic). In both cases, nature does not have value in itself, but only in its relationship to humans (Birnbacher 2002). In the case of pathocentrism,[8] we are also committed to direct moral responsibility toward all other sentient animals, while in biocentrism,[9] all living beings are afforded direct moral consideration (where the decisive criterion is life itself). Lastly, in holism,[10] it is argued that all living and inanimate entities, including superorganismic entities (i.e., entire ecosystems such as mountain chains or entire

6 This does not mean that in the respective accounts the human being is always at the center of the argumentation. It is precisely with regard to those on whom the focus is placed that the accounts differ from each other—in this respect, the scheme could be misleading at first glance. The "central" position of the human being at the starting point of all circles results only from the equally neutral and trivial fact that it is about *her* view of nature surrounding her. A justified point of criticism of Gorke's scheme is that the possibility of a collision of interests or obligations is not apparent. However, this is precisely where ethical problems often arise. The scheme is only used here to show the basic types of nature ethics, with their respective areas and criteria, at a glance.

7 Representatives of anthropocentric approaches in ethics, for example, include those of Jürgen Habermas (1997) and Ernst Tugendhat (1994), for whom the ability to communicate and cooperate (which they grant only to humans to a sufficient extent) are central criteria in the attribution of direct moral consideration. Another classic example of an anthropocentric approach to moral consideration of humans vs. animals is provided by Immanuel Kant in the second section of his *Groundwork of the Metaphysics of Morals* (Kant [1785] 2011, AA IV).

8 Pathocentrism can be traced back to Jeremy Bentham, who in 1789 emphasized in his *Introduction to the Principles of Morals and Legislation* that it was not the ability to reason or the ability to speak but the ability to suffer that was decisive for who is to be given moral consideration (Bentham [1789] 1970, 283). Arthur Schopenhauer carried out pathocentric considerations in 1840 in *On the Basis of Morality* in his so-called ethics of compassion, and Peter Singer developed pathocentrism further in the 1970s in *Animal Liberation* and *Practical Ethics* in the guise of a preference-utilitarianism (Schopenhauer [1841] 2007; Singer 1976).

9 Biocentrism is represented, for example, in the work of Albert Schweitzer (Schweitzer [1923] 1971) and Paul W. Taylor (1986). In his work *Civilization and Ethics* of 1923, the former placed what he called a "reverence for life" at the center of his ethics and conceived an egalitarian biocentrism.

10 In Arne Naess's (1973, 2008) holistic approach, the unity of humans and nature is expressed in so-called deep ecology. In German-speaking countries, Martin Gorke (1999, 2000, 2010) is one of the main proponents of holism in ethics.

species) fall within the realm of direct moral responsibility on the criterion of existence.

All basic types of nature ethics that include more than a single group of entities in the moral community are in principle conceivable in both *egalitarian* and *hierarchical* terms, depending on whether *all* suffering, *every* form of life, and *every* type of existence must be treated equally or whether there are hierarchies between suffering, living, and existing entities. This weighing problem applies equally to pathocentric, biocentric, and holistic approaches. However, it is important to note that all basic types of environmental ethics in their pure form postulate in principle, and above all, equality with regard to the claim of moral consideration of all members of their defined moral community. If strict equality is linked to an outstanding moral status, and if strong protective rights are derived from this status, then there is the possibility of treating all designated beings equally among themselves and in a specific manner compared to non-designated beings. For example, if faced with conflicts of interests, we could demand they be excluded from any weighing procedures of interests *from the outset*. The equality of all human beings with regard to their dignity means that their *status as persons with human rights* is postulated. Human rights guarantee inalienable, absolute and universal rights for every human person.[11] While in animal experiments the suffering and killing of animals is weighed against the scientific knowledge of humans, experiments on human persons who did not give their free and informed consent are (rightfully) unthinkable. People are generally excluded from such a practice. No matter how great the possibility of scientific progress, the life and the *inviolable dignity of the person* should not be balanced against it. This is one of the strengths of anthropocentric legal concepts, one that is usually not achieved as easily by the other three basic positions. For anthropocentrists, basically only the human being deserves direct moral consideration and thus, in principle, is given preference over all other living beings and things. The rights and interests of people are in principle often not recognized as equivalent to other goods, interests, and rights, such as the integrity or the life of animals and plants.[12]

11 The origins, foundations and problems of the human rights idea are discussed in more detail in Section 8.4.1.

12 However, there are also tensions among anthropocentrists when it comes to the balancing of rights or interests of *two different individuals*: if mother and embryo are both considered as persons with the same strong moral status and protective rights, such as a right to life that cannot be revoked, or if the human rights of the mother cannot be weighed against the human rights of the embryo, then a termination of pregnancy cannot be defended ethically *solely under these premises*. Whoever wanted to make such an argument in favor of such a point of view could do three things: First, he could argue that the

In essence, the discussion about extending the idea of "rights" in general and the concept of personhood in particular to nonhuman animals revolves around the dissatisfaction with the idea that such a unique status and the subsequent protection should be afforded to human persons alone.

If in the following the strength of rights or even the strength of personhood rights is addressed or demanded for animals, then this refers to the dissatisfaction with a two class ethics that allows for the weighting of important goods and interests in the case of some beings (animals) but, rightfully so, not in the case of others (humans).

Living beings who are to be morally considered in themselves (who deserve direct moral consideration) are *moral patients* (Regan 2004, e.g. 151–56; Blumer 1997). I would like to make it clear that, while the term as I use it refers to living beings who are to be considered morally, it is not necessary that they be able to act morally themselves. In this sense, children can also be spoken of as moral patients. In principle, the concept of moral patients is first and foremost synonymous with the idea of members of our moral community who have intrinsic value and are to be morally considered for their own sake. Connected to what has been said so far, and according to the circles of moral consideration provided by Gorke, only humans are clearly moral patients in all basic positions of environmental ethics. But pathocentric positions, for example, argue that animals capable of suffering also fall under this category.

Within this schema, the opposite of the moral patient is the *moral agent*. Being a moral agent involves membership in the moral community, but it goes beyond that. Moral agents are not only affected by our actions and have to be considered morally; they are themselves morally active, possess moral judgement, and can act morally. There is a strong connection between the status of moral agents and their capacity for moral autonomy. But Tom Regan rightly makes it clear that there can be different interpretations of what autonomy means. Regan contrasts Kant's definition of an autonomous person, who can

embryo does not yet have personhood status, and therefore its interests cannot outweigh the mother's interests. But secondly, he could also maintain the embryo's personhood status and still accept abortion. He could say that, at least in the case of medical indications, there is a kind of self-defense situation in which it is permissible to save the mother's life. Thirdly, he could observe that if the mother's life is seriously threatened, then two equally strong fundamental rights of two people actually collide. So we are faced here with two options: either the mother loses her life, or the embryo loses her life. This could mean simply that no morally correct decision can be made at all. Provided one considers it ethically justifiable that one intervenes decisively (even doing nothing in this case is an intervention, as it could lead to the death of the mother), then performing an abortion could be morally tolerated.

dictate for herself and follow reason-induced moral principles, with a more basic definition of autonomy. At the center of his minimal idea of autonomy are intentional actions and the pursuit of one's own preferences (desires and beliefs). Kant's understanding of autonomy merely leads to criteria for moral agency—and therefore full moral responsibility. While most mammals are clearly not morally responsible for their behavior, they could still be considered as living beings with *preference autonomy* (Regan 2004, 84ff., 151ff.). Regan therefore endeavors to develop a de-intellectualized notion of autonomy. He does well to try. Those who pursue human rights-analogous animal rights can only borrow from Kant to a limited extent. One might deny from the outset that Kant has an empirical approach to reason and autonomy. Animal rights theorists who want to argue from the limited reason and autonomy of animals must, however, presuppose that both can be proven in the behavior of living beings. My book also makes this prerequisite: its goal from the outset has been to search for definitions of culture, language, and theory of mind applicable to animals, and to seek evidence for this in ethological studies. But if reason and autonomy are understood as empirically provable capabilities, then completely new (and no longer originally Kantian) ways unfold to define both. Animal ethics moves away from elaborated, autonomous moral judgement that puts emphasis on a reason-based function of distinguishing good and evil without contradiction and following the categorical imperative. Instead, the perspective opens up to understanding autonomy in a less complex but also less abstract way. Regan rightly makes it clear that being autonomous in the Kantian sense presupposes being able to act fully morally. To be able to act morally, however, does not have to be synonymous with autonomy in the Kantian sense. Children, for example, are not moral agents in the Kantian sense, but nevertheless possess preference autonomy in Regan's terms. (For further aspects of Regan's theory, see Section 8.2).

Whether being a moral agent should result in a stronger intrinsic value is generally disputed today. It is clear that all moral agents are also moral patients. But it is possible that not all moral patients must also be moral agents. This could apply not only to nonhuman animals but also to certain groups of people. Not all people possess the full rationality of a normally developed, morally capable and responsible person. This is the case, for example, with severely mentally disabled or senile people, infants, babies, fetuses, and patients in a coma. People who, among other things, are no longer or not yet oriented toward ethical norms and are not able to act in accordance with these norms are often referred to as *human marginal cases*.[13] What is not being

13 The term is considered ableist and discriminatory today. I will avoid using it (which is not always possible when referring to debates that are circling around this term). If I use the

questioned is whether or not these people are human beings. Instead, it is unclear on which grounds we can identify them as *persons* holding the same strong moral status as "paradigm" persons even if they do not possess the (full range of) personal qualities which are often considered as characteristic for the status of a person.[14] If an ethical theory were to attribute intrinsic value to moral agents alone, and count only them among the moral community, then cases such as children or sick, disabled, and elderly people would be excluded. So one can either attribute to them the status of a person, although they do not possess all personal qualities or not all of them to the full degree, or one can pursue a different concept of personhood from the outset, which then also includes these people and other animals.

The distinction between *moral rights* and *moral duties* is closely connected with the distinction between moral agents and moral patients. It is obvious that not all members of a moral community have to fulfill all duties. Some moral patients, such as children, have strong protective rights (such as basic human rights) without having any of the duties of the adult citizen. To be a bearer of (some) rights, one does not have to be a bearer of (all) duties and obligations, at least not when one is human. To a certain extent, *the exceptions* to our claim of personhood and status are particularly interesting when it comes to animals. It is the question as to why and when human persons qualify as having personhood rights, *without* (all) personhood duties and obligations being imposed on them, which spurs animal ethicists' interest in the argument from "marginal" cases.

8.2 Basic Positions in Animal Ethics

8.2.1 *Peter Singer's Utilitarianism*

Besides the question of who is to be directly morally considerable and on the basis of which criteria, the question arises of *to what extent* every member of the moral community must be considered. In addition to a differentiation

---

terms "borderline," "marginal," or "paradigm" person I will put them in scare quotes to make clear that I distance myself from these labels.

14 If we assume that (some) nonhuman animals could be nonhuman persons then we can speak about "borderline" cases here as much as we speak about "borderline" cases with respect to some "marginal" human cases. But since animals as nonhuman persons are always handled per se as "borderline" cases for personhood, their "borderline" status is already included in the concept of nonhuman persons to a certain extent and does not always have to be specifically identified. I will come back to the ethical problems that come with the assumption of "borderline" or "marginal" cases later when discussing the concept of personhood.

between criteria for moral consideration (e.g., suffering or life), a distinction needs to be made as to how strong or absolute the intrinsic value of a moral patient or agent is, especially if it collides with the intrinsic value of others that are morally considerable. The two predominant traditions of animal ethics, utilitarianism (represented by Peter Singer, for example) and the animal rights approach (represented by Tom Regan, for example), differ on precisely this point.

Both approaches employ the argument from "marginal" cases and refer to comparable characteristics of (some) animals and (some) humans. Singer, however, shrouds this argument in the language of interests and preferences to be weighed up against each other, whereas Regan speaks of rights. Beside the idea of rights, he emphasizes the value of living beings beyond any considerations of utility. This idea, which is at the center of the attribution of rights to animals, will be examined below by comparing the positions of Singer and Regan.

Peter Singer is one of the best-known proponents of a modern utilitarian argument in animal ethics, in which the *interests and preferences* of moral patients form the core. Here, acting morally means "starting with a broadly utilitarian position" and taking into equal consideration the interests of all those affected by a decision (Singer 2008, 11ff.). All interests must be weighed, and then the course of action that is most likely to satisfy the interests of as many of those affected as possible, or which can be said to have the best consequences for all those affected on balance, must be chosen (26). Singer's preference utilitarianism differs from classical utilitarianism in that it is not oriented toward the greatest happiness of the greatest number, measured by a maximization of pleasure and a minimization of pain, but more specifically toward the promotion of the preferences of those affected (this is the case for the kind of utilitarianism Singer has developed in his books on animal ethics and defended for the major part of his career, see Singer 2008, 11ff.; 2009c, 155ff. Only recently, he changed his position into hedonistic utilitarianism in Lazari-Radek & Singer 2016).[15] Singer himself coined the term "preference utilitarianism" to

15 In a discussion with his critics, Singer writes: "An intrinsic preference is something that is desired for its own sake. An instrumental preference is a preference for something as a means to satisfy either another instrumental preference, or an intrinsic preference" (2009c, 192–93). In his pursuit of utilitarianism which focuses on inter-personal behavior, *intrinsic* preferences are important. A patient in pain who wants to kill herself because she is depressed may have an instrumental preference (the desire to kill herself) to achieve an intrinsic preference (a painless condition). According to Singer, her intrinsic preference must be observed and followed, which is not necessarily her desire for suicide (2009c, 154ff.).

refer to the version of utilitarianism he espouses. Under this theory, an action is judged not by its tendency to maximize pleasure and minimize suffering but by the degree to which it corresponds to the preferences of the beings affected by the action, including the consequences of the action (Singer 2008, 11 ff.). According to Singer, the consideration of preferences is the first step in applying a universal aspect to the ethical decision-making process.

Anyone who wants to go beyond this step by including other universal ethical ideals (like the ideal of individual rights) carries the burden of arguing why this is necessary. He or she must provide reasons then, why non-utilitarian rules and ideals should be accepted instead, or additionally (Singer 2008, 27).

"Equality" in Singer's utilitarianism therefore means "equal consideration of interests"—and this consideration is not limited to members of our own kind. It commits us to accept "that it extends to some nonhuman animals" (2008, 28). The effort to extend the principle of equality to other species runs up against the prejudice that the interests of animals should not be taken seriously. But this prejudice is "no better founded than the prejudice of white slave owners against taking seriously the interests of their African slaves" (2008, 29). People who hold the latter belief are racists, while representatives of the same attitude with reference to animals are "speciesists." The latter discriminate in favor of their own species by attaching greater weight to the interests of their own species when their own interests collide with those of others: "Human speciesists do not accept that pain is as bad when it is felt by pigs or mice as when it is felt by human beings" (2008, 32; similarly, see Singer 2009a, 6ff.). According to Singer, antidiscrimination movements fighting for the rights of animals, slaves, or women follow a common logic of argumentation. They reject the preference of the interests of a single group on the basis of arbitrarily chosen, supposedly higher-status characteristics of this group (2009a, 6, 9). Singer argues that

> it is on this basis that we are able to say that the fact that some people are not members of our race does not entitle us to exploit them, and the fact that some people are less intelligent than others does not mean that their interests may be discounted or disregarded. The principle also implies that the fact that beings are not members of our species does not entitle us to exploit them, and it similarly implies that the fact that other animals are less intelligent than we are does not mean that their interests may be discounted or disregarded. (2008, 29)

Consideration for the interests of others, be it humans or animals, should not "depend on what they are like or what abilities they possess." But Singer adds:

"although precisely what this concern requires us to do may vary according to the characteristics of those affected by what we do)" (2008, 49). It is important not to overlook this addition and to understand in general the characterization of the principle of equality in Singer: it is not a description of a *currently existing equality*—not all humans and certainly not all living beings *are* actually the same—but it is a *normative principle* that informs us of how we should treat living beings. If this principle is extended from one group (humans) to other groups (nonhuman animals), this does not mean that we must treat all these groups equally or grant both groups exactly the same rights:

> Whether we should do so will depend on the nature of the members of the two groups. The basic principle of equality does not require equal or identical *treatment*; it requires equal consideration. Equal consideration for different beings may lead to different treatment and different rights. (2009a, 2)

While Singer emphasizes the principle of equality, certain characteristics play a role in the way in which living beings and their rights are treated. Much then seems to depend on which "treatment" and which "rights" different living beings receive due to their "nature."

What criterion does Singer ultimately use to attribute direct moral consideration? He says that all humans as well as many animals are actually "*the same*" with regard to their ability to suffer. Since Bentham, the ability to suffer has widely been regarded as "the vital characteristic that entitles a being to equal consideration" (2008, 50). It is not arbitrarily chosen but is "the only defensible boundary of concern for the interests of others," since it is not simply another skill (such as the ability to speak or the ability to do higher mathematics). Instead, the ability to suffer (or its opposite, to feel joy and happiness) is "a prerequisite for having interests at all, a condition that must be satisfied before we can speak of interests in any meaningful way." (Singer 2008, 50; 2009a, 7). The ability to suffer points to a fundamental and significant interest shared by humans and all other animals capable of suffering: the interest not to suffer (Singer 2009a, 8). For Singer, this interest is so significant that it almost seems to lead to a "right" not to suffer. Although Bentham originally speaks of the "right" of animals not to suffer, Singer wants to avoid this term, since he views rights as something secondary, derived from the interest not to suffer. In Singer's and Bentham's utilitarian argumentation, to speak of "rights" means nothing more and nothing less than the following: a creature capable of suffering has the right for its own interest not to suffer to be taken into consideration equally *in relation to the same interests of other living beings*. However, no right to life or the like is established by this. Singer certainly appropriately describes the role of the ability to suffer as a basic prerequisite for the possession of interests. The importance of taking preferences into account for ethical decisions and actions is also becoming clear. However, his theory goes beyond these fundamental ethical considerations on the principle of equality and must do so because important questions remain open: Who has interests and intrinsic preferences? What different types of preferences are relevant here? How much do they weigh in each case in the balancing of interests? And what role do the different characteristics of living beings—including their different socio-cognitive characteristics—play when it comes to the evaluation of preferences?

Singer (2009b, 74) describes the following four propositions as the important premises of his theory. The last one especially links with the questions I asked:

1. (A factual [not moral] premise): "Humans are not the only beings capable of feeling pain or of suffering."
2. "Pain is bad, and similar amounts of pain are equally bad, no matter whose pain it might be."
3. "We should consider ourselves responsible both for what we do, and for what we refrain from doing."
4. "The seriousness of taking a life depends, not on the race, sex, or species of the being killed, but on its individual characteristics, such as its own desires about continuing to live or the kind of life it is capable of leading."

The last premise offers in my view the greatest scope for criticism as it stands against the backdrop of the utilitarian approach, in which individuals don't have an absolute right to life. Instead, we are faced with weighing procedures. In fact, Singer opens up hierarchies between living beings—not as far as the validity of the principle of equality is concerned, nor with respect to the basic ability to suffer, but rather as far as the *right to life* of individuals and the *value of different lives* is concerned. It is here, among other points, that Singer's theory becomes controversial for critics. The differences that he recognizes in the treatment of living beings and with reference to their rights are not based on species boundaries, and this does not exclude humans. In the case of some people—e.g., when it comes to severe disabilities—it allows euthanasia or infanticide, which is intolerable not only for representatives of the principle of the sanctity of life (i.e., anti-abortionists) but also for associations for and of disabled people.[16] Regarding the question of killing another human being,

16 Singer has met with great recognition and sometimes with strong criticism for complex reasons, which cannot be examined in detail here. See the compilation of critical voices together with counterreplies by Singer in Schaler (2009).

Singer writes: "The account I have favoured for many years is that the most serious reason against killing arises when the being has an awareness of his or her existence over time, and wishes to continue to live" (2009c, 153). If a living being is not capable of suffering, has no consciousness of its own existence over time, or does not wish to live on, then, according to Singer, there is no immediate reason to save her life at all costs.[17] The human fetus, for example, clearly has no intrinsic preferences according to Singer (see, e.g., Singer 2009c, 157). When referring to the case of a newborn baby with massive and irreparable brain damage, he writes that its vegetative state could mean that it could neither speak nor recognize other people, act independently of others, or develop a kind of self-consciousness. It is conceivable that parents who know that the living conditions for the child will not improve and who may not be able to pay the cost of treatment for the child would ask doctors to end the child's life painlessly (see Singer 2009a, 17ff.). Singer emphasizes that he does not say that severely disabled babies should have their lives ended, but that the parents of these children, if they think it would be better if the child died, should have the option either to let the child die, or to have the doctors end the child's life should the latter mean a more humane death (2009d, 189). According to Singer, critics such as Harry J. Gensler would usually not deny that decisions as to whether such a severely impaired life is worth living are made anyways. They would also advocate the retention of life-sustaining activities and simply "let some children die." Singer, however, does not want his critics to get away with the causal reference to death occurring "naturally" in such cases. He wants people to take responsibility for such ethical decisions: "I see little difference between causing death by withdrawing a respirator, when one knows that the infant is unable to breathe unaided, and causing death by a lethal injection. (In practice, the difference between what Gensler recommends and what I recommend may be even less than this—among health care professionals, talk of "easing the pain" is often a euphemism for a life-shortening dose of morphine)" (2009d, 190–91).

What Singer seems to be interested in is the fact that the active killing of patients is highly controversial and not legally permitted (in his view because legal systems follow the Christian principle of the sanctity of life).[18] But at the same time, the same people who would not allow the killing of the fetus or

---

17 Singer, however, recognizes some social (relational) arguments that refer to the emotional state of the parents or possible adoptive parents (see, e.g., Singer 2009c, 2009d).

18 Singer does not specify here which legal systems he means; he speaks generally of "the law" (2009a, 189).

disabled babies would agree to the killing of many animals without any hesitation: "How can they justify their different judgements? Adult chimpanzees, dogs, pigs, and members of many other species far surpass the brain-damaged infant in their ability to relate to others, act independently, be self-aware, and any other capacity that could reasonably be said to give value to life" (2009a, 18). Here it becomes clear that, on the one hand, Singer argues on the basis of the argument from "marginal" cases and, on the other hand, he cites classical personhood abilities when it comes to a right to life. In that case, socio-cognitive abilities make a difference. We consider ourselves persons, and this self-conception is central when it comes to the problem of killing. The decisive factor is not our coming to the world as biological beings but the beginning of our own perception of ourselves as persons:

> I can identify myself either with the late fetus, when consciousness begins, or with the developing infant, when self-awareness develops, or perhaps even with the young child I can remember being. Since I have no psychological connection with the fetus I once was, and that fetus had no desires for the future, we are not compelled to say it was in the interests of the fetus to live.... Once there is a being conscious of its own future, with desires about that future, which we can satisfy or frustrate, the consequences of ending its life are different. (2009c, 162)

However, according to Singer, personal identity and the concrete indication of the point in time of its emergence are not of central importance for his theory. He claims that his theory can account for the graduality in the development or existence of skills. The intrinsic problem of killing, then, grows parallel to the increase in a child's self-consciousness (2009c, 162). According to Singer, killing a fetus in a late abortion seems a "lesser evil" than killing a severely disabled newborn, which would in turn be a lesser evil than killing a healthy newborn child, and this would in turn be a lesser evil than killing a healthy adult—a hierarchy that can be severely criticized. Here, certain socio-cognitive abilities become *direct criteria* of moral consideration within an ethical scheme that does not provide any basic protection for individuals. A utilitarian balancing procedure may also tempt us to apply a hierarchy to the value of living beings instead of demanding an absolute limit of socio-cognitive complexity that defines what is ethically appropriate *for all* who fall within this limit. This, however, would require the acceptance of inalienable and fundamental rights, which do not fall below the standard of human rights. Conceptions of such rights reject violations of human rights in the form of killing, torture, and the loss of freedom *generally, for all* moral and legal bearers of such rights. Once it is determined that disabled, elderly, young, and newborn people belong to this community, then any hierarchization and ranking within this group can be rejected. It would not be the lesser evil to kill a healthy newborn child than it is to kill an adult human being; both acts would be equally morally reprehensible. Torture would also be something that we *on principle, under no circumstances, should conduct on any* human being.

In nonhuman animals, too, apart from the capacity for sentience, the question of consciousness and self-consciousness, for future-oriented desires and interests plays an important role in Singer's theory. In the treatment of animals, especially when it comes to the ethical justifiability of their (painless) killing, the degree of such abilities similarly comes to bear.[19] With reference to the contradictory position of those who would not kill the infant but would have no problem killing animals, Singer comments: "This does not mean that to avoid speciesism we must hold that it is wrong to kill a dog as it is to kill a human being in full possession of his or her faculties." In order to avoid a therefore contradictory, speciesist position, it is only necessary to allow living beings with similar characteristics ("who are similar in all relevant respects") to have the same right to life ("have a similar right to life"). Within these limits, it would still be possible to say that it is worse to kill an adult, healthy person (who, for example, has self-consciousness, plans his future, and enters into meaningful relationships with others) than a mouse (which probably cannot do any of this) (Singer 2009a, 18–19). Whichever abilities one would present here, however, according to Singer, in his typical reasoning, they would certainly not follow our species boundary. If we want to protect a severely disabled or extremely senile person, we must also protect some animals. Singer recognizes the double-edged nature of the argument:

> This argument cuts both ways. It could be taken as showing that chimpanzees, dogs, pigs, along with some other species, have a right to life and we commit a grave moral offense whenever we kill them, even when they are old and suffering and our intention is to put them out of their misery. Alternatively one could take the argument as showing that the severely retarded and hopelessly senile have no right to life and may be killed for quite trivial reasons, as we now kill animals. (2009a, 19–20)

---

19 Their ability to suffer, in this account, is irrelevant: along these lines, it seems the "lesser evil" to kill a mouse in an experiment than to kill a chimpanzee if it is assumed that the former has no self-awareness in comparison to the latter.

Singer himself considers both positions, although they avoid speciesism, to be unsatisfactory: "What we need is some middle position that would avoid speciesism but would not make the lives of the retarded and senile as cheap as the lives of pigs and dogs now are, or make the lives of pigs and dogs so sacrosanct that we think it wrong to put them out of hopeless misery" (2009a, 20). Singer wants to leave open the way to infanticide in extreme cases, such as that of severely disabled newborns with irreversible brain damage, or euthanasia in the case of sick or elderly people, if they express the wish to die (2009d, 189–90). He also wants animals that suffer and cannot be cured to be put to sleep. To Singer, it seems incompatible with the assumption of an absolute right to life to leave such exceptions open. In his opinion, killing a person is not always morally wrong. It must be considered whether there is a "prospect of a meaningful life" and whether existence without terrible pain is possible (2009a, 20).

Interestingly, Singer's rejection of speciesism does not imply that all lives have the same value. Self-consciousness, thinking with foresight and planning, a sense of the future, meaningful relationships with others, and other sociocognitive abilities (2009a, 19) are not relevant to the question of how much a living being suffers, since pain is simply pain and is perceived as such, no matter what other abilities a living being possesses. However, these skills are important when it comes to the killing problem: "It is not arbitrary to hold that the life of a self-aware being, capable of abstract thought, of planning for the future, of complex acts of communication, and so on, is more valuable than the life of a being without these capacities" (2009a, 20). According to Singer, a self-conscious living being is aware of herself as a distinct entity. She knows about her past and future, and she also has desires for her future. In killing such a living being, you then "thwart that person's most significant desires. Killing a snail does not thwart any desires of this kind, because snails are incapable of having such desires" (2008, 77). If a decision has to be made between the life of a healthy adult and that of an animal, then, according to Singer, the life of the human being should be saved. What is decisive here, however, are the characteristics of the human being, not its species. At the same time, he believes, this is the reason why we can no longer say that, in principle, the life of every human being is always preferable to the life of every animal. Members of our species that lack the corresponding characteristics are thus not always entitled to strong rights of protection: "When we consider members of our own species who lack the characteristics of normal humans we can no longer say that their lives are always to be preferred to those of other animals" (2009a, 21).

To assess the ethical justifiability of animal experiments, Singer draws on this parallel again. An animal experiment can only justifiably admit suffering and the death of animals if it is of such great importance that we would carry it out in the same way on a severely disabled person. This sets—benevolently interpreted—extremely high demands on the benefit of experiments. Less charitably interpreted, Singer could be accused of not being able to enhance the moral status of the animal without lowering the moral status of human beings. Alternatively, an attempt could be made to depict painful or even lethal experiments with all animals, both human and nonhuman, as ethically unacceptable. The fact is that Singer really believes that it is sometimes morally justifiable to research on disabled people or people in general. If it were true that the life of one subject could save several other human lives, then conducting the experiment would be the right thing to do in his opinion (2009a, 85).

It becomes apparent that Singer's utilitarian heritage has far-reaching consequences: neither the life of a human being (or animal) nor the physical integrity of a person can be absolutely guaranteed, as one would expect from a right to life or a right not to be tortured. According to Singer, torture would "almost always be wrong, but it is not absolutely wrong" (Singer 2009a, 85). If it were the only way to determine the location of an atomic bomb in New York City, then torture would be necessary. If a single animal experiment could cure leukemia, it would have to be carried out. At least he admits, that these might be, after all, "extremely rare" cases. In real life, no such *direct benefits* arise from such methods where there really is *no alternative* (Singer 2009a, 85). However, the dictates of utilitarianism would present clear answers to these cases. In addition to the problems of utilitarian reasoning in extreme situations, their answers in everyday life are sometimes unsatisfactory. Benefits, costs, suffering, interests, and preferences are all difficult to determine. Above all, however, the result of a weighing scenario focusing on safeguarding the interests of the largest number of people affected can lead to results that run counter to a strong moral intuition: the intuition that the individual (human and nonhuman) should count for more in its intrinsic value.

8.2.2 *Tom Regan's Animal Rights Approach*

Tom Regan, who defends an animal rights approach, accuses Singer of degrading those to be considered morally (whether human or animal) as mere containers of interests and preferences. Attributing someone a real intrinsic value means to consider the *individual herself* as valuable, not only to consider her as a (worthless) container of interests which count. Regan therefore speaks of *inherent value* to emphasize the weight of the intrinsic value of a certain group of living beings (2004, xxi et seq.). Individuals with inherent value and rights should never be treated as if they were resources for others: "In particular, harms intentionally done to any one subject cannot be justified by aggregating benefits derived by others. In this respect, my position is anti-utilitarian, a theory in the Kantian, not the Millian, tradition" (2004, xvii). However, Regan's path deviates from that of Kant when it comes to who must be regarded as the bearers of rights (with a view to the concept of autonomy, this has already been clarified).

While Kant considers moral agents exclusively for their own sake, Regan seeks to provide a larger circle of individuals with basic rights. Each *subject-of-a-life* has an inherent value. These are all those living beings who are conscious beings whose lives have a quality of experience for them, and for whom therefore it is important what happens to them, no matter whether secondary usefulness in relation to others exists or not:

> individuals are subjects-of-a-life if they have beliefs and desires; perception, memory, and a sense of the future, including their own future; an emotional life together with feelings of pleasure and pain; preference- and welfare-interests; the ability to initiate action in pursuit of their desires and goals; a psychophysical identity over time; and an individual welfare in the sense that their experiential life fares well or ill for them, logically independently of their utility for others and logically independently of their being the object of anyone else's interests. Those who satisfy the subject-of-a-life criterion themselves have a distinctive kind of value—inherent value—and are not to be viewed or treated as mere receptacles. (2004, 243)

The focal point of Regan's definition is the ability to consciously suffer, and the ability to act according to (or orientate toward) desires, preferences, intentions, beliefs, memories, anticipations, and expectations. The satisfaction of needs, frustration when any such satisfaction is not achieved, survival or death, all this constitutes the individual quality of life and is important for the individual and its way of life. Not explicitly mentioned here are self-consciousness or meta-cognition (i.e., a theory of mind), nor does Regan mention in his definition the term of personhood. Still, in this passage as well as in other passages in Regan's work it becomes clear that his criteria for being a subject-of-a-life are nevertheless demanding. They are still primarily found in socio-cognitive abilities. He speaks of the presence of psychological states and dispositions and of an inner mental life (2004, xvi) as well as an understanding of the future (2004, 81). Subjects-of-a-life are individuals with beliefs, desires, and intentions who initiate actions because they want to achieve something or satisfy their wishes. Some of their beliefs, such as beliefs about preferences, are "general beliefs about the connection between choice and the satisfaction of desire.... And since many of the beliefs they have involve their having expectations about the future, these animals must also have a sense of the future—indeed, a sense of their own future." This notion of their own future entitles us, according to Regan, to assume that animals not only possess consciousness, but also self-consciousness (2004, 80–81). Regan's characterization of the subject-of-a-life seems basic at first glance, but in fact it involves a range of quite sophisticated socio-cognitive abilities. Thus, it is probably often underestimated. Although it comprises more living beings and applies to more basic socio-cognitive abilities than a demanding concept of personhood does, it can—like any theory that sets criteria—only include living beings that fulfill them. Regan also sets limits in extreme "borderline" cases, for example in the case of very severe mental disabilities or babies under one year of age (2004, 78). Because they are not subjects-of-a-life, they cannot be protected by his arguments, which of course does not mean that they cannot be protected by other arguments or indirect duties.

The core and the strength of Regan's theory lies, on the one hand, in the fact that he includes some very basic criteria and, above all, in the fact that he avoids the balancing scenario and instead wants to establish real rights tied to the inherent value of the subject-of-a-life. The aim is a strong moral rights status that is very similar in content to the status of the human person. Regan writes that he wants to argue for more than just animal rights: "*The case* does more than argue for animal rights. It seeks to describe and ground a family of basic human rights, especially for the most vulnerable members of the extended human family, for example young children.... I never would have become an animal rights advocate if I had not been a human rights advocate" (2004, xiii).

Regan seems to have solved the problem of (most) "borderline" cases better than Singer. As long as they are subjects-of-a-life, they have the same inherent value and rights as indisputable persons. In relation to Singer, Regan has shifted the boundary between moral patients and those who are not morally considerable for their own sake "downward." He can thus also accommodate many animals without problems. In his opinion, "mentally normally developed mammals from the age of one year" and a number of other animals, such as birds and fish, should be regarded as subjects-of-a-life (2004, xvi, also p. 78). However, the relationship between all subjects-of-a-life is not completely clear. Do all subjects-of-a-life (human persons and "marginal" cases, other mammals, and possibly many other vertebrates) really have the same moral status and the same, equally strong rights? The criticism could be made that the rights approach would ultimately lead to a relativization of differently complex living beings. He levels differences in the complexity of living beings by assigning the same intrinsic value to all, in deliberate ignorance of these differences per se. Jean-Claude Wolf (1992, 73, transl. MK) says that Regan's criteria are also "empirically variable variables." He asks: "Why would value, which is linked to graduable properties, not itself be graduable? Why shouldn't some beings have more intrinsic value than others?"

In fact, Regan reaches some limits of his broad definition of the subject-of-a-life with inherent value and rights. In the discussion about so-called lifeboat cases, he cites considerations that indicate that he also accepts differences within the subjects-of-a-life. If only four places in a lifeboat and four people and a dog were ready to board, then the dog would have to remain behind (Regan 2004, 324–25, xxvii–xxxii). This is so because death is a greater loss and greater harm to every single person than to every single dog: "Why? Because the loss that death represents is a function of the number and variety of possible sources of satisfaction it forecloses" (2004, xxix). It could not be assumed that the loss that death meant for a human being was less than the loss for a dog. However, the severity of the loss is the only thing that needs to be determined and compared here. Even if one million dogs would have to be sacrificed and the four people alone would be allowed to get in, this would be the morally correct decision, because it still applies that the loss by death would be more significant for the individual human beings than the loss by death for each of the individual dogs (Regan 2004, 325, xxix).

While it makes sense under a rights approach that the perspective on the right to life of every subject-of-a-life should be preserved here, it is not sufficiently clear why death represents a loss of varying degrees for different subjects-of-a-life, or why the life of a human being contains a higher level of total satisfaction, which, in addition to this, also has a moral weight. In other places, however, Regan makes it clear that he actually believes that "human life contains within it the possibility of a richness not to be found in the life of other animals." The reasons for this are "our advanced cognitive, aesthetic, moral, and spiritual capacities" (Regan 2004, xvii–xviii).[20] Here, therefore, all the other, more complex personhood characteristics more or less come into the moral arena through the back door—although they were initially left out of the definition of the subject-of-a-life and the naming of morally relevant criteria. It is one thing to acknowledge that human beings possess these "higher" (in the sense of more complex) abilities; it is a completely different matter if their existence makes a difference for moral treatment. Even if Regan repeatedly

20 According to Regan, the value of a life must always be considered on a case-by-case basis. He says that he would never say that the life of *every* person is in principle *always* richer than the life of every dog. However, his statement about the million dogs to be sacrificed seems to demonstrate the opposite. Regan assumes that the adult person has *the same possibilities plus many more than the adult dog* (2004, xxxiv).

emphasizes that this does not detract from the same inherent value of the four people and the dog, one wonders whether the inherent value of the human being—even if only under the pressure of the emergency situation of a lifeboat scenario—is "enriched" by certain characteristics that go beyond that of a subject-of-a-life. Can the absolute, inherent value itself, as Wolf (1992, 73) asked, be something that is thought out gradually? This would actually fundamentally contradict Regan's rights idea.

There are two ways of dealing with the dilemma that has arisen here. First, it could be said that an inherent value, which is tied to strong protective rights, which are absolutely essential even in extreme situations, is actually only possessed by those subjects-of-a-life who have these "richer" characteristics. Then, however, one has to deal with a concept that, although it is based on rights in contrast to preferences that have to be weighed, only strongly enforces these rights for subjects-of-a-life with "richer" experiences in the sense of human persons. Or, secondly, one can maintain that there is an inherent value and equally strong rights for all subjects-of-a-life. Then one has arrived at a serious moral status and strong conception of rights—for quite a large group of animals, certainly for many animals we use on farms, in labs, as zoo or companion animals and so on. Regan obviously prefers a strong conception of rights, but allows for a special argumentation in extreme emergencies. This seems to be inconsistent, but the only alternative to that would be to assume a moral stalemate between the stakeholders in the lifeboat. One could only escape from this by "rolling the dice" in favor of one or the other, or by then (and only then) consciously citing other (possibly also anthropocentric) arguments. That one actually prefers one's own brother to a stranger in a lifeboat may not imply that one's brother has different human rights than the stranger. It could be similar among all the subjects-of-a-life: why I prefer humans to dogs would then be a question of species membership or social closeness and no longer of the moral status of those affected. The difficulty here lies in the fact that such worst-case argumentation based on genus or social proximity must be reserved just as strictly for extreme lifeboat scenarios as the validity of Regan's additional assumption of a richer human life. Besides, according to this argumentation, it might even be morally correct to take the dog on board. Someone might feel more socially connected with their long-time pet than with a stranger. Regan presumably wants to avoid situations like these; he probably does not believe that arguments of belonging to a species or arguments of personal or social closeness are morally decisive, even in exceptional situations. It seems to remain an unsolvable problem: lifeboat scenarios are not comparable to everyday scenarios. Regan makes the mistake of not clearly naming his possibly anthropocentric special approach to the richer life of humans as such. Instead, he pretends that it fits seamlessly into his reflections on the moral status of the subject-of-a-life.[21] But he doesn't fail to point out the exceptional character of lifeboat scenarios themselves. These are indeed quite clearly moral exceptions and not transferable to our everyday interaction with animals. Animal experiments, for example, are not an immediate emergency in the sense of a lifeboat situation, but everyday research practice. Regan therefore rejects them and stresses the rights of animals as subjects of life.

Other philosophers have rightly pointed out that the importance of lifeboat scenarios for the question of rights and inherent value of living beings should not be overestimated. Mark Bernstein (2002) writes, for example, that it may well be the case that in extreme, hopeless cases it is justified to assume otherwise secondary differences with regard to additional qualities or characteristics of persons. One could prefer a virtuous person to a less virtuous one. However, this does not mean that we are entitled to *create* a dangerous or even life-threatening situation for the less virtuous person in order to benefit the virtuous. Bernstein uses the following example to illustrate what he means: If both he and Mahatma Gandhi needed a heart transplant, and only one organ was available for the procedure, then Gandhi should get it. If, however, this organ was in his, Bernstein's, body, then no one in the world—not even Gandhi—would have the right to take it from him and have it transplanted in him: "Likewise my (purportedly) superior virtue to that of a mouse may allow my illness to be treated before his; it does not allow a healthy mouse to be infected (or a mouse to be bred for scientific purposes) so that a search for a cure for my illness may be expedited" (Bernstein 2002, 536–37).

Similarly, it can be argued that, although under the pressure of choice one may favor humans over animals in a lifeboat, the targeted killing of animals for the benefit of humans is still far from justified. All in all, an ethics that starts from rights and does not regulate the extreme case, but regulates the normal case by applying those rights, is ultimately still a different ethics than one that is fundamentally utilitarian from the outset. Regan can therefore say with some conviction that the richer life of some subjects (probably, in his case,

21 Unfortunately, it is not quite clear whether the criterion of a richer life in Regan applies only to humans. First, the criterion is not described as an additional criterion applicable to all subjects-of-a-life. Second, it is not systematically introduced and explained in its function, its scope and its significance within Regan's overall theory, but seems a bit appended. Third, this is done in a discussion that deals specifically with the issue of dog vs. human. Fourth, the criterion is not discussed for other collisions of rights (especially not for the collision between the rights of two nonhuman subjects-of-a-life, e.g., between cat and chimpanzee). Due to these ambiguities, it is unclear whether the accusation of anthropocentrism is appropriate here.

humans) is *exclusively* decisive under the collision of equally strong rights (such as the right to life). Otherwise, when it comes to the richer life-form of the human being, he sees no reason to fundamentally grant us more rights to exploit animals or rights to dispose of them: “Although I believe that human life contains within it the possibility of a richness not to be found in the life of other animals—because for example, of our advanced cognitive, aesthetic, moral, and spiritual capacities—this difference provides absolutely no basis whatsoever for our exploitation of other animals” (Regan 2004, xvii–xviii). Furthermore, he strictly distinguishes between the *inherent value of a living being* (which he always wants to assume to be the same and absolute) and *the value of a life in the sense of the quality of a life*. The quality and value of a life can vary dramatically, up to a life that seems to be barely worth living. The value of the individual who leads this life, however, can still be the same inherent, strong value that also belongs to any other subject-of-a-life (Regan 2004, xxiii). This distinction is important for Regan when it comes to understanding and evaluating the lifeboat situation. Again and again he has to defend himself against the fact that critics do not take this distinction into account. Thus it is not the inherent value of subjects-of-a-life that is graduated, but the value of different lives.

Astonishingly similar problems remain with Singer and Regan: Both philosophical accounts assume an initially very strong basic demand for equality (with regard to the consideration of preferences and the inherent value of living beings). But in Singer’s case—in every ethical decision, not only in extreme cases—interests are weighed against interests, and the interests of (human and nonhuman) persons are then more important than those of all other living beings. Regan reasons from the viewpoint of the same inherent value of all subjects-of-a-life; the value of different lives, however, is then graduated with a view to more complex skills—which, however, only comes into effect in extreme situations. Nevertheless, there remain two controversial points in Regan’s thinking: Is it not a weak argument that—at least in extreme situations—the proper treatment of a subject-of-a-life is suddenly no longer oriented to her inherent value, but to the value of her life? And in this case—even if one recognizes a graded value of a life and accepts this “solution” in extreme situations—is the value of a life of animals fairly assessed here? Is the value of their lives actually so distinguishable from the value of a person’s life with regard to them supposedly lacking “advanced cognitive, aesthetic, moral, and spiritual capacities”? With regard to the graduality of socio-cognitive abilities, there may be minimal levels even for such abilities, which some animals also fulfill. The question then is no longer whether humans have the ability X plus the “richer” ability Y (and therefore their life offers more possibilities and represents a higher value), whereas animals only have ability X (and therefore their life offers fewer possibilities and represents a lower value). The question is: Do animals also have a degree of ability X and, possibly, degrees of ability Y that are sufficient for their life to offer enough possibilities (for them) to represent the same value as the life of a human being? What if animals not only have the same inherent value as individuals, but also the value of their lives cannot be underestimated to such an extent that it is so clearly below the value of a person's "rich" life, as Regan believes? This would be followed by the arguments of the first half of this book.

It is clear that different positions in animal ethics diverge with regard to the question of how to handle the valuation and weighting of interests and preferences of living beings (in Singer) and the valuation and weighting of the rights of living beings (in Regan). Great challenges arise for both approaches, rights and utilitarian, when it comes to how much animals ultimately count in relation to humans as persons and, above all, in relation to human "marginal" cases. What is important is that it is always the consideration and weighing of the morally relevant characteristics—in a utilitarian or a rights perspective—that, at least theoretically, has the power to make a difference. It is not the case that certain morally relevant characteristics or characteristics of personhood would be of importance in only one of the approaches. Often arguments with respect to the ability to suffer are attributed to pathocentrism and a pathocentric argumentation above all to utilitarianism. The argument from cognitive similarity, on the other hand, is attributed to animal rights approaches. However, this ends in a very general and artificially delimited juxtaposition of these two positions. The impression is created that the ability to suffer plays no (significant) role in animal rights approaches and that utilitarian approaches, conversely, are not about certain characteristics and there are no "rights elements." It should be made clear that this is not the case. Singer's utilitarian approach, in particular, which strongly emphasizes the ability to suffer, ultimately arrives at a point where the distinction between persons and nonpersons is made by espousing a concept of interests. Furthermore, his approach is also based on something like a "right" not to suffer. This holds if we presuppose that it is a preference of the being in question not to suffer and that she has, from a utilitarian perspective, something like a right to have her preferences taken into account. Regan, on the other hand, speaks of rights, but avoids talking about persons. He sticks to an idea of the subject-of-a-life that seems to leave out classical personhood abilities (such as a theory of mind). This gives the impression that his account might be less demanding than personhood accounts. Furthermore, scholars following him might sometimes underestimate how intellectually demanding his criteria are. The price for this is that it also remains unclear how strongly animal rights can be enforced vis-à-vis personhood or human rights and whether, in extreme situations, the "richer" life of humans should play a role after all.

Socio-cognitive abilities and their role in the definition of the moral status of human and non-human animals pose challenges to both approaches. In order to further avoid a direct comparison of animal rights vs. utilitarianism, I will now present the three main arguments of animal ethics in detail, without their assignment, at the outset, to a particular current of thought. The following sections first present the argument from suffering and the argument from endangerment (meaning the threat of extinction many species face) in more detail and then come separately to the argument from cognitive similarity. The first, two are the "classical arguments" in animal welfare ethics and conservation ethics. However, our reasoning about the human-animal relationship can gain in strength if we add the argument from cognitive similarity. My discussion will lead us to an investigation of *human rights-analogous animal rights* (*or personhood rights for animals*), which targets the aforementioned moral status of the person and tries to establish basic human-rights-like rights for animals.

8.3 "Classical" Arguments in Animal Ethics and Animal Welfare

8.3.1 *The Argument from Sentience and Suffering*

Within animal ethics, arguments are used that are developed from all four basic types of environmental ethics (anthropocentrism, pathocentrism, biocentrism, and holism) and from the two most prominent basic positions in animal ethics (utilitarianism and animal rights). Pathocentric arguments have prevailed to the greatest extent. A first and certainly the most prominent answer to the question of why we should protect a given group of animals, let's say rats (a typical laboratory animal species), pigs (a typical farm animal species), or apes (a typical zoo animal species) is therefore: they are capable of suffering. There is a lot behind this answer, and there are also very different forms of pathocentrism (see Mayr 2003). What follows are only a few important points.

Pathocentric arguments are initially based on the basic assumption that the infliction of suffering, especially unnecessary suffering, is ethically unjustifiable. This *principle of nonmaleficence* is a broad principle that also guides human interaction with peers and is already expressed in the Hippocratic oath. To this day, it obliges the physician to look after the well-being of the patient and adhere to a prevention of loss (which is expressed in the attitude *primum nil nocere*: "first, to do no harm"). In contemporary bioethics, this precept is understood under the terms "beneficence" and "nonmaleficence" (Wiesing 2000, 22).

Beauchamp and Childress included the principle of nonmaleficence among the central rules of biomedical ethics in their book from 1979; the rules also include respect for autonomy, beneficence, and justice (Beauchamp and Childress 2008). Schicktanz (2002) and others argued that these basic moral principles of biomedical ethics should also be modified and applied to today's animal ethics debate. As a basic principle of pathocentrism, the principle of nonmaleficence in utilitarianism, especially in the case of Jeremy Bentham, is highlighted in animals: since they too have the capacity for sentience, that is, for suffering and happiness, they must be included in considerations of interests and within the utilitarian goal of the greatest happiness of the greatest number. For utilitarians, any being that feels pleasure and displeasure, that can be satisfied (happy, comfortable) or dissatisfied (unhappy, distressed), is particularly worthy of consideration. Thus, at the center of the pathocentric argument is not the *fact* that animals can suffer (this would represent a naturalistic fallacy) but the *utilitarian norm* of promotion of the well-being of as many beings as possible who are capable of suffering (Irrgang 1998, 834).

For David DeGrazia the principle of nonmaleficence is one of the basic principles of all ethics. He formulates the principle as follows: "It is wrong to cause extensive unnecessary harm to others without their consent" (DeGrazia 2005, similarly in 1996a). DeGrazia believes that the principle of nonmaleficence is a moral principle that is beyond any serious doubt. Any system of thought without this principle would hardly be recognizable as a moral system (DeGrazia 2005). On this, most people would probably agree. DeGrazia applies the principle of nonmaleficence, but no longer in its general, pure form. He does not speak of the general imperative of nonmaleficence, as the Hippocratic oath does, but of avoiding "extensive unnecessary harm." This certainly expresses the concern of utilitarian pathocentrism. If it is assumed that there is unnecessary injury, then this also implies that the possibility of unavoidable harm (i.e., necessary harm) exists. Similarly, the talk of significant harm indicates that harm is being differentiated according to severity. Both refer to the recognition of decisions made on the basis of weighting.[22]

Pathocentrism is based not only on the principle of nonmaleficence but also on a number of other basic assumptions. It is also assumed, for example, that sensations as subject-object relations are always assigned with a value.

---

22 It is significant however, that a *quantification of suffering* alone does not provide us with a *hierarchization of living beings* (on the problems of applying hierarchies, see Mayr 2003, 263ff.).

They are always authentic, usually immediate and do often not allow much reflective distancing at the moment they are perceived—especially in the case of severe pain (Mayr 2003, 120). The feelings of animals are also generally authentic in this sense and they also perceive pain directly as negative sensation. Due to their *comparability* of the ability to suffer, and due to the *principle of equality*, many animal species we use and consume, experiment on and kill must be morally considered from a pathocentric perspective (Mayr 2003, 121ff.). In this sense, as has already become apparent, Peter Singer, in particular, demands that the same interests of not suffering pain and enjoying life be morally taken into account in the same way. The same applies to Tom Regan, who also requires application of the same intrinsic value for human and nonhuman beings.

How strong the life of an animal is being protected in a pathocentric position can vary, however. In hypothetical conflict situations in which either the life of an animal or that of a human has to be sacrificed, Regan and Singer both decide to kill the animal. Regan justifies this, as explained, by the fact that the death of a human being would deny many more possibilities for having satisfying experiences than the death of an animal. Singer argues that the life of living beings with outstanding cognitive abilities has a higher value than the life of those lacking these characteristics. Thus, normally the human being is preferred. Still, one should certainly admit in Regan's case that the "exception" in the lifeboat is not "the rule" of his animal rights ethics. Birnbacher (2002), too, concludes that, from the recognition of an intrinsic value of animals being capable of suffering, the majority of pathocentrics solely derive a requirement to avoid or reduce actively induced suffering. Only some of them—probably those who include discussion of rights in their pathocentric approaches—formulate more far-reaching demands. These demands include, for example, a requirement to prevent and alleviate suffering not caused by oneself or a prohibition on killing (as is evident in vegetarianism), in the rejection of lethal animal experiments, and in the rejection of keeping animals for further use of their bodies after killing (see Birnbacher 2002). The problem here is that within a utilitarian framework with pathocentric assumptions alone, a prohibition on killing cannot actually be formulated. A really painless killing (if it exists at all) for the benefit of the greatest happiness of the greatest number as well as the sacrifice of an individual capable of suffering is theoretically permissible in utilitarian pathocentrism—but for many, this becomes counterintuitive when it comes to the sacrifice of a human being. In the end, a prohibition on *painful* killing remains justifiable in a utilitarian-pathocentric framework, but a ban on *pain-free* killing does not necessarily. Problems such as these give the impression that pathocentric arguments have limits and might not always be far-reaching or strong enough without certain additional elements (cf. Irrgang 1998, 835; Mayr 2003, 263ff.). The criticism is thus asserted that although sensitivity to pain is an ethically relevant criterion for assessing harm, it is probably not an ethical principle that is equivalent to other principles, such as the principle of respect for autonomy or the principle of justice (Irrgang 1998, 835).

No matter which precepts are derived from pathocentric approaches, the greatest challenge for pathocentrism already follows from the principle of nonmaleficence alone: namely, that harm and suffering must be something which we can *assess* or *evaluate*. For practical implementation, especially when it is often ultimately a matter of weighing suffering, we even need to *quantify* and *measure* harm and suffering. After all, the aim of avoiding suffering is often to see when the same animal suffers more or less, or whether an animal would suffer less than another animal in a given situation. This results in the need to assess well-being (*welfare assessment*). Well-being can, in fact, be judged in pathocentrically founded utilitarian welfare considerations. In this instance, ethics and animal welfare depend on biology and veterinary medicine. They tell us what animals need, when their well-being is at risk, and when they are suffering. With respect to this, Würbel (2007) states that, from a biological point of view as well as from the advocatory position of humans, a clear distinction must be made between our personal opinion about animals and their truly objective well-being. The assessment of the latter is the target of any ethically and biologically founded animal protection and animal welfare law.[23] Consequently, the animal herself justifies what she needs for protection.

One means of making suffering in animals measurable is by analogy. The suffering of animals may not be *directly* measurable, but it is at least *indirectly* inducible and can be plausibly demonstrated (Würbel 2007, 19; with reference to Sambraus 1995). The conclusion by analogy refers to the comparison between animal and human and is based, according to Würbel (2007), on three criteria: the similarity of the nervous systems, the similarity of the situations, and the similarity of the reactions (19), whereby the decisive criterion is the homology of structures that have developed in the course of evolution from the same precursors (20). However, species that are not evolutionarily closely related to us can develop similar adaptations through convergent evolution. So

23 The animal ethics underpinning (German) animal welfare law—is, according to Würbel (2007, p. 15 f.), a "hybrid ethics" that includes principles (such as "the animal as a fellow creature" or "protection of life and well-being") and utilitarian considerations. What we then weigh up in a utilitarian matter is the suffering of the animals against the benefit for humans.

if the analogy conclusion is particularly plausible with homologous structures, it is not necessarily implausible with analogous structures. The decisive factor here is whether these species are confronted with similar or similarly complex challenges in their ecological niches, which have led to the development of comparable or comparably complex adaptations. Not only our closest relatives, but also species that are evolutionarily distant must therefore be taken into account when it comes to analogies. Although the similarity of the nervous system in the sense of a homology may then not exist, they can show similar or similarly complex reactions in similar situations and also be capable of suffering.

The basis of the assessment of suffering and well-being in animals is that animals have *needs* that are so important for their welfare that their nonperformance leads to suffering. Animals may not be able to communicate their needs and preferences verbally, but are very much capable of doing so by way of their behavior. To assess preferences, ethology uses a microeconomic approach which was developed to assess people's demand for consumer goods: "Luxury goods ... are typically in demand when they are cheap. If the price rises, the demand for them also decreases rapidly. Such a demand is called "elastic." On the other hand, goods that are vitally needed are also in demand at the expense of luxury goods when the price rises" (Würbel 2007, 24; transl. MK). Würbel continues with reference to Dawkins (1990): if we can vary the time or energy an animal has to spend on certain activities or on the use of a certain resource, the conditions are in place to quantify the demand for that activity or resource. The different elasticity of the demand curves for different activities and resources then reflects "the hierarchy of importance of these activities and resources from the perspective of the animals" (Würbel 2007, 24).

How are preferences and well-being or the opposite (negation) of well-being, namely suffering, connected? Here, too, evolutionary theoretical considerations can help: Systematic preferences have emerged as a result of natural selection. Animals should therefore prefer the conditions that best maximize their reproductive success (fitness). This can be explained by the fact that subjective sensations only make sense in evolutionary biology if they help animals to select the most favorable conditions for themselves or to avoid disadvantageous conditions. According to Würbel, this means that beings with subjective perception feel more comfortable under pleasant circumstances (maximization of reproductive success) or suffer less than they do under circumstances that they find unpleasant (2007, 25). Of course, not only physical but also psychological needs related to the social, cognitive, and emotional abilities of a living being are to be included in such considerations. This will be discussed in further detail below.

For the time being, the question of why we should protect a given group of animals has been answered many times, at the latest since Jeremy Bentham: because they might be capable of suffering. This answer also contains analogies, because we consider our own ability to suffer to be central in our assessment of ethically justifiable behavior. Different pathocentric approaches centered on the well-being of animals have developed, up to and including the attempt to determine preferences, needs, and suffering. A successful assessment of well-being appropriate to the needs and behavior of an animal is certainly both a direct consequence of and, at the same time, the greatest challenge to pathocentrically-based animal protection.

8.3.2 *The Argument from Endangerment and Conservation*

A second possible answer to the question of why we should protect a given group of animals, like for example great apes or dolphins, is that they might be endangered. The foundational precept of this answer is that the survival of whole species, not only of individuals, is important. The argument for the protection of the individual animal, therefore, does not primarily refer to a claim to protection for its own sake, but to the animal's role in and importance to the ecosystem. The aim is to protect species in their natural and evolved diversity, making *species diversity* and *biodiversity* the goal of conservation efforts. "Red Lists" have been used since the 1960s to estimate the level of extinction or degree of endangerment of species. The more endangered a species is, the more intensive species conservation programs should ideally be. Species protection is thus part of nature conservation, which deals, on the one hand, with the protection of populations and, on the other, with the protection of entire habitats (biotopes, ecotopes, etc.). Species protection as part of nature conservation aims at habitats, because species are a necessary, functional component of them. When species die out, the habitat is affected because each species influences and maintains the function of an ecosystem in its own way, especially when it plays a key role—for example, as predator. Conversely, if habitats disappear, species become extinct. To be sure, intact ecosystems and biodiversity are also of central importance for the survival of humans: without biological-filter functions, without securing food production, without ecosystem stability or the many other functions that intact ecosystems fulfill, even our own species cannot survive in the end. This fact is now taking on increased importance, especially with a view to future generations. Here it becomes clear that arguments for species protection are often anthropocentric arguments at the same time. According to Birnbacher (2002), anthropocentric approaches to justify nature conservation point to the resource character of the inherited stock of landscapes, biotopes and biological species. In view of the complexity of the ecological interactions and the limited assessability of the risks, we cannot afford to allow or even accelerate the current—in human history unprecedented—extinction of biological species. This holds at least if we are indeed interested in our own survival (Birnbacher 2002). There are other anthropocentric arguments, such as the claim that in principle it is impossible to estimate which previously unknown economic, scientific, educational, and aesthetic value (some) species might have. Accordingly, the endangerment and conservation argument can be combined alternatively or additionally with other anthropocentric arguments. We could argue, for example, that it is desirable for aesthetic reasons to preserve the beauty of diversity. If we destroy it, we thereby deprive future generations of a valuable experience, and this constitutes an aesthetic loss. As far as the aesthetic functions of nature are concerned, according to Birnbacher, there is much to suggest that "the need for nature—in the sense of nature not yet completely overhauled by human civilization—will increase rather than decrease in the future." In an overly structured world oriented toward the satisfaction of human needs, the need for "the self-sufficient, self-awakened, restfulness of nature" would certainly increase (Birnbacher 2002, transl. MK).

Whereas anthropocentric principles might be sufficient to argue theoretically in favor of *environmental protection*[24] whether or not anthropocentric and pathocentric approaches are also sufficient to justify the goals of *species conservation* is controversial. Finally, biocentric and holistic arguments, which attribute intrinsic value to entire ecosystems and also to inanimate nature, are conceivable for the preservation of species diversity (Birnbacher 2002). However, while biocentric and holistic foundations of nature conservation can, like pathocentrism, do without the detour via human needs, they are also confronted with the weighing problems mentioned in the case of pathocentrism. They are difficult or even impossible to implement on an egalitarian basis, while in hierarchical conceptions, human needs are often given priority over the preservation of nature (Birnbacher 2002).

From the argument from endangerment and conservation, which usually ascribes an anthropocentrically determined value to species diversity, species

24 "It is undisputed that anthropocentric principles are sufficient for the theoretical justification (although possibly not for motivational support) of standards of environmental protection, as environmental protection is about the avoidance, prevention and reduction of present and future impairments of human survival, human health, and human well-being. In this respect, an environmental problem is a particular type of social problem" (Birnbacher 2002; transl. MK).

protection measures are to be derived that primarily have the survival of species or subspecies in mind rather than the welfare of individuals. Later I will discuss the fact that modern zoos are adopting the species protection argument more than ever before. This is problematic from many angles. Among other things, zoos argue that compromising the welfare of some individual ambassador animals (by confinement) is justified because it serves the greater good to preserve endangered species in the zoo. Thus, their argument exemplifies to what extent individual welfare and species conservation can collide.

8.4 A "New" Argument: Personhood Rights for Animals

My previous remarks have already prepared a third answer to the question of why we should protect a given group of animals (let's continue to assume we are talking about apes). This answer could be: we are supposed to protect animals because they are similar to us in their social and cognitive abilities. Of course, this answer only picks out an important aspect of similarity: proponents of this argument, for example, usually also emphasize similarity with regard to complex emotions.

Both socio-cognitive and emotional similarities with humans are subcategories or varieties of a more comprehensive argument that refers to the *evolutionary kinship* between humans and animals. It is important that arguments based on socio-cognitive or emotional similarity do no adopt the naïve and actually wrong claim that only species that are *evolutionarily close* to humans have similarities to humans. Even *evolutionarily distant* species, such as birds, can resemble us in certain aspects of their cognitive or social abilities. (Remember, for example, the tool behavior of the New Caledonian crows). In this sense, the argument from socio-cognitive similarity goes beyond that of evolutionary proximity in a decisive way. Animals that are cognitively similar to us, so goes the argument, should share with us a number of basic and strong moral rights. The basis of this demand is that it is assumed that we grant ourselves these rights also or above all because we are living beings with complex social and cognitive abilities, with consciousness and self-consciousness. As persons, we are entitled to these increased moral rights in the form of certain personhood rights or human rights. To use the argument from cognitive similarity in animal ethics, therefore, means to represent a rights approach or, more precisely, an animal rights approach analogous to human rights. The proponents of the argument from cognitive similarity do not claim that some animals *are in fact* human persons. The argument states rather that some animals have a sufficient number of abilities relevant for the status of personhood and that they share these abilities to a sufficient degree with us to be able to say that they are *like* human persons. Therefore, it makes sense to consider them as *nonhuman persons.*

I will first look at philosophical encyclopedia articles[25] to provide an overview of the content, the origin, and justification of human rights and/or personhood rights, whereby, when it comes to their application to humans, problems with the explanatory strategies become clear. Here I return to the problem of "marginal" cases. Afterward, I will address demands for basic human rights for apes and dolphins and conclude with some pros and cons for the argument from cognitive similarity. As in the previous sections, the aim is to present the argument from cognitive similarity and to introduce it into the arguments of the proponents of personhood rights for apes and dolphins. The resulting strong ethical claims are then first contrasted with how we actually deal with these animals in captivity and in the wild before they are finally discussed in their meaning, scope, and interdependence.

8.4.1 *On the Origin, Justification, and Problems of the Concept of Human Rights*

We grant special protective rights in the sense of human rights to members of our kind. These are to be understood as inherent, indispensable rights that humans possess (Ganslandt 1995, 846). In particular, rights and freedoms such as the right to life, the right to a fair trial, the freedom of movement, the freedom from torture and enslavement, or the freedom of thought and speech are considered human rights. In this book it is assumed that human persons—including so-called human "marginal" or "borderline" cases—have human rights and that the advent of the guarantee of these human rights represents an indispensable moment for modern civilizations. Based on this assumption, and without giving up on it, it should be discussed whether some animals could also be granted some human rights-analogous animal rights.

25 The following German encyclopedia articles are referred to here (as the philosophical debate on these topics is an international one, English encyclopedias and handbooks will not differ significantly in terms of content): the entry on "Human Dignity" by Bayertz (1999) in the *Encyclopedia of Philosophy* (*Enzyklopädie Philosophie*) and the entries for "Human Rights" by Klenner (1990) in the *European Encyclopedia on Philosophy and Sciences* (*Europäischen Enzyklopädie zu Philosophie und Wissenschaften*), by Ganslandt (1995) in the *Encyclopedia of Philosophy and Theory of Science* (*Enzyklopädie Philosophie und Wissenschaftstheorie*) and by Hilpert (1998) in the *Lexicon of Bioethics* (*Lexikon der Bioethik*), as well as the entry "Human" by Rentsch (1999) in the *Encyclopedia of Philosophy* (*Enzyklopädie Philosophie*).

The idea of human rights has a centuries-old tradition in the history of thought. The most important human rights movements, their catalogues, and their predecessors are the Magna Carta Libertatum (1215), the Petition of Rights (1628), the Habeas Corpus act (1679), the Bill of Rights (1689), the American Declaration of Independence and Virginia Bill of Rights (1776), the French Revolution, with its *Déclaration des droits de l'homme et du citoyen* (1789), the Universal Declaration of Human Rights, which was adopted by the United Nations General Assembly in 1948, and the Charter of Fundamental Rights of the European Union, which has been legally binding since 2009. The constitutions of several states, such as the Basic Law of the Federal Republic of Germany (Deutsches Grundgesetz) of 1949, also profess inviolable, inalienable human rights as the basis of every human community (§. 1, Para. 2) (Ganslandt 1995, 847; Bayertz 1999; Klenner 1990).

Human rights are basic, fundamental rights that apply in an egalitarian manner to everyone everywhere and at every time. They are considered *inalienable* and *universal*. Their absolute validity is independent from state determination or refusal. It is either derived from superseding, affirming sources of divine or natural law, or from the quality of humanity without any reference to metaphysical justifications (Ganslandt 1995, 846; Klenner 1990, 367). If we follow the secular branch of argumentation here, then there is something unique about the "human" in the concept of human rights. In this reasoning it becomes clear that, for whatever reason, we ascribe a value to the human person. According to Hilpert (1998), the normative core of human rights is the idea that the human being as such is accorded an end in itself—an idea that is captured by the term "dignity" in Kant and his followers, but also in key human rights documents. This value is ascribed to humans beyond their empirical appearance, and it means that "there are conditions of existence and expressions of personhood, which must under no circumstances be taken away" (675, transl. MK). The concept of human dignity is thus an anthropological and ethical category at the same time, which in the wake of the Enlightenment and the French Revolution was increasingly transferred to legal, political, and social areas and thus enters into a close relationship to the idea of human rights (see Bayertz 1999, 824). According to Bayertz, the concept of human dignity plays a dual role as an ethical-anthropological category and as a constitutional concept and thus serves as a bridging idea (Bayertz 1999, 824). It is, so to speak, the link between morality and law, between the human being and her human rights.

Despite the clarity of the function of the term in its double role, the content and foundation of "human rights" remain blurred and controversial (see Bayertz 1999, 824). What is clear, however, is that, first of all, its content has the anthropological dimension mentioned and that whoever wants to define human dignity must include the question of what constitutes the specific nature of human beings (see Bayertz 1999, 824–25). Above all, modern philosophy, according to Bayertz, emphasizes three "moments" of humans in this respect: first, the nonfixedness of humans, meaning that all other beings are prescribed in their way of being by nature or by God, while humans are free in the choice of their way of life. Humans are capable of creative self-determination. Second, the rational nature of humans, their ability to think and act rationally, is emphasized. The third "moment" is their autonomy (Bayertz 1999, 824).

Regardless of its postulation as value or legal norm, the concept of human dignity also requires justification. Here, Bayertz mentions three different but not necessarily mutually exclusive justification strategies, which are to be distinguished. The first is based on religious convictions and interprets the reason why human beings have dignity, for example, in Christian terms, namely, that humans have been made (or exist) in God's image. According to Bayertz, such religious justification strategies have the privilege of "being able to justify the unconditional and absolute validity of human norms strictly by referring to God and the order of being created by him, i.e., its impenetrability, indispensability, and unrestrictibility (Bayertz 1999, 825, transl. MK)." However, according to Bayertz, religious reasons can only convince those who share the corresponding faith: "if human dignity is to serve not only as a category of (private) theological morality, but as a (publicly) binding constitutional concept in a pluralistic society with a state that is ideologically neutral, religious reasons cannot lead to success" (825, transl. MK). For this reason, secular philosophical strategies of justification usually refer to the tradition of natural law or philosophy of reason (825). An empirically and historically based third strategy consists in deriving the content of the concept of human dignity *ex negativo* from recognized acts of violation. Because one does not want human rights violations, such as the atrocities of the Nazi regime, to ever be repeated, the need to uphold human rights is pointed out: "This strategy thus resorts to a respective consensus on a historical experience of degradation. However, this is also its limit: it is successful where it applies an existing consensus to the concept with regard to a transgression (e.g., torture, slavery, racial discrimination); however, it encounters difficulties where such a consensus does not exist" (Bayertz 1999, 825, transl. MK).[26]

26 Whether, for example, artificial insemination constitutes an offense against human dignity, is controversial and, according to Bayertz (1999), cannot be clarified on the basis of the concept of human dignity alone (825).

The idea of *human rights*, which is linked to the idea of human dignity, can also be traced back to traditions ranging from antiquity to the teaching of international law in the philosophers of the Stoa and later in Cicero, as well as to the philosophy of the High Middle Ages and social contract theory.[27] Together they form the basis of the modern concept of civil society and human rights emerging in the Enlightenment.

In the modern human rights tradition, human rights are no longer explicated in the context of an order of creation or as elements of an objectified moral law, "but are interpreted as subjective rights found in human reason and to be justifiable to any reasonable being" (Hilpert 1998, 675, transl. MK).[28] In modern and contemporary philosophy, human rights are also linked to the status of the person, and the person endowed with human dignity is attributed an unconditional meaning or an absolute value (Sturma 2002; Höffe 1997, 226). The modern concept of personhood originates with John Locke: those possessing personhood are intelligent and thinking beings who have reason, reflectiveness, and self-consciousness, who can lead their lives deliberately and who therefore have to deal with future courses of action in a special way (Sturma 2002).[29] The effective further development of the concept of person, however, is carried out by Immanuel Kant, who develops a direct connection between the person and human rights in his ethics: "The metaphysical dignity of the person is categorically determined within the framework of an elaborate moral-philosophical foundation procedure. As a free being who is endowed with reason and independent of mere causality of nature, the human being thus sets herself apart from the rest of nature, by setting nature itself a purpose" (Sturma 2002, transl. MK; with reference to the *Metaphysics of Morals* (Kant, I. [1797] 2017 and the *Groundwork of the Metaphysics of Morals* (Kant [1785] 2011). Kant stipulated the requirement that the person should be ascribed an unconditional meaning or an absolute value in the second formula of his

27 Modern explanatory approaches formulate, for example, discourse theory and political theories. For representatives of development politics, the basic needs of individuals are immediate reference points for human rights claims. Such claims are then being addressed to society, state, and international community alike. They pose, however, the question of what needs are linked to human rights and how they should be weighted in relation to each other (Hilpert 1998, 675).

28 Similarly, Ganslandt (1995), Klenner (1990), and Bayertz (1999) all stress that the modern idea of human rights has developed particularly as a moment of Enlightenment philosophy (see Ganslandt 1995, p. 847; Klenner 1990, 367; and Bayertz 1999, 824).

29 But other philosophers before Kant, such as Boethius and Rawls, already characterized the person as being determined by reason (Sturma 2002).

categorical imperative, the formula of humanity, which asks us to treat all persons as ends in themselves (Kant [1785] 2011, AA IV, 429).[30]

Since Kant, persons have been characterized by the fact that they can act morally and are moral agents. More important, according to Höffe (1997), however, is the fact that persons are at least the primary (and perhaps the only) *objects* of ethical action (i.e., they are moral patients) (Höffe 1997, 226). This means that every person who has human rights or personhood rights has strong protective rights. What exactly makes a person? Among the essential conditions of personhood, according to Sturma (1997, 2002), are self-consciousness, epistemic differentiation capability, emotive expression, communication, education, awareness of time, situation-independent language, and emotional and social ties to other persons. Above all, under the influence of the newer Anglo-American philosophy, "person" in present-day philosophy in this sense is "applied as a systematic, basic term that usually refers to a subject of epistemic states of consciousness and moral attitudes or beliefs" (Sturma 2002; transl. MK). In everyday language we often interchangeably use the expressions "person" and "human." But even if all humans (and only humans) may be persons, they are not such, according to Höffe, because they are humans: a criterion independent of this is sought (Höffe 1997, 227). The search for the central definitional features of the person, Höffe continues, will therefore also be controversial in the future, because an exclusion criterion emerges from them. In modern bioethical debates, it is therefore asked whether fertilized ova, fetuses, individuals who are in a coma, or the severely disabled are also persons (Höffe 1997, 227). Moreover, Sturma (2002) thinks that one of the habits of social space is not necessarily assuming the convergence of human and person. This does not mean that the "semantic divide" (Sturma 2002, transl. MK) between human and person is unproblematic in practical terms. On the contrary, two kinds of reactions can be observed here: on the one hand, it is denied that the lives of one human and one person can diverge in one and the same individual. In this sense, a stronger accentuation of the anthropological power regarding the concept of person is asked for. On the other hand, due to "discriminatory uncertainties" of the concept of the person, its scope has been broadened and there are attempts to impose a species-neutral application (Sturma 2002, transl. MK). The latter strategy includes attempts to extend personhood status to some animals.

The overall picture is that in the secular determination of what constitutes the quality of the human being and ensures its specific human dignity with

---

30 The central explanations can be found in Section II ("Transition from popular moral philosophy to the metaphysics of morals") of the *Groundwork of the Metaphysics of Morals*.

human rights deriving from it, socio-cognitive abilities in the form of humans' ability to reason are cited. The same is true for social capacities and social bonds. In this instance, the human being appears as a person who possesses reason, is free to act, and leads an autonomous life in the community (see also Rentsch 1999, 814). Nevertheless, all the concepts mentioned here—that of the human, that of her specific human dignity and human rights, as well as that of the person—are multivalent terms, each dating back to several traditions of thought in the history of philosophy, law and political science. In terms of their content and validity, therefore, they are anything but clear. They often refer to ideas of the human being that are metaphysical and sometimes religiously charged. In the case of the concept of human dignity, Bayertz points out that its main difficulty lies in the fact that it is based on metaphysical conditions, the difficulty of which would become clear from the "difference between the level of justification and the level of application of human dignity": on the one hand, human dignity is reduced to certain characteristics such as freedom, rationality and autonomy, "but on the other hand, individuals (zygotes, embryos, infants, the mentally ill, the comatose) who do not possess these characteristics should also fall under their protection" (Bayertz 1999, 825; transl. MK). In view of the difficulties of determining the content of and justifying human dignity, Bayertz concludes that the scope of this concept for the solution of ethical controversies must not be overestimated (826). In principle, "it should be stated that the reasons must be given for what human dignity is and why it has been violated in a given case....In an ethical context ... this cannot be conducted in any other arena than in the realm of discourse through rational arguments" (826; transl. MK). It should be noted that these difficulties already occur in discussions about the application of personhood status for *human* embryos, fetuses, and babies or severely mentally disabled, highly demented, and comatose *people,* and not only when it comes to the applicability of personhood to nonhuman animals. Certain cognitive or mental achievements that are constitutive for personhood often lead to cointerintuitive consequences, according to Höffe (1997). Two ways to avoid these difficulties are mentioned here.[31]

According to Höffe (1997), one could avoid the possibility of counterintuitive consequences in the *distinction between current and potential persons* (the latter applies, for example, to fetuses) or between *categorical and social persons* (meaning persons in a strict sense and persons in a social sense, whereas

31 It is not so important here, whether these attempts lead to real solutions. However, they certainly had and still have a decisive influence, e.g. on the debate on the status of the embryo. Both suggested lines of arguments have supporters as well as detractors.

the latter category applies, for example, to children) (Höffe 1997, 228). Such a distinction does justice to the empirically ascertainable otherness of fetuses, children, etc., and recognizes that these are potential (as opposed to current) or social (as opposed to categorical) persons. However, this distinction alone does not yet make clear whether they should nevertheless be granted full personhood status as such (by which both embryo and mother, for example, are full persons with full personhood status and whereby abortion would become ethically problematic) or whether a potential person only has a graded personhood status and/or graded rights (with which the current person and her full rights would be given priority in ethical conflict situations).

Irrespective of this, one may also see differences between the moral worthiness of potential persons (embryos) and social persons (such as children) who are already among us and interact with us, although they are not yet fully morally responsible.

A second way to avoid counterintuitive consequences in the application of the concept of person follows the first proposal, but adds the important differentiation between *being a person* and *having the status of a person*. Wimmer (1999), who proposes this distinction, would like to deal with situations "in which it is unclear or controversial whether human life that can be referred to as 'personal' exists." These can be situations in which personhood "does not yet exist," "no longer exists," or possibly "cannot be assumed at any point in human existence" (336; transl. MK).[32] There are cases, he says, "in which human personhood is clearly not, is no longer, or is not yet assigned, as in the case of deceased people or human fetuses. Yet they are nevertheless assigned personhood status, with the corresponding practical consequences." For example, we expect that everyone pays last respects to a corpse, and we have legal obligations toward a final will and testament. We assume strong moral duties regarding the prohibition of experiments on humans, and we assume that parents have an obligation to refrain from doing anything that endangers the unborn

32 Wimmer (1999) defines "person" quite broadly (but also with cognitive criteria) as "a living being that is able to reflect on and behave toward itself and its own behavior." This concerns self-reflection with theoretical components (based on perception and cognition) and practical elements (based on the formation of opinions and value judgements), in which the possibility of every morality and ethics lies. Of course, we could also employ other basic principles, such as the principle of balancing the needs and interests of living beings. "If it includes the idea of personhood, it could be understood as its extension: not only persons and their interests, but also the interests of other living beings (nonpersons) must be morally considered. However, in cases of conflict, the principle should then be understood in such a way that it commands the dignity of affected persons in every case" (Wimmer 1999, 330, 335–36; transl. MK).

child (Wimmer 1999, 340). Here it becomes clear, first of all, that "although the personhood of a person has a temporal beginning and a temporal end, its personhood status does not," and, secondly, that "a personhood status can only be meaningfully spoken of in cases where beings are basically and generally, by their nature, capable of becoming persons" (Wimmer 1999, 341; transl. MK). If this is the case, then on the basis of the distinction between being a person and holding the status of a person, we can say that "with the death of her brain, a human being loses her characteristic of being a person because a functioning brain is a necessary condition of being able to act like a person; yet she does not thereby lose her status to be treated as a person" (343; transl. MK).[33]

Obviously, Wimmer wants to uphold the idea of a person and a strong notion of personhood related to certain (socio-)cognitive abilities. All members of the moral community of persons (even those who are only persons in a limited way) enjoy full personhood status and all personhood rights, since by their nature they basically belong to the group of persons. Wimmer does not fall into the trap of arguing whether or not "borderline" cases *are in fact* persons. From the outset the question is not: Are these living beings persons? It is: Do they have personhood status even though they are no longer persons, though they are not yet persons, or, though they will never be persons? Wimmer restricts the scope of his distinction: he does not believe "that with the proposed distinction between being a person and having personhood status, all the moral problems tormenting us here can be solved" (1999, 341). The tensions in the concept of person and human dignity itself continue to exist—especially in view of the novelty, confusion, and ambiguity of our scope for action extended by science and technology. In Wimmer's arguments, it appears that the concept of the person as a cognitively determined term is difficult, but not obsolete, and that there are possibilities of maintaining it. In the following sections, we will continue to explore the extent to which we can and should stick to it in relation to animals.

Among those authors who do not approach the subject matter in terms of the concept of person or dignity, but in terms of the concept of human rights, similar problems as the ones discussed by Bayertz and Wimmer are referred to.

33 In contrast to Wimmer, I assume that to be alive and possess minimal perception and reaction functions is a basic, necessary requirement for being a person *and* for having personhood status. I consider deceased and comatose patients with irreversible consciousness or irreversible brain damage as people who *were persons* and *had* personhood status. Please note that this can still give rise to specific duties towards them (like a duty to treat their bodies in a respectful manner). Furthermore, Wimmer's point that being a person and having personhood status can differ remains important for all other "borderline" cases where the individuals are still alive.

For Klenner (1990), human rights are justifiably "one of the encyclopedic problems for which there is no comprehensively accepted solution, if only because antagonistic interests (within a society as well as between societies) also articulate themselves within them." "Above all," Klenner says, "the societal causes and the theoretical justifications for human rights are controversial" (371; transl. MK). In view of the large number of justification approaches concerning human rights, Hilpert (1998) notes that these should not lead to resignation before the aim of justification. Rather, it should be an encouragement to assume "that the potential of the human rights idea to connect to complex cultural, religious, and moral traditions is not yet exhausted" (675; transl. MK).

Ultimately, this also opens up the opportunity for discussions about extending personhood status to nonhuman animals.

In the case of the concept of the human being, which is inscribed both in the ideas of human dignity and human rights, Rentsch (1999), too, concludes that, despite the presence of a recognizable philosophical trajectory for the purpose of determining the human being, the human being cannot clearly be identified as such: "One can exaggeratedly speak of an absence of human beings in philosophy. Instead of speaking about the human being, philosophy tells us about the mind, body and soul, freedom, determinism and indeterminism, the individual, person, subject, and self-consciousness. It leaves human beings to be dealt with by empirical anthropologies. In place of an explicit reflection on humans, classification in life's realms takes place in far-exceeding transhuman orders." In this context, it is also true that the determination of the human being by reason "only partially and selectively reflects her everyday life" (816; transl. MK). The same could be true, of course, for nonhuman animals.

Some of the issues raised by the extension of human rights to animals are undoubtedly related to the problems outlined above, which are already included in the justification and attribution of human rights to human beings. At a minimum, it should have become clear by now that if animals are nonhuman persons and possess dignity, then there are conditions of existence and expressions of personhood, the possibility of which must under no circumstances be taken from them. How far advanced ethical claims of the application of personhood to animals are, on the one hand, and how far away we are from providing animals with such conditions of existence and expressions of personhood, on the other, will be discussed in the following.

8.4.2 *The Great Ape Project and the Hiasl Case: Human Rights for Great Apes*

The Great Ape Project, promoted by Paola Cavalieri and Peter Singer, is a paradigmatic case in which the moral status of humans and animals, in general, and of humans and their closest relative—the great ape—in particular, is being reconsidered against the background of their cognitive similarity. The project's claims are supported by many renowned philosophers and scientists and include extending some human rights to great apes in order to protect them more effectively. A declaration has been drafted that calls first for the right to life, secondly for the protection of individual freedom, and thirdly for the prohibition of torture for these animals (see Cavalieri and Singer 1993). It is notable that this initiative is not so much about such human rights as the right to work or freedom of opinion and belief. These correspond to the human person as a citizen and a participant in his or her specific, politically constituted society. They have no relation (at least no direct relation) to the welfare and interests of great apes. This differs with regard to the right to life, freedom, and physical and psychological integrity, since these are fundamental rights that are also of direct importance for the welfare of animals. The goal of the Great Ape Project is to extend the moral circle of those who are to be treated equal to all great apes, since they are our closest relatives and most closely resemble us in their characteristics and lifestyles. Granting them these human rights would have far-reaching consequences. It would mean a ban on animal testing on apes and a ban on their captivity in zoological gardens if both are not in the interests and welfare of the animals.

Following the Great Ape Project's demands would not mean that biomedical research would no longer be possible. Instead, the demand is to either replace the comparatively small proportion of animal experiments on great apes with alternative methods or to deliberately dispense with such experiments. One could argue that the abolition of experiments with great apes could lead to an increase in the number of animal experiments overall. However, this would only be the case if the experiments on great apes could be replaced by experiments with other animals. It is conceivable that instead of such replacement experiments, clinical studies with human patients could be introduced or methods applied in which no animals are used. Concerns that the abolition of animal experiments with great apes could prevent high-level research already presuppose that experiments with apes generally pursue high-level research objectives and produce important scientific findings. This presupposition must, of course, be scrutinized. I will point out later on that some experts doubt it, and with good reason (see Section 9.1.4). The impact of the Great Ape Project's demands on the quality of biomedical research could be far less than many suspect. Peter Singer also tries to clarify further misunderstandings concerning the scope of the consequences of the Great Ape Project:

> Some of the opposition stems from misunderstandings. Recognizing the rights of great apes does not mean that they all must be set free, even those born and bred in zoos, who would be unable to survive in the wild.
>
> SINGER 2006

The proponents of the project certainly do not want to release zoo and lab animals into nature, where they might be exposed to a variety of dangers, without a plan. This is not about a naive paradisiacal idea of an intact home in the wild, but about long-term, complicated, and scientifically informed management plans for the conservation of species. If habitats no longer exist or can no longer be secured, controllable reserves represent a possibility of providing animals with a new environment that comes as close as possible to their natural habitat and leaves them with the greatest possible freedom without endangering their survival at the same time.

The demands of the Great Ape Project have made it to the political stage several times: in 1999, for example, New Zealand followed suit in legislation, and in June 2008 the Spanish parliament signed a party-spanning declaration granting apes fundamental rights, calling on the Zapatero government to join the Great Ape Project (Roberts 2008). Most prominently, the US-based Nonhuman Rights Project works through litigation, public policy advocacy, and education to secure legally recognized fundamental rights for nonhuman animals. Their strategy is to file petitions for writs of habeas corpus, working through the common law on behalf of animal clients such as great apes and elephants.[34]

While the work of the Nonhuman Rights Project is widely recognized and also supported by distinguished philosophers such as Kristin Andrews, Bernard Rollin, and Will Kymlicka (Mezzina 2018, Andrews et al. 2019), similar endeavors in other countries, while probably less prominent, are equally interesting. For instance, in Austria, where there is a very active animal welfare and animal rights movement. A few years ago, a single chimpanzee attracted attention due to his lack of fundamental rights. I will use this case to illustrate how animal rights advocates typically argue and how courts typically react in litigation on behalf of nonhuman persons. Because it was about the future of an individual animal named Matthias (nickname Hiasl) the “Hiasl case” was a mixture of a concrete animal welfare concern and a process of strategic interest as developed by animal rights activists; its aim was to persuade national and European authorities to make a paradigmatic statement on the moral and legal status of great apes. In the following, I will present the history, the process

34 See www.nonhumanrights.org.

flow and the questions of the case on the basis of the article of the initiators of the process published in 2007 in the journal *Alternatives to Animal Experimentation* (ALTEX). I will then explain and discuss some of the legal foundations (for more on this, see Balluch and Theuer 2007).

The chimpanzee Hiasl was born in 1981 in the jungle of Sierra Leone. When he was ten months old, his mother was shot dead in front of him and, together with eleven other young chimpanzees, he was sold by an Austrian animal dealer to the pharmaceutical company Immuno. At Hiasl's destination near Vienna, his fate was to become a subject in hepatitis and AIDS research together with a female offspring from the group, Rosi (Balluch and Theuer 2007, 335). Just one day before the arrival of the animals in Vienna on April 28, 1982, however, Austria signed the Convention on International Trade in Endangered Species of Wild Fauna and Flora, (CITES).[35] The import of the chimpanzees thus became illegal due to the lack of required CITES documents, and the confiscated animals ended up in the care of the Vienna Society for the Protection of Animals. Hiasl and Rosi grew up in the care of a human family: "Hiasl, hence, is socialized like a human and considers himself to be part of the human species. Until today, he reacts to other humans as being his social partners, his rivals or his sexual mates" (Balluch and Theuer 2007, 336). Immuno was sued for violating the CITES Agreement. However, in an appeal (and later in another case, in which it sued for its ownership rights to the animals) it was ordered that Hiasl and Rosi should be returned to the company's laboratory. The animals were now six years old. The court justified its decision with the fact that, according to the Austrian Civil Code (ABGB), the animals counted as "things" and thus had no interests. The only existing interests in this case were the property interests of the owners (Balluch and Theuer 2007, 336). In 1989, however, the ABGB received an important addition: while §285 did, in fact, declare that any entity that is not a person is to be considered a thing, the addendum 285a made it clear that animals are expressly not objects and are protected by special laws. However, the provisions applicable to goods would apply to them insofar as there were no deviating provisions. On the basis of this amendment, the Vienna Society for the Protection of Animals filed an appeal against the return of the animals. It understood the addendum in such a way that animals, in principle, were not objects and thus had a value in and of themselves that was higher than the owners' interest in the animals as experimental tools (Balluch and Theuer 2007, 336). The court rejected this objection in September 1989 and ruled in favor of Immuno, but no steps were taken to remove Hiasl and Rosi from the care of the animal welfare association. The animals were

35 For more information on CITES, see also Section 9.1.

now eight and a half years old. They remained in the care of the animal protectionists for many years to come.

Immuno was eventually taken over by Baxter, and the company stopped chimpanzee experiments in 1999, at which point custody of Hiasl and Rosi was officially transferred to the Vienna Animal Welfare Society. In 2005, the Austrian parliament passed a ban on testing on apes, which came into force on January 1, 2006. Hiasl and Rosi moved into a specially built, four-hundred-square-meter facility at the Vienna Society for the Protection of Animals [Wiener Tierschutzverein], where they were cared for intensively and have been offered an enrichment program (Balluch and Theuer 2007, 336).

In 2006, however, a new problem arose. The Viennese animal protection association found itself in financial difficulties, and the chimpanzees, whose maintenance costs amounted to five thousand euros per animal each month, ran the danger of being counted among the assets in a bankruptcy proceeding. Legally the animals belonged to the association's assets and, in the worst-case scenario, they could have been sold to the highest bidder. They were again facing the possibility of ending up in an experimental laboratory (Balluch and Theuer 2007, 337).

It was decided, on the one hand, that the case should be brought to the attention of the public and, on the other, that it should go before the court with the help of legal and animal ethics experts. A group of animal protectionists and scientists led by Martin Balluch (president and campaign leader of the Association against Animal Factories [VGT], Vienna) and Eberhart Theuer (Austrian lawyer and legal adviser in the proceedings) brought the case to the European Court of Human Rights, with the help of a legal trick.

In 2006, an anonymous person donated a large amount of money to Balluch on the condition that the latter would only be able to gain possession of the money if the chimpanzee Hiasl were to be given a legal trustee (a kind of caretaker), who would also have access to the money and who would have a say in deciding what to do with it. The donation was a strategic move by the animal rights activists to create a financial interest for Balluch and thus a (legally valid) incentive for applying for a trustee for the animal. The application was filed in February 2007 at the District Court of Mödling. The proposal was supported by four scientific expert opinions. The argument to be put forward was that Matthias "Hiasl" Pan, in accordance with Austrian law, should be considered as a person in court (specifically, a person who, like a disabled child, is entitled to a trustee) (Balluch and Theuer 2007, 337). This was the beginning of a legal tug-of-war, in which a statement on the central question of the case has been consistently avoided right up to the end.

The district court in Mödling initially ended the trial because Hiasl was neither mentally disabled nor threatened, both of which would have been prerequisites for appointing a trustee. In May 2007, the opponent appealed this first ruling and argued that as a young animal Hiasl was traumatized by the wearing of a muzzle and the cruel killing of his mother. In addition, he had been locked up for the rest of his life and was unable to take care of himself. His interests were seriously endangered due to the threatened bankruptcy of the animal protection association. As a person with a trustee, he would also be able to accept the donation; otherwise he would lose it (Balluch and Theuer 2007, 339). The case was once again rejected, now with reference to procedural reasons: the applicant did not formally have the legal standing to submit the application for trusteeship. Thus the question of whether or not Hiasl is a person remained unanswered. The case next went to a higher court, the Regional Court of Wiener Neustadt, which rejected the appeal in September 2007: only the legal administrator or the person himself could file an objection. A third attempt by the animal rights activists at the Austrian Supreme Court argued that the law cited by the regional court should only be applied if a legal administrator *already existed*. However, his appointment was still under discussion. In other cases, close relatives could also have appealed to the court if the applicant was unable to do so. Furthermore, in addition to Hiasl's interests, the interests of the applicants were also affected here (because Balluch would lose his part of the donation, if no administrator should be appointed. This was a condition the donor had tied to the donation). The Supreme Court once again denied the appeal and the controversial question of the case with a formal justification: a judicial review was not a legal remedy. On the grounds that Hiasl had not received a fair trial and that the question of the status of a person had not been investigated by withdrawing to procedural arguments, the case was to be continued and would be heard at the European Court of Human Rights (ECtHR). The appeal was registered there at the beginning of 2008, but the case was ultimately rejected.

Obviously, the applicants would not pursue such a case in this way if, in their opinion, Austria's civil law would not allow for a decision in their favor: article 16 of the ABGB, which dates from 1811—i.e., from the pre-Darwinian period—states that every human being is also a person. Balluch and Theuer, who see the term "human" as biologically (genetically) defined, also recognize this.[36] In the case of beings such as robots and computers, who behave in the same way as humans but are not human beings, we have no problem

36 They understand the term "person," however, as a philosophical category.

excluding them from the human species. However, it may become more difficult if one is confronted with creatures that are obviously similar to humans in terms of appearance and behavior and are also genetically very close to them. Civil law, according to Balluch and Theuer (2007), rightly includes beings who are genetically human but cannot act as persons because of mental disabilities or restrictions. But what about those animals that are genetically and cognitively similar to us, but that are not genetically considered human beings? Balluch and Theuer consider it important that civil law, with the statement "Every human being is a person," states explicitly and rightly that every human being is a person, but not that every person is also simultaneously a human being (see Balluch and Theuer 2007, 337ff.). Would it therefore be possible to recognize chimpanzees as persons under the Austrian Civil Code?

> Until today, there is no judicial literature on this subject, since everybody apparently assumed that they knew which creatures are human and which are not. This might have been a reasonable assumption to make in the pre-Darwinian days of 1811, when the law was written. However, since the onset of taking evolution seriously, a number of different species or subspecies of humans have been identified. Take, for example, Neanderthals. Would they count as humans according with Article 16 ABGB, if, say, they suddenly appeared alive in a remote Himalayan valley? Or *Homo habilis*, or *Homo erectus*, or the recently discovered *Homo florensis*, who apparently still lived only 12,000 years ago?
>
> BALLUCH and THEUER 2007, 337

Three terms, "human," "person," and "thing," are therefore under debate. First of all, the authors turn to the concept of the human being, which they try to grasp biologically. In their opinion, this is a first possible method for classifying apes differently than we have done so far:

> In human rights charters...basic rights are recognized for "members of the human family." To provide this phrase a scientific meaning, it must be interpreted as the term "family" of the Linnean classification. The biological family today's *Homo sapiens* belongs to is the family of Great Apes, which includes chimps. If the term "human" has to be considered in its more narrow sense, then it must refer to the genus "*Homo*," which, after all, is Latin for human. Now, while the question [of] which species belong to the genus *Homo* might be decided upon controversially, a well based

> scientific argument can be made that chimps (and bonobos) must be part of it as "*Homo pan.*"
>
> BALLUCH and THEUER 2007, 337

According to Balluch and Theuer, our understanding of the genus *Homo*, and thus our understanding of who belongs to the "human family," must be rethought.

Interestingly, in the tenth edition of his *Systema Naturae*, from 1758, whose appearance is considered to be the beginning of modern zoological nomenclature, Karl von Linné (Linnaeus) first described humans among the primates and then described a second human species, the *Homo nocturnus*, an anthropomorphic description of the chimpanzee. It was only later that chimpanzees and bonobos were allocated a separate place in their own genus, the genus *Pan* (Marks 2000).

Today, the taxonomic classification of humans and other great apes falls within a common family, but in different genera. Humans belong to the order of primates, which is divided into two large groups, the Strepsirhini (wet-nosed primates) and the Haplorhini (dry-nosed primates). Within the suborder of Haplorhini, humans fall under the infraorder of Simiiformes. These, in turn, are divided into two large groups, the Platyrrhini (New World monkeys) and the Catarrhini (containing Old World monkeys and apes). These are identified by morphological and molecular genetic findings as well as by their geographical isolation as separate communities of descent. The apes (Hominoidea), as a superfamily within the Catarrhini, comprises two extant branches: the gibbons, or lesser apes, and the great apes. Within the family of great apes (Hominidae) the human belongs to the subfamily Homininae, which includes the Hominini tribe (the genera *Homo* and *Pan*) and the Gorillini tribe. The Ponginae (orangutans) are not included here, since chimpanzees and gorillas are more closely related to humans than to orangutans (Schmidt 2010, 574ff; Wehner and Gehring 2007, 659; Hickman et al. 2008, 932–33). From here on, the human being's path separates from that of other apes: under the great apes, there are assumed to exist different genera: *Homo*, *Pan*, *Gorilla*, and *Pongo*. Within the genus *Homo*, the single surviving species are those of *Homo sapiens*. *Homo neanderthalensis* is considered as its own, extinct species here. Regarding the human relationship with other great apes, taxonomy currently assumes a separation *within the family* in *different genera*.[37]

---

37 From a molecular and morphological point of view, if one wants to assign the systematic units (taxa) to the traditional categories "class," "order," "family," or "subfamily," then one

However, it would also be conceivable to separate them into different species *within a common genus*, as was Linnaeus's intuition. Accordingly, based on the latest findings on parentage and genetic similarity, the following structure could be applied as an alternative taxonomic classification, which Balluch and Theuer (2007) also consider: the family of great apes might include the common genus *Homo* with the species *Homo sapiens* as well as *Homo pan* (*Homo pan troglodytes* and *Homo pan paniscus* would then be subspecies of *Homo pan*). A reclassification in this sense is justified in view of the great genetic similarity, according to Balluch and Theuer (2007). With respect to humans and chimpanzees, this is approximately comparable to that we assume for the Przewalski horse (*Equus przewalski*) and the domestic horse (*Equus caballus*). Both belong to the genus of horses (Equidae), a group in which we also place donkeys and zebras. Other pairs, such as brown bear and polar bear or lion and tiger, are also taxonomically grouped as subspecies of a common genus on the basis of comparable genetically similarity. In cases of comparable degrees of kinship, according to Balluch and Theuer (2007), we should proceed in the same way for humans and chimpanzees (337–38). Other biologists agree in principle to this argumentation on the basis of genetic similarity (see Elango et al. 2006; Wildman et al. 2003).

Thus, biologically seen, there is much to be said for the fact that chimpanzees could also be included as "humans" (in the sense of being members of the genus *Homo*) and are to be regarded as persons according to § 16 ABGB. On the other hand, it would be redundant to say that all human beings are persons if the categories "human" and "person" were intentionally and extensionally equivalent.[38] If the terms "human being" and "person" are not to be equated, then, in addition to the biological definition of the family, subfamily, and genus of the human being it is also necessary to clarify what is understood under civil law as a "person." Balluch and Theuer, like other representatives

would most commonly refer to the Hominoidea as a superfamily, to the Hominidae as a family, the Homininae as a subfamily, and the Hominini as a tribe, should one want to reserve a separate designation for *Pan* and *Homo*. However, modern zoology textbooks are increasingly avoiding these classical categorizations, as they might be confusing. The family of great apes has been determined differently throughout the history of zoology. In contrast to today, only the human being and his extinct ancestors were traditionally described as Hominidae, and all other great apes were separated from them in the family Pongidae. A modern phylogenetic system and a modern classification of the family of apes includes all great apes including humans as monophyletic taxa (Wehner and Gehring 2007, 659).

38 This would mean that the expressions "human being" and "person" were of equal meaning and denoted the same individuals, i.e. they were identical in content and scope (cf. to this argument also Wimmer 1999, p. 337).

of this argumentation, believe that the distinction between a person and a thing on which civil law is usually based is problematic. They deduce therefrom that animals such as chimpanzees can be more accurately described as persons than as things within this distinction, if it is made.

In civil law, natural persons (people) as legal entities and bearers of rights and obligations are opposed to "legal entities" or "things." According to § 90 of the German BGB, for example, the latter are physical objects, which means that they are "perceptible, spatially delimited and physically controllable" (Zerres 2005, 35). Animals were classified as things under the Austrian Civil Code until 1988 and under the German Civil Code until 1990. Although they have been expressly excluded from this category since then, their situation under civil law has remained virtually unchanged. Conceptually, precursors of things in the sense of § 90 of the German Civil Code, according to Schmoeckel et al. (2003), are the *res* of Roman law. The definition of the Roman jurist Gaius, whose writings presumably originated between AD 130 and 180, is largely the basis for today's legal applications. His final decisions were transferred to Justinian's legislative work and thus had a major influence on European civil law (Schmoeckel et al. 2003, 308). Gaius, for example, distinguished the physical *res* (*res corporales*) from the nonphysical *res* (*res incorporales*), whereby, for example, a plot of land, a garment, or even a slave belonged to the physical *res* (Schmoeckel et al. 2003, 308). If animals today, in contrast to slaves and other people who have long been legally discriminated against on the basis of the classifications of Roman law, are considered as "physical things," the question arises whether animal rights would not have to be enforced analogously to the abolition of slavery—which, as has already been shown, is an idea that many animal rights activists have promoted for some time (e.g., Singer 1976).

According to Schmoeckel et al. (2003), stimulus for the further terminological and conceptual development of the term *res corporales* came not from further legal definitions but from the factual definition that Kant offered at the beginning of his *Metaphysics of Morals*, according to which "any object of free choice which itself lacks freedom is therefore called a thing (*res corporalis*)" (Kant [1797] 2017; AA VI, 223).[39] In the discussions about the concept of things since the beginning of the nineteenth century, there have been several exacerbations and restrictions, especially with respect to how things were to be understood as components of the physical world or the external world of goods, which humans legally control, where they see possibilities for use and

39 The central explanations for the distinction between persons and things can be found in the section "Preliminary Concepts of the Metaphysics of Morals (*Philosophia practica universalis*)," in Kant's *Metaphysics of Morals*.

exploitation and which they ultimately consider to be an asset component (Schmoeckel et al. 2003, 310ff.). Instrumentalization concepts such as these are usually considered incompatible with the concept of intrinsic value or a form of dignity in animals. As we see more clearly today, according to Barta (2007), the stringency of the conceptual separation between a thing and a person also has disadvantages: namely, crossovers between legal subject and legal object probably exist.[40] Animals are obviously difficult to classify here. They could be beings who embody such a crossover on the basis of evolutionary and cognitive kinship.

Discussions about the allocation of animals amongst persons or things represent a continuation of a philosophical discussion that has been going on for centuries about whether an animal can be assigned to the category of objects, subjected to humans, or whether animals must be conceded an intrinsic value that those of an inanimate nature (and also plants) cannot have (Schmoeckel et al. 2003, 4). With further reference to philosophy, in antiquity, according to Schmoeckel et al., Aristotle (and later, in the Middle Ages, Thomas Aquinas) assigned animals an intermediate position between the rational beings and inanimate things. Descartes, in describing animals as mere automatons, set in motion, especially in England in the eighteenth century, countermovements emphasizing the relationship between humans and animals and the difference between animals and inanimate things. Lord Erskine was also influenced by this: according to Schmoeckel et al., he was probably the first to transfer the idea that animals are not only property from the philosophical to the legal sphere. In 1809, he introduced the draft of an animal welfare law in the English House of Lords and thus initiated the Cruelty to Animals Act in 1822. However, this was still a penal law that left the private legal position of the animals largely unchanged (Schmoeckel et al. 2003, 323ff.).

It was not until the twentieth century that animal rights activists demanded that the civil status of animals be improved, and the German and Austrian laws were supplemented by addendums. However, these addendums can also be seen as more or less strong, symbolic legislative acts (Schmoeckel et al. 2003, 325). The criticism is that, despite the clarification that animals are not things, they have not been given a truly outstanding legal status. The following applies in Germany as well as Austria. The addendum says that, unless otherwise specified, the provisions applicable to goods in civil law shall apply mutatis mutandis to animals (§ 90a, sentence 3, BGB) (Deutscher Tierschutzbund 2019).

---

40 The human corpse, according to Barta (2007, 199), represents such a crossover, for example: it would nowadays be understood as a thing, but as one that has to be met with particular reverence.

Among other things, animals can be the object of all criminal offences relating to the protection of property: for example, theft (§§ 242 et seq. of the Criminal Code), embezzlement (§§ 246 et seq. of the Criminal Code), robbery (§§ 249 et seq. of the Criminal Code), receipt of stolen goods (§§ 259 et seq. of the Criminal Code), or damage to property (§§ 303 et seq. of the Criminal Code). The Swiss Foundation "Stiftung für das Tier im Recht" (a center of excellence for animals in law, ethics, and society) also states that animals in Germany, like in Austria, have no legal status comparable to that of humans. Throughout Europe as well as worldwide, it generally applies that animals legally still count predominantly as things. Although this is no longer theoretically the case in some countries, such as Switzerland, Austria, and Germany, according to the center they have "no actual rights and obligations that are legally enforceable." They are not on the same legal level as people, and they have no legal capacity. Instead, they remain "assets" in the ownership or possession of persons. They are therefore also subject to the power of disposal of their owner, whereby the owner, of course, has to observe animal welfare legislation and other mandatory legal regulations. The term "animal rights" is often used, but from a legal point of view this terminology is not correct. This is vividly demonstrated by the fact that animals do not have even the most basic of all rights, the right to life. Animal welfare legislation protects, more or less successfully, against unjustified pain, suffering, damage, and fear as well as other interventions in the dignity of the animal, but not against being killed. Still, animals are at least granted the right to animal-friendly husbandry (Stiftung für das Tier im Recht 2019). Despite the many critical voices and the as-yet modest number of practical cases of application, however, one can see a trend throughout Europe for introducing a special civil legal status for animals (Schmoeckel et al. 2003, 325–26).

The parties in the Hiasl case and similarly arguing philosophers and jurists rightly point out a questionable and still-influential basic distinction and its effects on animals in civil law. They draw the following conclusions from the recognition of the questionability of this distinction and from the fact that it nevertheless forms the basis of global civil law: according to the current legal situation, we are forced to classify animals in this rough categorization, and if we have to do so, then at least some animals are clearly more accurately included in the category of persons than in the category of things. According to Hiasl's representative, in order to say who the ABGB is referring to as a "person," the philosophical background at the time when the law was passed (around 1811) must be taken into account. Back then, against the background of Enlightenment philosophy, the ability to reason was established as the decisive criterion for personhood. Accordingly, following Kant, § 16 of the Austrian Civil Code ABGB formulates that every person has innate rights, which are already evident through the ability to reason. Today, against the background of research into cognitive abilities in animals, this definition should include, at a minimum, apes, especially in view of the fact that they appear to possess a theory of mind. Having a theory of mind would imply that other individuals are recognized as having interests, which means recognizing them as other persons and holders of rights (Balluch and Theuer 2007, 338). In asserting this, Balluch and Theuer certainly go far. The role of theory of mind for recognizing others as persons with rights is anything but clear. While theory of mind can surely be regarded as the foundation and also precondition for moral judgment, it has been questioned whether moral judgment is the only capacity that can motivate truly moral behavior. Animals who can act on the basis of moral emotions could also already act morally (Monsó 2017; Rowlands 2012). To support their claim that at least some animals, such as great apes, can act morally (maybe even without having a theory of mind), Balluch and Theuer could refer to the conceptual work of scholars like Monsó and Rowlands. They could also provide initial indications for moral capabilities in animals by referring to their prosocial behavior, adherence to fair play rules, cooperation and altruistic helping, and similar behavior patterns that are currently being discussed as moral by biologists as well (see, e.g., Bekoff 2007a; Warneken and Tomasello 2006; Broom 2003; de Waal 2006).

Hiasl's legal representatives and the supporters of the Great Ape Project rightly question the category of the person and its scope. It follows from the determination of the inadequacy of the legal distinction between person and thing that there is a lack of a suitable category for animals. Excluding them from the group of goods or things (as they are classified in supplements to the German and Austrian Civil Codes) but then in fact only treating them legally as things seems to be unsatisfactory. There are two possibilities: first of all, one could be prepared to include some animals in the category of the person. Then one would have to treat them legally as persons who are not capable of carrying out legal or criminal acts themselves. Similar to children, they would be persons with partial legal capacity, whose interests are, for example, being represented before the law by trustees. But we could also try to find another solution: it may be that ethical problems in dealing with the moral status of animals, especially animals cognitively similar to us, may be coming to the fore today and are recognized as such, problems that were not yet clear at the time of the creation of existing civil codes. Therefore, a fundamental redefinition might be necessary, one that could be expressed in the establishment of a third category between human persons and things or in the establishment of a second subcategory under the upper category of the person. It would be the category of a nonhuman person. Humans would then represent human persons and (some) other animals would be considered nonhuman persons.

Certainly apes, just like human children, are not fully legal entities who can be held morally accountable and are capable of committing criminal offences (for that, a person needs to have moral judgement in the way a healthy, human adult has). Therefore, it would certainly be necessary to discuss which concrete, probably restricted rights the status of the nonhuman person would entail. There is much to be said on this matter; however, it seems likely that we would have to grant at least some fundamental, strong rights of protection in the sense of the three rights of life, freedom, and the absence of torture. The criticism on the Great Ape Project and of the Hiasl case—that anyone who talks about these basic rights for apes would weaken the basic rights of persons such as children or the disabled—can also be easily reversed: *due to the fact that* someone is demanding basic rights for nonhuman persons, they strengthen the position of those people who have also had to fight for their rights as human persons for a long time. According to Balluch and Theuer (2007), "in a world where legal rights for all great apes are accepted, no one could question the existence of rights for certain ethnicities, minority groups, or genders of *Homo sapiens*" (341). In principle, it is not a matter of reducing the rights of human beings (persons), but of strengthening the moral status of nonhuman persons. Much seems to depend on whether the latter can be achieved without the former. It could be more appropriate to think about a separate, yet-to-be-determined category of the nonhuman person than to deviate from the unranked legal capacity and equality of all human beings within the category of the human person. Graduated, partial legal capacity must either be conceivable without the risk of instrumentalization or abandoned. The protection of all human beings, irrespective of their mental abilities, must not be allowed to fall behind—but this is also not inevitable as soon as animal rights are promoted.

Granting all people, irrespective of their intellectual abilities, the status of personhood, which affords them equal dignity and human rights, is also a moral decision of principle, which has been taken from the experience of injustice. The example of the protection of prenatal life shows that *actually being a person* or *bearing the qualifications of a person* are not the only criteria for granting an entity personhood rights. We help each other out, for example, with arguments from potentiality, saying that such entities should be afforded the status of (quasi) persons holding a right to life simply because they will eventually be human persons. We might not even want to argue, in some cases, and simply adopt the stance that a "marginal" person is to be afforded full personhood status because we are all human beings. Why we protect people with human rights is thus the result of a whole group of arguments that, although centered on the idea of personhood (which does not necessarily have to be up to date), also substantially supplements it. It is a historically developed decision that all people, regardless of their mental endowment, gender, or skin color, are persons with the same universal human rights. This moral decision is at the very heart of human rights. Wimmer (1999) rightly points out that possessing dignity is not a quality or ability alongside other qualities or abilities of humans. The concept of the person is, furthermore, not the summary of such characteristics and abilities. Neither does its application stand or fall with the fact that certain mental abilities are present or exist after a certain degree. Such assignments and classifications "are morally prohibited," Wimmer says, "even from the principle of the person itself, since one of the main aspects of the ethos of respect for human dignity lies precisely in the protection of the personhood rights of the weak, the physically and mentally handicapped, or those who have not yet developed" (339; transl. M.K.). Thus, Wimmer believes that we can consider such qualities and abilities that we *generally associate* with being a person also (though only in a certain breadth and indeterminacy) as *criteria for personhood status*. This could become "significant" when it comes to the question of "whether we have to recognize other species of living creatures—e.g., higher animals … as persons." Then, too, we cannot label some individuals of a species as individuals to be respected while not doing so for others. Rather, it is a question of "deciding whether the whole species is to be regarded and treated as equal to us in this essential respect and therefore granting them equal moral consideration." If it were about our fellow human beings, we would not be dependent on these criteria in the same sense, "because we are (anthropologically) always already familiar with the personhood or with the becoming-a-person of everyone who carries human identity" (Wimmer 1999, 339; transl. M.K.).

The reference to personal characteristics in all their breadth and inconsistency, therefore, makes a certain sense in its application to arguments in animal ethics. However, as has been shown, the determination of personhood is already difficult in its relation to the human being—and, according to Wimmer and other authors, there is only a general, broad, and indeterminate characteristic of its application here. This should temper our expectations. In the animal ethics discourse, the scope and power of the argument could also prove to be less far-reaching than we hope. What nonhuman persons are would need to be determined independently, and we should certainly be wary of claiming that something can only be achieved for the welfare of animals if we manage to include them as persons. And finally, it is also true in the case of great apes that although classical criteria for personhood can be cited at the core of the argument, they are not full-fledged persons in comparison to human beings. In the case of apes, we also have complementary arguments (such as their imminent extinction) to tilt the moral scales in favor of a moral status that could be as strong (or nearly as strong) an argument as personhood status. The fact that the Hiasl case was considered worthy of dispute, and the fact that the demands of the Great Ape Project are increasingly being heard worldwide, indicates that we could be on the way to a paradigm shift, despite all the theoretical, legal, and philosophical difficulties. It should be clear: as long as the required third category between things and human persons does not yet exist, it must be decided in which of the two existing categories apes are described more aptly. In the current global legal situation, they are probably more accurately categorized as persons before the law than as things.

In conclusion, the number of philosophers who demand that at least great apes be accepted into the community of persons is growing, just as an increasing number of biologists now argue for a common *Homo* genus due to genetic similarities. For example, the representatives of the Great Ape Project agree in similar arguments with Diamond (2006) and Wise (2000). As misplaced as a chimpanzee may seem in the Court of Human Rights in Strasbourg today, the bioethical and political significance of the issue should not be underestimated. And we should be aware that the group of living beings that we cannot accurately describe either as human beings or as things will continue to exist and grow in size in the future. It will possibly include *hybrid beings* such as chimeras, clones, and robots, whose moral and legal status we should clarify—it is better to do so before we create them. *Currently existing hybrid beings* include the great apes discussed here, but also other species such as dolphins and elephants, who might similarly fulfill criteria for personhood. This group must generally be open to other species, not only in terms of genetic kinship but, more importantly, in terms of cognitive kinship. Lastly, one might also pose the question as to whether or not *previously existing hybrid beings* such as *Homo neandertalensis*, *Homo habilis* or other extinct species of the genus *Homo* were persons. They, too, are not modern humans, and they certainly did not yet have our pronounced cognitive abilities. Still, we usually perceive them as human persons.

We will now turn to Thomas I. White's argument in favor of a personhood status for dolphins. Similar arguments are employed here, but this author is more fundamentally skeptical toward the concept of personhood than the representatives of the Great Ape Project have been so far.

8.4.3 *The Demands of Thomas I. White: Personhood Status for Dolphins*

In the 1980s, the American philosopher Thomas I. White wrote a chapter about persons, or personhood status, in an introductory book on philosophy. The question of what a person is and whether a dolphin could be a nonhuman person marked his first foray into the subject. In the course of further study of the animals and their abilities, as well as the scientists researching them, he came across a wealth of anecdotes and research results on the intelligence of these animals. The centuries-old claim that humans were the only living creatures gifted with thoughts and emotions began to come into question for him too, and he began to wonder seriously whether dolphins were "someone" instead of "something," "indeed, a someone who perceives the world and makes decisions in a way similar to how we humans do" (White 2007, 4). At the same time, right from the start, he noticed a fundamental otherness and complexity, a tension, which he wrote about in his book *In Defense of Dolphins: The New Moral Frontier* in 2007. Two questions guide his research interests: "What kind of beings are dolphins?" and "What does our answer to the first question say about the ethical character of human/dolphin contact?" (White 2007, 13). My book follows a very similar approach to that of Thomas White. White presents in detail scientific results on the intelligence and social behavior of dolphins and discusses their anatomy and physiology as mammals adapted to the water; the structure and potential of their brains; and whether they think and feel, solve problems, and understand language. Particularly with regard to their social intelligence, he shows that dolphins possess outstanding intellectual and emotional abilities. He assumes that even skeptics at least have to admit that there is a remarkable likelihood that dolphins are nonhuman persons because of these abilities and that some forms of our dealings with them may not be ethically justifiable.

Within his reflections on the status of dolphins as nonhuman persons, White, similar to Balluch and Theuer (2007), initially refers to two important distinctions: he separates the terms "human being" and "person" and refers to the former as a scientific or biological concept, while the latter is a philosophical category: "*Human* refers to any member of the biological category *homo-sapiens*. *Person* refers more to the combination of advanced traits by which we define ourselves—things like self-consciousness, intelligence and free will" (White 2007, 8). The distinction between human and person and their definition follows a further distinction that we already know, namely the distinction between things and persons: according to White, persons—in contrast to tables, chairs and amoebas—have rights, such as the right to life, the right to be free, and the right to be treated equally. We consider it wrong to violate a person's rights. Thus, "what starts as a theoretical discussion in metaphysics quickly migrates to the applied world of ethics" (White 2007, 8). White already refers to ethically difficult cases here: Is every human being also a person? He refers to the bioethical discussions about switching off life-sustaining measures in brain-dead patients and abortion. In both discussions, the relevant question is whether patients or embryos and fetuses (still or already) have personhood status. In brain death, according to White, some people would allow for the switching off of life-prolonging measures precisely because what made the brain-dead individual unique is no longer there. He believes that switching off is not the same as murder for many people because the person they knew is already dead (White 2007, 8).[41] But not everyone would agree with that, and so White finds it difficult to say what can really be included in the category of a "person." Even when it comes to people, the certainty we hope for is often lacking, especially when we turn to "borderline" cases (e.g., at the beginning and end of life). It is therefore unlikely that the problem will diminish when species boundaries are crossed.

After White has already asked whether every person is a person, he then concludes with the question of whether there can also be nonhuman persons:

> After all, in theory, any being with self-consciousness, intelligence and free will, for example—no matter what the species—would qualify as a person. It's this question that led me to study dolphins, because many of the stories we hear about them suggest that we've come upon a person who isn't human. (2007, 9)

In answering this question, he summarizes what philosophers consider to be essential components of a happy life. First of all, basic physical needs (including food, protection, safety, health, etc.) have to be fulfilled. Secondly, we would have to be treated in a certain way. By this he means that something that we feel uneasy when individuals are treated unfairly, discriminated against, manipulated, or lied to. Accordingly, from an ethical point of view, something is called "right" when it increases the likelihood that the conditions and needs mentioned above are met. Something can be described as "bad" if it prevents us from fulfilling them. These statements also presuppose a minimal ethics, which, according to the author, corresponds to our general understanding of

41 Whether the matter is not more complicated in the status of the embryo or fetus, however, must remain open. Both bioethical "borderline" cases are only touched upon by White and are not discussed in terms of their significance and scope, which is why it is not entirely clear what position he would take within the debates.

human rights (which, in turn, are based on respect for the necessary conditions for a happy and dignified life) (White 2007, 10). Accordingly, in the preceding section on the justification of human rights, I stated that there seem to be conditions of existence and expressions of personhood which must not be denied to human persons or possible nonhuman persons.

From here White moves on to dolphins as one of several species of highly developed mammals whose outstanding social, cognitive, and emotional abilities we have learned more and more about in recent decades. But in the case of dolphins, he believes, and for the same reasons, it would be necessary to reconsider whether primates and elephants would also have to be understood and treated more appropriately as nonhuman persons. These living beings would probably also have to be given a different moral status, and their interests would have to be given greater consideration in moral decisions (White 2007, 11).

An example in which the moral status of nonhuman persons would benefit dolphins as soon as their interests are balanced against human interests is the tuna industry. This industry benefits mankind while simultaneously endangering dolphins, because they are victims of bycatch. Other cases to be discussed could be the captivity of dolphins for entertainment, research, therapies, or military purposes. Such examples raise the question as to whether advantages for humans can outweigh the damage done to dolphins when it comes to ethical considerations (White 2007, 11) and whether it is justified to subject dolphins to such cost-benefit considerations at all. I will elaborate on the problematic treatment of dolphins in captivity below.

In order to avoid anthropocentric and speciesist perspectives, White introduces what he believes to be a promising view of dolphins as *alien beings*, a proposal originating from the marine biologist Diana Reiss and added to the arguments already known from the Hiasl case and the Great Ape Project in White (2007). Understanding nonhuman animals in this way proves to be a particularly effective conceptual model because, on the one hand, it underpins the idea that these animals could be nonhuman persons and, on the other, it reckons that important concepts, such as the concept of the self or intelligence, can develop very differently in nonhuman and human persons. It also opens up a new field of questions with which we can approach animals and their moral status: How can we avoid the pitfall that our evaluation of their intellectual and emotional abilities is interpreted as inferiority from the outset? Don't properties such as intelligence have to be specifically defined or filled with content? Is the classical concept of a person most suitable for such an investigation? Can we apply other concepts in which we can be more confident that our characterization is free of unintentional species prejudice? Is it possible, for example, that humans and dolphins are so different that what can do them harm could also differ from one species to the next? According to White, the understanding of dolphins as fundamentally different beings makes us more sensitive to the differences between species and makes us more aware of their philosophical and ethical significance (2007, 12).

In a very clear and plausible way, White demonstrates why we want to understand and treat ourselves as people:

> The basic reason that humans expect other members of our own species to treat us in particular ways (to respect our dignity, for example) comes from the kind of consciousness our brains produce. We are self-conscious, unique individuals who are vulnerable to a wide range of physical and emotional pain and harm, and who have the power to reflect upon and choose our actions. *Because we can experience pain so intensely and because we value our ability to choose our own actions so deeply*, we neither accept nor minimize it when other people hurt, coerce, threaten or manipulate us. We object to such actions so strongly that we label them not just "inconvenient" or "unpleasant," but as wrong. Ethics—our labelling actions as "right" or "wrong"—is grounded in the idea *that the type of consciousness that we have gives us special capacities and vulnerabilities*. When we label something as "wrong," then, we're saying that *it crosses the line with regard to not respecting some fundamental features that makes us human*." (2007, 155; italics added)

This ties in with the already described definition and justification of human rights. Despite all the divergences about what distinguishes a person, White (2007) believes that there is still a consensus in the literature that a person is a living being with a certain kind of highly developed consciousness and an inner world (156). This, too, is in line with the core statements that have been given priority in the history and tradition of human rights. White draws up a list of eight criteria for personhood status, which he has taken from the literature.[42] It defines persons (1) as living beings, (2) with consciousness, (3) with

42 He refers here, for example, to Daniel Dennett in his book *Brainstorms: Philosophical Essays on Mind and Psychology* and Joseph Fletcher's ideas in *Humanhood: Essays in Biomedical Ethics* (Dennett 1988; Fletcher 1979; and White 2007, 157n1). However, these criteria can also be found in other authors (see the introduction to the concept of human rights and the concept of the person in Section 8.4.1). Certainly it is practically impossible to establish an *exhaustive* list of such criteria. However, this does not preclude the assumption of a *representative* list suitable for a working definition. This must, just as White's list attempts to do, summarize the most frequently cited criteria. Bernstein (2002,

positive and negative feelings, and (4) endowed with emotions, who (5) possess self-consciousness (6) control their behavior and (7) are able to recognize other persons as such and can treat them appropriately. Criterion (8) adds that persons possess a number of cognitive skills (are able to learn, store, and retrieve information; to solve complex problems with analytical thinking; and can communicate in a way that indicates thinking) (White 2007, 156–57). All these criteria, he believes, are partially or wholly applicable to dolphins (2007, 157ff.).[43]

Further interesting ideas follow these considerations. Under the heading "Problems with personhood," White asks himself whether the term of the nonhuman person actually covers all significant aspects of the dolphin research results. If we take a close look at the skills and qualities contained in the aforementioned personhood concept, it becomes clear that this concept is not nearly as neutral as we may have assumed or as White himself had hoped. What we actually need would be a species-neutral definition of person; only this could do justice to nonhuman persons:

> If we take the way that dolphins move in the water as the standard for "swimming," even the performance of the best human swimmer doesn't come close. Yet surely we wouldn't say that this means that humans were unable to swim. In other words, when we specify the criteria for complex abilities, it's important to recognize how differently things might look from the perspective of other species.
>
> WHITE 2007, 165–66

In the case of his investigation, he believes, it would be important to recognize the importance of the dramatic difference between the habitats in which

525) concludes that a representative list of morally relevant characteristics proposed to distinguish humans from nonhuman animals includes above all rationality, autonomy, and linguistic and moral ability (cf. Amber 2002, 525). White's list is more comprehensive and much more detailed.

43 Some of these criteria are certainly present—even in complex form—in all (vertebrate) animals. This applies to criteria 1, 2, 3, and possibly 4. In the latter case, philosophy and psychology make complicated distinctions between feelings and emotions, or between primary and secondary emotions, and so on. Here it would be necessary to check what kind of complex emotions animals have. However, some of the above-mentioned eight criteria for personhood status have not yet been found in all animals and are probably not present (in a complex form) in all animals. This could apply to criteria 5 and 7 (which refer to a theory of mind) and parts of criterion 8. The fact that these abilities may not be present in all animals does not mean, of course, that some building blocks and preforms of these abilities are not to be found in many animals.

humans and dolphins have developed and now live. Standard definitions of personhood would fail at precisely this point. The problem here lies both in what classical definitions of personhood most often emphasize and in what they ignore (White 2007, 166). White sees an unjustified overemphasis above all with regard to the ability that we ourselves possess in an outstanding way: the ability to speak. He sees the danger of underestimation especially in the case of other social skills. With regard to both areas, it is equally important that it is inappropriate to measure the skills of other species that have adapted to very different niches—such as life in the water—against human competencies. For example, White says, it would be far from clear that dolphins communicate with each other in the same way as humans do. In the ocean, communication and social interaction had to be shaped differently due to ecological conditions. White wants to understand intelligence in a species-specific way[44] and considers it possible that two species as different as dolphins and humans, both of which are endowed with large brains, have adapted to their respective environments in very different ways under clearly different conditions (this applies to their inner/mental as well as to their outer/physical endowment). In such an understanding of cognition and intelligence, the ability to speak is a biological adaptation that has arisen and developed in humans because it is useful to them in coping with their lives in their niche. At least in the form in which it has developed in humans, it does not have to be equally useful under other ecological conditions. A spoken language, as we can think of it in terms of orientation, may not be a particularly helpful tool for life in the ocean because of transmission conditions of sound in the water. However, complex social interaction can take place in a different way.

White believes that the only competencies of persons on the previously presented list for which the existence of other individuals is necessary at all concern the ability to recognize others as persons and to communicate with others. However, the most important competence of human persons would usually be self-consciousness, a far-reaching, solitary process in which an independently and autonomously conceived living being reflects on herself (White 2007, 175). All other social skills have often been neglected in the history of philosophy.[45] The overemphasis on this self-conscious, autonomous, Kantian individual is, according to White, typical of European philosophy. However,

44 He uses a definition of intelligence referring to Gardner (1999) (White 2007, 167n3).

45 One could say, however, that feminist theories and also certain currents in moral philosophy have been tackling this gap in the history of thought. Instead of the overemphasis on autonomy and reason, ecofeminist theory since the late twentieth century has emphasized the importance of emotions and the role of dependency and cooperation with increasing importance across many disciplines. For example, Alasdair MacIntyre (1999)

the individual, and thus the person, can also be thought of in a completely different way. This is the case, for example, in ancient philosophy as well as in other cultures (such as some Indian cultures). But even in Western society, psychologists assume that people describe themselves differently in terms of how they see themselves. Some of us have a "separate self," a more detached self-conception, while others would stress their social bonds, their attachment to social responsibility, and thus promote a socially dependent self-conception (a "connected self"). Neither this last group of people nor, in White's opinion, dolphins could be described correctly when it comes to personhood concepts that overemphasize the independent and autonomous individual (2007, 175–76).

Based on these considerations, White calls for the establishment of a *socially* oriented concept of personhood. Such a concept, if it is to be applicable to dolphins, must do justice to the strong involvement in a school. Intensive communal experiences, according to Kenneth Norris (1991), as well, could shift the limits of self-understanding and include other individuals as an integral part of one's own self-image, even leading to the suppression of individuality and the prioritization of group interests (Norris 1991, 186, 137ff.; White 2007, 179ff.). In addition to shared echolocation and the ability for joint attention (see Pack and Herman 2006; Gregg et al. 2007), many other observations suggest that the dolphin's life is particularly intense in terms of social interaction and group interests. This includes behaviors that are linked to their physiology, such as the fact that they have to swim consciously to the surface of the water in order to breathe. It is assumed that they support each other in this process, take care that conspecifics do not lose consciousness, and in an emergency push others to the surface of the water (White 2007, 20, 130–31; with reference to Norris 1994, 340; Connor and Norris 1982; Caldwell and Caldwell 1977). Furthermore, dolphin schools are not "kin groups" but are described as "remarkable cooperative societies" in which there are no strict hierarchies (White 2007, 126; with reference to Norris 1994, 342; Reynolds III et al. 2000, 124–25). In schools, different animals fulfill different roles (White 2007, 131–32), and the coordination in the group becomes particularly obvious within cooperative hunting strategies. Here it was observed that dolphins synchronously throw themselves onto flat stretches of beach to catch fish that they have driven before them on the bow wave (so-called hydroplaning), that they form mud rings, or that they gather fish swarms into balls (herding) and then push forward in ordered formations into the gathered swarms (White 2007, 77ff.). All this

called for the *recognition of dependency*, and Carol Gilligan (1999) referred to *the other voice*, and thus to values arising from relationships.

follows a strict distribution of roles and demands precisely synchronized, alternating formations. Up to three hundred animals are involved in such cooperative hunting strategies and coordinate their behavior (White 2007, 78; Würsig et al. 1994; Würsig 1983; Benoit-Bird and Au 2003; Benoit-Bird and Au 2009b; Benoit-Bird and Au 2009a). Bernd Würsig (1983) comments on this interaction as follows:

> It is a dynamic, ever-changing system, which may require organization by these large-brained and communicative social animals. Differential role-playing and pre-meditation (such as a decision that certain members do particular things in order to meet various contingencies) may be important in this kind of cooperation. The degree of behavioral flexibility to encompass novel situations appears well developed. (5–6)

Benoit-Bird and Au (2009a) have followed the nightly hunting behavior of East Pacific dolphins (*Stenella longirostris*) via 3D sonar recordings and also speak of a remarkable degree of coordination, of very strict geometry, precise timing, and orderly alternation. Four clearly distinguishable geometric formations were recorded. This behavior represents a decisive strategy of energy production in a relatively low energy environment in which individual hunting strategies would only provide insufficient energy production through food (Benoit-Bird and Au 2009a, 539, 133ff.). A complex synchronization with simultaneous flexibility of the entire situation requires, in fact, that the animals communicate with each other. It is assumed that the dolphins communicate with each other via clicks during such coordinated hunting phases, not via whistles, and perhaps also via other, additional visual means (they could also be guided by the bioluminescence of their prey caused by their own movements). In the phases between the different geometric formations, Benoit-Bird and her colleagues were able to detect the exchange of clicks (Benoit-Bird and Au 2009b, 543, 545). They state: "Whatever the mechanism by which clicks are involved in group coordination, the results highlight the potential flexibility of delphinid acoustic signals, contrary to the existing assumption that echolocation and communication are separate and compartmentalized" (545). Such behavior, as White also states, presupposes the socio-cognitive competences mentioned here, an understanding of mutual dependency, and knowledge of mutual competences and roles. This implies that the individuals involved show patience, restraint, and reliability. Managing such skills within a group would certainly be a complex task (see White 2007, 129–30).

But social intelligence is not only evident in the cooperative behavior of dolphins. The same is true of their competitive behavior. Male dolphins in particular are known to form alliances of two or three, especially for sexual dominance over certain females. Even the formation of an alliance with up to fourteen animals, was investigated. This alliance comprised different, changing sub-alliances and lasted for more than two years (see, e.g., Connor et al. 2000). Smolker (2001) thinks that these animals are not only faced with the challenge of understanding and cultivating relationships within their own alliance, but also of cooperating with other males outside the alliance with whom second-order alliances, so to speak, are formed:

> Cooperation among male mammals even on an occasional basis is rare enough, and alliances formation, where males form long-term cooperative bonds, is even rarer. But nothing as complex as long-term alliances among pairs and triplets forming long-term alliances with other pairs and triplets has been found in any other mammal besides dolphins and humans.
>
> SMOLKER 2001, 158–59

White stresses that according to Connor et al. (2000) the formation of alliances is also known in chimpanzees, but not in the multiple nested complexity of the first- and second-order alliances we find in dolphins (see also Connor and Peterson 1994, 126).

Although there is much to suggest that the competitive and aggressive behavior of dolphins is also a sign of their particularly complex social cognition, marine biologists seem to agree, according to White, that among dolphins there is a relatively rare escalation of aggression. Despite intensive observations, there are few reports or scientific observations of dolphins attacking other, smaller dolphins or individuals of their own school.[46] All in all, the social life of dolphin schools seems to be more focused on cooperation than on competition and aggression. White thinks this can be explained by the strong dependence of the animals on each other: "In such a group situation, the best strategy is to use just enough aggression to get one's own desired result without doing any major damage—killing someone, making them feel so alienated that they want to leave the group or to stop pulling their weight, or fracturing the group into factions" (White 2007, 140). If aggression occurs, it is a limited level of violence, which is clearly controlled. In this sense, aggressive behavior

46 A few cases, however, including cases of infanticide, are known (see White 2007, 139–40, especially footnote 42).

in dolphins can also be regarded as a sign of social intelligence (White 2007, 139ff.).

According to White, just how far dolphins follow the school's interest can be seen in their escape behavior. So far it is considered an unexplained phenomenon that swarms caught in purse seines could easily jump out of the nets at the top, but do not do so. Physiologically, dolphins would be able to escape, but will not do so until the rest of the school has the same opportunity. Kenneth Norris investigated this behavior by catching Hawaiian spinner dolphin schools in a ring net. Even if a gap of twenty feet was left in the net, the animals stayed together. In such stress situations, according to Norris, individual interests would be reduced to zero (Norris 1991, 186).[47] Based on this and the results of many other researchers, White says, dolphins might be said to possess a *social self*, regarding their own self in very strong connection with others (having a "sense of I-connected-with-others") (White 2007, 176, 182–83). From this point of view, the classical standard of the independent and autonomous human person could be inappropriate or deficient when it comes to dolphins (White 2007, 182–83).

When applying our definitions of intelligence and personhood, White's credo is to be cautious and to consider alternatives. Nevertheless, the fact that we measure dolphins against criteria as deficient in terms of their actual abilities, and the fact that they do well even when measured against these inadequate criteria, would underline the need to take them into greater account morally (White 2007, p. 182). It may seem contradictory at first that White makes both claims here: that these animals possess outstanding abilities according to our concepts of intelligence coined on humans, and also that these concepts need to be revised. He seems to acknowledge that in a certain sense we cannot help but start from our own abilities when we look at the abilities of other animals. All knowledge about other minds must always be based on our own experience and knowledge, or on the terms we have for our abilities. Thinking is thus always "subjective" in the very trivial and general sense of epistemic anthropocentrism. This basically applies to everything that a human thinks. However, this does not mean that our perception perspective or our concepts of mental abilities are in principle insufficient or inadequate when it comes to abilities in other animals or when it comes to the whole spectrum of these abilities in other humans. This largely depends on the *content* of the concepts we apply. In this respect, the starting point is a hermeneutic and comparative approach,

47 However, according to White, the animals are certainly also acoustically blind due to water turbulence and the air vortex of the hunt, and otherwise effective defense strategies do not work in confined spaces (White 2007, 179–80, 195–96, 197–98).

but it can lead to an insight that goes beyond this and calls us to develop species-specific, nonanthropocentric concepts. The decisive factor then is whether one's own human knowledge and ability can be left behind *as a yardstick*. To understand the knowledge and skills of animals, we have to rely on conclusions drawn through analogy and dependent on comparison with one's own abilities up to a certain point, but caution is needed in assessing what animals know and are able to do. To think, for example, that the highly technical and institutionalized form of culture we find in (some) humans is the only possible form of culture at all fails to recognize both species-specific and local forms. It also neglects the possibly much more basal core of what is common to different cultures. It is therefore not irrelevant, despite all skepticism with regard to anthropocentric, monolithic, and exclusive definitions of cognitive competencies, that a being knows or can do something. However, this is irrespective of whether a being *knows exactly the same thing* and *can do exactly the same thing* as we ourselves do it. White has exemplified this with the communication and social interaction of dolphins. Other animals may have a different knowledge and (gradually) different abilities than we do. In fact, it is very likely that this is the case. They live in their niches and do not possess many abilities in the same way that humans do. On the other hand, however, they also have abilities that humans do not possess or have in a less pronounced form. The dolphin's echolocation with its kind of perception and communication function may be an example of this. The fact that they possess certain knowledge and certain abilities can therefore be relevant, both epistemologically and ethically, without the need for a direct comparison of content with human abilities.

From a biological perspective, echolocation is only one sense of perception among others. Philosophically, however, it is particularly interesting because one is dealing with an area of perception to which the human being has hardly any epistemological access. It is no coincidence that Thomas Nagel's (1974) much-praised essay *What Is It Like to Be a Bat?* selected the bat and her echolocation to show that different perceptual abilities and different worlds of perception that fundamentally surpass the sensual experiences that we ourselves are aware of lead to experiences that are fundamentally concealed from us. Of course, the problem of limited access to other minds also applies to other people. But at least we share the same biological channels of perception with other people. In a way, we can put ourselves in the minds of other mammals whose sensory organs function similarly to ours. Even with species that have perception skills that are relatively underdeveloped in humans (because they hear much better or see more sharply than we do, for example), we still have a basic understanding of what it means to hear and to see. But when a dolphin or a bat actively records her environment via echolocation, we reach the limits of what is fundamentally conceivable and comprehensible for us in terms of "seeing" or "hearing." Traditionally (often with reference to Nagel), this leads to a resignation of philosophy before the assessment of other minds. In White, on the other hand, it becomes clear that the otherness of the animal is not fundamentally subordinate to the way in which humans construct their environment. Otherness, before which we must be at least partially resigned in epistemological terms, does not mean the end of recognizability.

Fundamental human rights have already been called for by philosophers and scientist for chimpanzees, bonobos, orangutans, and gorillas within the framework of the Great Ape Project and initiatives like the Nonhuman Rights Project and the Hiasl case. However, White takes a decisive and interesting step beyond this demand. In his concluding chapters, what he indicated at the beginning of the book becomes crystal clear: he considers the traditionally defined status of a person neither the only nor even, possibly, the best criterion by which to justify animal rights. On the other hand, he does not think this criterion can be dispensed with, either. Rather, he chooses the concept of the person in his investigation for strategic reasons first, assuming that even the most radical anthropocentrist would have difficulties justifying the way people treat dolphins if it could be shown that we need to give serious thought to whether these animals are more appropriately been described as nonhuman persons (White 2007, X).

White shares this tactical argument via personhood status with the authors of the Great Ape Project. He also seems to assume, like Hiasl's lawyers, that we should classify these animals more accurately as nonhuman persons than as things against the background of the existing legal situation, which forces us to differentiate between things and persons. But he does not seem to think this solution is optimal. In this respect, he exposes the recourse to the concept of the person as a necessity that is at the same time an emergency solution and, in doing so, points to weaknesses in classical concepts of the person and their limited scope. On the one hand, he is dependent on the concept of the person, because it can easily be connected to the concept of dignity and rights and, on the other hand, he tries to fill it differently for dolphins than for humans and thus to do justice to the differences in species. In humans, the main reasons why we believe we deserve moral consideration are our particular form of consciousness, our self-consciousness, and our uniqueness as individuals with a different and clearly discernible personality. In addition, there is our lifelong memory and personal history, the wide range of physical and psychological damage we may be exposed to, and the fact that we reflect on our actions and choose to take them (White 2007, 186). That is why we have—in Kant's sense—inherent value and dignity; a human person does not only have an instrumental value or a price, but is an end in itself. As individuals, we are to be morally considered. Since there is more evidence that dolphins are also self-aware creatures, they should—from an ethical perspective—share a form of special moral status and protection with us: "Because dolphins appear to have a consciousness that makes them vulnerable to a wide range of physical and emotional pain, and that gives them the power to reflect upon and choose their actions, they should be seen as having intrinsic worth and moral standing as individuals" (White 2007, 186).

In cases where there are reasonable and comprehensible doubts that one of our actions is harmful to another person, we are obliged to reconsider and stop this action. This is the strength of having status as a person. If it is awarded, then benefits for human beings will no longer be weighed against the right to life of a nonhuman person, but fundamental personhood rights will be cited, going beyond any weighing of interests.

Those who argue that in cases in which we are faced with a person's basic right to life against that of a dolphin, we should not prefer the latter, misjudge the situation in two ways: First, as has already been said, when the right to life of two human beings clash, in principle no morally "right" decision can be made at all. To prefer the dolphin to the human would fail to be the "correct" decision, as would its opposite. Furthermore, legislators have left ways out of such dilemmas, for example, by clarifying that killing in self-defense in a life-threatening situation is not murder. This, too, does not mean that the killed person is not recognized as a person. Moreover, in such existential situations (and only in these situations), it is questionable if it is morally wrong to give priority to our peers or family members, and to employ additional arguments that lie beyond basic human rights.[48] Whoever thinks that the demand for human rights for dolphins means that decisions such as "me or the animal" could counterintuitively lead to one's own death is also, above all, ignoring that, practically speaking, such choices are extremely unlikely. Normally, *extreme* situations in which animals are *fighting for life* do not come into conflict with *alternativeless* human needs and their basic rights to life, individual freedom, or bodily integrity. Instead, the rights and interests that collide are most often of a very different quality. For example, with the right to life of endangered dolphins, we are talking about a collision of the right of wealthy industrial nations to cheap fish caught by drift nets. In this case, the basic right to life of dolphins as nonhuman persons would then have to prevail in principle. This would apply equally to human and nonhuman persons, and even if we could not be quite sure whether our action actually caused the presumed damage. The maxim

48 This refers to the above-mentioned lifeboat scenarios.

that actions that substantially endanger human and nonhuman persons are to be stopped also applies when we cannot be completely sure whether our counterpart is a being with personhood status: "In a questionable situation," White believes, the principle of "benefit of the doubt" applies, i.e., the decision in favor of the individual to whom damage would be caused in a doubtful case (White 2007, X, 187). So if there is a possibility that our behavior could harm a human or nonhuman person, we are asked to stop the behavior (even if we cannot say with absolute certainty whether personhood is at stake). The danger that, first, our actions could cause serious harm and, second, that the injured beings could be persons is sufficient to have to refrain from these actions (White 2007, 188). White calls for a genuine "interspecies ethics" that not only denounces clearly inhumane ways of dealing with animals (for example, when it comes to unnecessarily inflicted pain, extremely brutal cruelty, neglect, and the like) and strongly condemns the extinction of species (White 2007, 185). A real interspecies ethics would also have the power to include animals if they qualify as nonhuman persons and are, as such, in the circle of those deserving direct moral consideration. The mere suspicion that dolphins could be nonhuman persons gives rise to justified and justifiable doubts as to whether certain practical measures can be ethically right:

> In assessing the ethical acceptability of human practices that affect dolphins, we need to make sure that we use "species-appropriate" standards. That is, I believe that whether something is right or wrong is in large measure grounded in the defining, biological traits of species. And this is where the fundamental differences between humans and dolphins come to play.
>
> WHITE 2007, 186

This means, among other things, that a behavior X toward a dolphin can be ethically problematic while the same behavior toward a human being might be acceptable. This is because the intrinsic value or dignity—regardless of whether a human or nonhuman person is involved—is bound to the personhood status and can be satisfied differently in different species. The species-specific basic needs—the conditions and expressions of personhood of different species—overlap, on the one hand, but, on the other hand, they differ, and these differences must be adequately understood and included in ethical considerations (White 2007, 188). We have already seen that profound ethical decisions have such a biological basis and that in other pathocentric approaches, too, we must determine what causes suffering and damage by focusing on the animal's biology (Würbel 2007).

Dolphins and humans share, according to White, a number of conditions necessary for their well-being, which are linked to basic needs. Life, security, and freedom from physical and psychological pain are indispensable. Children (human children and dolphin calves) need protection and access to someone who teaches them vital skills and ensures that they develop the skills necessary for life and functioning in the social group. In addition, both species depend on the necessary conditions for relationships to develop and persist (White 2007, 188–89). Life in the ocean may require a close network of social relationships and bonds, and the brains of dolphins may be specialized in social intelligence (see Norris 1991, Jerison 1986, White 2007, 189, 176ff.). The most fundamental condition for their emotional development and well-being would then be membership and involvement in a dolphin school. If relationships and group membership are so important to dolphins, then this leads to significant implications as to what harm and suffering means to these animals. Depriving them of the possibility of maintaining important relationships with other dolphins, cooperating and communicating with each other, and depriving them of the possibility of fulfilling the role of a useful group member could then be extremely problematic (White 2007, 189). We will come back to this when it comes to the conditions of dolphin captivity.

White concludes his interspecies ethics in a consistent fashion. The following seems clear to him: if, due to similar cognitive and emotional abilities, we grant human persons and nonhuman persons the same status, then we also have to consider the groups morally equal. For him, nonhuman persons have the same moral significance as human persons, and "any defense that ignores this is guilty of anthropocentric bias" (White 2007, 198).[49] In his evaluation of human practices affecting dolphins, it becomes apparent time and again in the ethical evaluation that he argues against these practices at several levels. He does so not only at the level of species protection (which, as we have seen, can also be justified anthropocentrically), but also at the individual level of the dolphins concerned. In pathocentric arguments he binds the cause of harm and suffering back to the cognitive and emotional nature of animals, and in Kant-oriented arguments he argues from the duty to regard persons as ends in themselves, never only as means. Here it is important to assess, respect, guarantee, and protect the conditions of well-being of dolphins and to evaluate them as high-ranking in any consideration of goods and interests.

49 He emphasizes this several times and writes: "The death of each individual dolphin in connection with purse seine fishing is as ethically unacceptable as if it were the death of an individual human" (White 2007, 195).

8.5 Problems and Limits of Personhood Rights

The argument from cognitive similarity, as seen in the Great Ape Project, the Hiasl case, and in Thomas White's work, can be summarized as follows, with loose reference to Kant: the categorical imperative calls on us to treat any person as an end in itself and never merely as a means of achieving other ends. This is derived from the way persons are: they are rational and autonomous, whereby autonomy is not to be understood as moral autonomy (hence the only loose reference to Kant), but it already begins with preference autonomy. If certain basic rights for persons are important to us humans because of the quality of certain characteristics and capabilities we have, then we cannot, in the sense of the categorical imperative, wish that they be ignored in the case of nonhuman animals should these animals have the same or comparable qualities. This is also about avoiding a self-contradiction. The argument from similarity with regard to personhood and basic rights that are analogous to human rights is thus a logically constructed, argument backed up by (empirical and normative) reasoning and not just an argument based on the emotional attachment of humans to certain animals. The determination of similarity (P1: humans have outstanding socio-cognitive abilities, and P2: some animals have important components of the same socio-cognitive abilities) is not directly followed by the derivation of rights (C: therefore, these animals should have basic personhood rights). Instead, two normative premises give the conclusion its normative content and allow the analogy (P3: certain socio-cognitive abilities are qualities constituting personhood and as such they are recognized as morally relevant values, and P4: all persons who [partly] have/will have these qualities should have basic human rights). P4 is required so as not to fall behind with human rights standards in "borderline" cases. Animal rights must not be bought at the expense of the rights of certain groups of people. The very fact that a society wants strong human rights for "marginal" cases, as well, shows that personhood status can be granted, even if being or becoming a person is a matter of graduality. This points to larger bioethical debates, which are to be discussed at the end of this book.

The advantages and disadvantages of the demand for basic rights for nonhuman persons have already become clear in the preceding sections. Finally, from a general perspective, a look at the problems and limits of argumentation with the help of cognitive or evolutionary kinship is to be confronted once again in order to absorb and present problems and opportunities of this argumentation that have not yet been addressed or have only been touched upon.

The first question is whether it is really necessary to demand *human rights* for apes. Gunther Nogge (1999) makes the provocative statement that if you asked gorillas, they would probably rather stay gorillas and be granted gorilla rights instead of human rights (449). One may easily agree with this. But this does not relieve us of the need to determine more precisely what gorilla rights are. They must also be justified, and the criteria for moral consideration must be stated. Anyone who rejects applying *human* rights to nonhuman animals, and would rather strengthen *animal* rights than extend human rights, does not avoid the main difficulty. Within the formulation of animal rights, one must think about which qualities and values one ties to which rights and become clear about whether other morally relevant abilities are to be considered in addition to the ability to suffer.

If one rejects the emphasis on cognitive similarity but wants to argue for a better protection of great apes, then one might ask whether the central question is, perhaps, that this is about our closest relatives. But why should evolutionary proximity per se play a special role in the moral community? According to Ingensiep (1999), the theory of evolution is and remains a (descriptive) theory, and as such it remains hypothetical and has no normative relevance (445). However, advocates of personhood rights for some animals do not simply infer normative requirements from biological facts. Instead, clearly identified, normative premises are being inserted. These are logically necessary to reach such a normative conclusion.

Humans and great apes are biologically and evolutionarily closely related and therefore share certain characteristics (e.g., cognitive traits). These properties are indicated as morally relevant qualities (see Ingensiep 1999). The argument, then, only states that we also recognize these qualities as such qualities in humans and bind the existence of them to the moral and legal status of the person. Consequently, we should do the same for apes (when we think these qualities are morally relevant, and humans and animals are significantly similar in terms of these abilities). Evolutionary proximity alone is therefore not the decisive premise. It is a descriptive fact that humans and animals are evolutionarily related. Derivation of a normative conclusion (how we should treat these animals) solely from such a descriptive premise (how they are) would constitute a naturalistic fallacy. We need at least one normative premise in between the descriptive ones to pass on some normative quality to the normative conclusion. However, this is the case in the argument we are discussing, the necessary normative premises (P3 and P4) are provided. Thus, whatever one might think about the argument, at least it does not constitute a naturalistic fallacy. In ethical discussions, the accusation of a naturalistic fallacy is often made lightly; a warning was already made in the case of the pathocentric argument discussed above, in which, of course, the moral requirement to avoid suffering is not directly inferred from the biological ability to suffer, either. The criticism of the naturalistic fallacy must therefore always be carefully examined with regard to background assumptions and the normative—or rather, merely descriptive—content of the premises (see also the explanations of Engels 2001, 2008).

If evolutionary proximity cannot be the key factor or the sole premise for establishing a special worthiness of protection, is it perhaps the fact that apes are seriously threatened with extinction? A special worthiness of protection can be derived from this fact, but this applies equally to all other species on the "Red List." If great apes are to be protected intensively, even more intensively than other species that are equally endangered, then our special *emotional* as well as *cultural and historical relationship* to these animals could be emphasized. The special worthiness of protection would then result primarily from these special relationships between apes and us, and would not be based on certain personal qualities that the animals themselves have. The disadvantage of this position lies above all in its arbitrariness, but it reflects a real problem in animal welfare: species with which we have difficulties in establishing an emotional relationship with are often unrecognized and forgotten.

The special worthiness of protection, therefore, seems to depend on the special mode of existence of some animals with simultaneous reference to similarities regarding characteristics of personhood in humans. What is clear is that higher socio-cognitive abilities such as language, culture, and theory of mind cannot simply be raised to a *sole* criterion for moral consideration and for belonging to the moral community. The term "higher" cognitive abilities, alone, could be criticized, as it already contains an anthropocentric evaluation. The attitude and the duties humans have toward the rest of nature, and toward animals which do not have such "higher" abilities, should not simply fall outside of an ethical view. At first glance, it might therefore appear that it would be better not to make these or other socio-cognitive abilities a central criterion within animal ethics, as otherwise the circle of those deserving direct moral considerations would be very small. So what happens to all the other animals at the moment when personhood rights are granted to a few of them? First, it is true that *this* debate excludes them, and that *these* rights cannot be granted to them. On the other hand, their material and legal position does not change in comparison to the moral consideration and rights they have on the basis of other established arguments. For the representatives of the argument from cognitive similarity discussed above, animals other than apes and dolphins certainly do not fall out of the ethical view. Although such arguments are limited in focus to individual species when it comes to the demand for personhood rights, this does not mean that they would not also demand moral consideration for other animals on the basis of their inherent value and even rights—but probably on another basis and for other (e.g., pathocentric) reasons. They cannot provide all animals with the required strong personhood rights, as their argumentation is per se only applicable to those species suspected of having personhood qualities, and as they do not emphasize the graduality of socio-cognitive abilities at the same time. But at least for this group they can make a stronger claim to protection without simultaneously putting all other animals in a *worse* position than before. The Great Ape Project, Balluch and Theuer (2007), and White (2007) do not demand that from now on only great apes or dolphins be protected and thus fall behind everything that the animal welfare and animal rights movement has achieved so far. Instead, they call for a small (currently limited but, in principle, unlimited) group of animals to be assigned *additional* rights. In doing so, they limit their research and their demands to a certain focus, but they do not thereby reduce the status of all other animals.

In order to fairly assess this rights approach, which moves along the lines of personhood and human rights, it should therefore be recognized that not all species enjoy the same rights or the same consideration in other ethical approaches, either. The boundary here may lie mainly between animals capable of suffering and animals that are not capable of suffering, which means that the established entitlements refer to a larger number of animals. But the entitlements themselves do not usually provide for inalienable, absolute rights. This means that many more animals may be considered overall, but with much weaker "rights." It should be acknowledged that the representatives of personhood rights for animals, such as Peter Singer, represent one of the strongest positions published to date in animal ethics. Even if, as in Singer's case, they do not otherwise assume rights for all animals, but represent utilitarianism instead, they nevertheless demand much greater consideration of all animals capable of suffering, from the laboratory rat to the egg-laying hen, than we have had up to now. So they cannot necessarily be accused of neglecting other species. If someone like Singer adopts an animal rights approach similar to human rights for apes, then this should be seen as a continuation or supplement to already quite strong demands for all animals capable of suffering.

The demands of personhood rights for animals, are certainly difficult within animal ethics, above all because pathocentric approaches are brought into the field as supposedly competing approaches. To build a dichotomy here goes back to the formative statement of Jeremy Bentham, already mentioned: "The question is not, Can they *reason*? nor, Can they *talk*? but, Can they *suffer?*" (Bentham [1789] 1970, 283). How can Bentham's criticism that cognitive abilities do not matter, unlike the ability to suffer, be met? Does this apply across the board? Is there perhaps a place for socio-cognitive abilities in ethical arguments? Based on the knowledge of his time, Bentham assumed that cognitive abilities were a very special and exclusive phenomenon in the animal kingdom and that the questions "Can they reason?" or "Can they talk?" would be answered with an explicit "No!" in the case of virtually all animals. This has changed fundamentally—although representatives of the argument from cognitive similarity do not seem to be quite aware of it. It must be assumed that the research results of the last decades have revealed something about the capacities of animals that Bentham could not yet have imagined, namely that astonishing social and cognitive abilities are much more widespread in the animal kingdom than they were traditionally thought to be before the 1960s. The idea that continuity and graduality between species also exists in the cognitive domain has also consolidated since Darwin (see also Engels 2007). In my opinion, the current empirical research situation supports a concept of the human-animal relationship that emphasizes a continuity between humans and animals in the area of socio-cognitive abilities, instead of repeatedly establishing new criteria of difference that prove unsustainable or are gradually perforated by empirical and philosophical research. Irrespective of this, Bentham's famous quote may contain a completely different difficulty. The question of what weighs more in animal ethics, the ability to suffer or cognitive abilities, could already be misleading in itself. It posits (or unconsciously reflects) a dichotomy where there may not be one, thereby disguising an important connection: the ability to suffer and cognitive abilities may not be considered separately. Sensations, feelings, and emotions, on the one hand, and cognitive abilities, on the other, might instead be connected. This means that a long tradition of pathocentric animal ethics must be complemented by new aspects of cognitive research that do not, however, compete with the criterion of suffering. A synthesis of sentience and cognitive abilities as criteria relevant to animal ethics can only be achieved in this way. The integration of cognitive abilities into pathocentric considerations, as opposed to the illustrated integration into rights considerations, will be discussed in Chapter 14.

In the end, the argument from cognitive similarity must be answered by the following: this is not about genetic-evolutionary proximity but about cognitive kinship, which could also exist in species that are more evolutionarily distant than the apes. As Rogers and Kaplan (2004) rightly note: "Only a handful of species have been researched in depth, and current findings would suggest that many more species might be found to have exceptional cognitive abilities, if we only looked" (193). The moral community would, therefore, be an open one, and it would have to reckon with new members all the time. This seems to me to be an important part of a "new" self-image of the human being and her position in nature, which is increasingly gaining in importance due to the increasing knowledge on animal cognition. In general, critics who argue that the Great Ape Project does not go far enough must be aware that its supporters should be open in principle to expanding the moral community even further (see Cavalieri and Singer 1993). The caution called for (i.e., to remain within the constraints of rigid demarcations due to empirical uncertainties) does not go far as a counterargument against an emphasis on cognitive similarity: just because we might later learn that we would also have had to include other animals in the circle of moral equals, this is by no means a reason not to accept the great apes at this point in time. Therefore, the argument that the Great Ape Project does not go far enough is not a valid argument countering this initiative.

Conversely, the criticism may be that the Great Ape Project and similar alternatives go too far. But anyone who wants to maintain that fundamental rights, such as the right to life or the right not to be tortured, should be restricted to members of a particular species must be able to give reasonable arguments for this. Should we not at least grant such fundamental rights to all living beings who have reason and consciousness, some aspects of self-consciousness, and corresponding emotional and social needs (cf. Singer 2006)? If the characteristics mentioned can be more reasonable described as gradual phenomena from an ethological perspective, this could mean that they should also find their way into ethical theories as gradual phenomena (see Chapter 13). The danger of underestimating the abilities of animals that are dissimilar to us, but still act in complex ways, seems to be great.

The graduality of socio-cognitive abilities therefore provides good (though not unproblematic) arguments for the justification of personhood rights that can be transferred to animals. On the other hand, it is precisely the graduality of cognition that cautions us not to be entirely satisfied with the derivation of personhood rights from the evidence of socio-cognitive similarity in some animals. This certainly does not exhaust the argumentative potential or importance of these abilities. A possible further instance in which one could emphasize the socio-cognitive abilities of animals additionally or instead would be within classical animal welfare considerations (see Chapter 14).

8.6 Opportunities for Personhood Rights: "Cognitive Relatives" as Ambassadors of Species Protection?

Apart from all the limitations of the argument from cognitive similarity, the perspective is, above all, associated with great hopes. The substantial upgrading of the moral status of great apes (the Great Ape Project and Hiasl case) and dolphins (Thomas I. White) should quickly and assertively lead to an improvement of the threat situation of these animals. The reasons for pursuing this goal are twofold: First, the demand for stronger protective rights for these families will certainly be pursued for *strategic* reasons. This applies both to philosophers and biologists who want to strengthen the rights of *all* (or all sentient) animals in the long term and to scientists who argue from the perspective of active species protection. A process is to be set in motion to protect the great apes, which in the long term will also lead to more effective protection of other species. Karl Ammann (2001) stresses the extraordinary potential of the great apes as ambassadors for species conservation: "I believe the apes above all other species have the potential to be ambassadors in the fight to save other endangered wildlife. If we cannot do anything for the great apes, what hope is there for the giant pangolin or the forests these creatures live in?" (72). In the case of the fight against the bushmeat crisis, the British government speaks of great apes as a "flagship species" with similar intent (UK Parliamentary Office of Science and Technology 2005).

Second, the reasons for the demand for stronger protection rights for apes and dolphins must also be understood in the light of the *particularly* concerning situation of these animals. Both in the wild and in captivity, these two families are heavily endangered by humans and their interests and, it seems, are inadequately protected by the moral and legal status they have to date. The demand for stronger individual protective rights is therefore made against the background of a hierarchy of interest that determines the status quo of human-animal interaction and is criticized by supporters—namely, the subordination of the fundamental interests of these animals with regard to their own well-being and survival to questionable human interests in utilizing these animals and/or their habitat.

The next chapter is intended to make clear how problematic these human interests of use actually are, in many cases, from an ethical point of view. They do not usually concern basic health or survival interests of humans, nor are they without an alternative. It should be shown that, regardless of the ethical position we may take, we are far from complying with the pathocentric precept to avoid unnecessary suffering in those animals capable of suffering. We are even farther away from granting great apes and dolphins the conditions of existence and expressions of personhood that should not be taken away from nonhuman persons under any circumstances. Sections 9.1 and 9.2 form an excursus within this book that highlights the contrast between the growing ethical demands for animal rights and the status quo in our dealings with our cognitive relatives. Only against the background of this excursus can the final discussion of the relevance of socio-cognitive abilities in animals for animal ethics and animal welfare be continued, in the last part of the book. The following sections concentrate solely on great apes and dolphins because the previously discussed philosophical positions have concentrated on these two families. Great apes and dolphins certainly entail those species to which all three main arguments of animal ethics and animal welfare (both the "classical" arguments from suffering and endangerment and the "new" argument from socio-cognitive kinship) *simultaneously* most clearly apply. However, the problems mentioned above arise in the same or very similar ways when dealing with other species. This is the case, for example, when it comes to other endangered species in the wild or the welfare of zoo animals in captivity.

CHAPTER 9

Kinship and Responsibility: the Discrepancy between Ethical Demands and the Status Quo

9.1 Case Study: How We Treat Great Apes

In the following sections I will discuss the situation of great apes: first those in the wild and then those in captivity.

Great apes still live in the wild in Africa and Southeast Asia. The question of the status of these last wild populations of our closest relatives is gaining importance, and not only with respect to the theoretical discourse on the legal and moral status of great apes. Above all, this is a socio-political issue and a question of species conservation. It is, moreover, a pressing one, irrespective of how species protection is ethically justified, be it with the similarity or evolutionary affinity of these animals to humans, with the general responsibility of humans for all other living beings, with human beings' special responsibility toward living beings capable of suffering, with the value of biodiversity, or with other reasons.

Many national and international conservation organizations and projects are committed to the survival of the great apes. Organizations such as Borneo Orangutan Survival (BOS) take care of orphaned animals and the protection of the animals in their habitats in the field.[1] The Great Apes Survival Partnership (GRASP) of the United Nations (UNEP/UNESCO), a strategic alliance of UN organizations, governments, nongovernmental organizations, foundations, and business partners, is particularly noteworthy at the international level.[2]

Increased efforts are now being made, for example within the GRASP, to involve the governments of the countries where these animals are found in protection efforts. All countries where great apes still live have signed the Washington Convention on International Trade in Endangered Species of Wild Fauna and Flora (CITES). CITES is a governmental agreement that covers internationally the trade in endangered living species and the trade in products derived from these species (such as meat, leather, wood, tourist curiosities, and medical products). The agreement covers more than thirty-five thousand species of animals and plants (see CITES d; c). However, only a framework program

1 See http://orangutan.or.id/bosf/, 2019/04/13.

2 See http://www.un-grasp.org/, 2019/04/13.

 | DOI:10.1163/9789004415072_010

has been defined, which must still be implemented at national level, including in the respective national laws. CITES is the result of an initiative of the International Union for the Conservation of Nature and Natural Resources (IUCN) and now has 183 members (see CITES d). The species protected by CITES are listed in three appendices, with great apes included in appendix 1, which lists the most endangered species (see CITES a; b). Engaging in trade with them is prohibited; exceptions for noncommercial purposes are only possible according to strict specifications (see CITES b; c). The 126 signatory states to the Convention on the Conservation of Migratory Species of Wild Animals (CMS), which was signed in 1979 and which also provides for the special protection of gorillas and chimpanzees, also include the countries where great apes spread (CMS/UNEP a). CMS, GRASP, and the World Association of Zoos and Aquariums (WAZA) declared 2009 the "Year of the Gorilla." At the ninth CMS Conference of the Parties in Rome, in 2008, resolutions on the protection of gorillas were adopted in this context. There is a special CMS Gorilla Agreement in place (CMS/UNEP b). The animals are listed in appendix I of CMS, which means that they belong to the migratory species threatened with extinction and that the contracting parties to the convention undertake to protect them strictly, to preserve or restore their habitats, and to remove obstacles on their migrations (CMS/UNEP a).

So there are national and international agreements and regulations to protect the great apes. Nevertheless, they are increasingly at risk and are therefore increasingly in the public eye.

However, this does not mean that we are actually aware of the fatal situation faced by our closest relatives in the wild. The issue comes to our mind rarely, although sometimes by ways of dramatic events. In 2007 photographer Brent Stirton won first prize at the World Press Photo Contest 2008 in the category "Contemporary Issues Singles" with a report for *Newsweek* about the killing of several mountain gorillas in Virunga National Park in the Democratic Republic of Congo (DRC) the year before.[3] Stirton's pictures and related reports, such as those of his colleagues Scott Johnson and Mark Jenkins, were published around the world.[4] The events they documented were summed up by *Newsweek* managing editor David Klaidman as follows:

3 The winning photo in particular, showing villagers carrying the carcasses of the apes to funeral, was taken up by the world press. The report was entitled "Evacuation of Dead Mountain Gorillas, Virunga National Park, Eastern Congo." The complete photo gallery can be viewed at www.brentstirton.com.

4 Scott Johnson's (2007) article appeared in *Newsweek*, and Jenkins's (2008) article appeared in the German edition of *National Geographic*.

> Four members of the same gorilla family had been shot at close range, possibly the worst such slaughter in 25 years. Both Johnson and Stirton are experienced war journalists. Yet they were still taken aback by the scene. Two things in particular struck them. One was how human the great animals seemed in death; a 600 pound male silverback lay with one hand across his chest, as if he had just been beating it. The other was the tenderness with which the rangers treated the bodies.
>
> Newsweek Staff 2007

More than one hundred villagers followed the rangers into the jungle, built stretchers out of branches, and carried the dead animals back to the camp in a three-hour march. quotes Stirton, an established Africa expert, as follows: "I've never seen a demonstration of compassion like that from the Congolese people before" (Newsweek Staff 2007). It seems significant that it is the concern about the last remaining great apes that is motivating the population to make such joint efforts. Animals are apparently also of interest to the various civil war parties in the region (mainly because of their economic value within ecotourism), and their protection could play a new role in the peace process in countries that have been plagued by conflict for decades (Benz and Benz-Schwarzburg 2010). The presentation of the World Press Award to Stirton was not only the recognition of an outstanding photographer; the event also highlighted the status of one of the most pressing contemporary problems.

Other renowned photographers, such as Karl Ammann and Jay Ullal, also report on the situation of great apes in the wild. However, despite the increasing awareness throughout the world, and despite the efforts of international species protection, the animals' numbers have declined drastically in recent years and decades. On the one hand, dangers are responsible for this which directly target the welfare and life of the animals. In addition, however, there are many other threat factors that indirectly affect the animals. Below, two long-known main threat factors are described with respect to their effects on great apes: the destruction of habitats and the trade and consumption of bushmeat. Afterward, a third factor that has so far been neglected is considered: armed conflicts. These conflicts seem to intensify both the deforestation problem and the bushmeat crisis even more drastically.

All great apes are fundamentally threatened by all of these and some other human influences, albeit to varying degrees. The bushmeat problem and the danger of armed conflicts for nature and animals have become particularly apparent in recent years for gorillas, bonobos, and chimpanzees in Africa, while the clearing of the rainforest, especially for oil palm plantations, most notably in Southeast Asia, is increasing massively and has negative effects, above all for orangutans in Indonesia. It will become clear in what follows that the critical situation of the animals is actually due to human influences. In dealing practically with our closest relatives, we lag far behind the ethical demands of theoretical considerations on the legal and moral status of these animals. Our relationship with our last closest relatives living in the wild is decisively influenced by this gap between ethical theory and practical treatment and thus presents itself as extremely problematic from an ethical point of view.

9.1.1 *The Red List Status of Great Apes*

According to the Red List 2019–1,[5] all great apes are endangered or critically endangered which means that they are clearly threatened with extinction.

The bonobo (*Pan paniscus*) is listed as endangered with a decreasing population trend (Fruth et al. 2016). So are all chimpanzee subspecies and varieties (Humle et al. 2016a) with the exception of the western chimpanzee (*Pan troglodytes ssp. verus*) who is listed as critically endangered, population trend decreasing (Humle et al. 2016b). The gorilla subspecies and varieties (Plumptre et al. 2019; Maisels et al. 2018) are all listed as critically endangered, trends decreasing, with the exception of the mountain gorilla (*Gorilla beringei beringei*) who was recently reevaluated as only endangered with an again increasing population trend (Hickey et al. 2018). All subspecies and varieties of orangutans are listed as critically endangered with a decreasing population trend (Nowak et al. 2017; Singleton et al. 2017; Ancrenaz et al. 2016).

The cited reports provide all arguments and reasons for the Red List assessments. I will only summarize a few interesting findings from them. The population trends of the great ape species usually point downward toward a further decline in population size. From the reports it becomes obvious that sometimes even Red List experts cannot really assess the trends, which is particularly worrying. All the great apes have seen a sharp decline in population sizes, the causes of which are either irreversible or unresolved. Fifty to eighty percent or more of the populations have been destroyed in the last three generations or are expected to be destroyed within the upcoming generations.[6] For all species, the decline in population is attributed to the decline in the habitat or quality of habitats and to the actual or expected exploitation of the animals.

5 The IUCN's annual Red List classifies species, with contributions from thousands of scientists from hundreds of different countries. The classification is based on strict scientific criteria that take into account very different aspects (such as the decline of population sizes in the past and their expected decline in the next generations, the question of whether populations typically only occur in isolated and scattered groups, and the number of individuals capable of reproduction) (IUCN 2001, 2019).

6 A generation of great apes is measured at twenty to twenty-five years.

The impairment and destruction of the habitat and its loss for the animals primarily results from human influences such as agriculture, road construction, and the exploitation of mineral resources (above all through the timber industry and mining). The number of animals hunted is alarming, and illegal poaching, massacres, animal trafficking, political instability, and diseases (including human-borne diseases) are all significantly contributing to the eradication of the great apes.

In the case of orangutans, it is particularly clear that individual animals or small groups only remain in a few rainforest islands, which endangers genetically healthy reproduction. Corridors for the migration of animals would be essential for food and reproduction, but often no longer exist (Caldecott and Milles 2005, 418; Nowak et al. 2017; Singleton et al. 2017; Ancrenaz et al. 2016).

Meanwhile, there are only few mountain gorillas that have survived in only two regions of Central Africa, in the Bwindi Impenetrable National Park in Uganda and in the Virunga Volcano region in the border area of Rwanda, Uganda, and the DRC. Experts estimate the number of individuals capable to reproduce on only 600 animals, but their population trend is at least pointing upwards again (Hickey et al. 2018). In the case of the cross river gorilla, the number of reproductive individuals is only between one hundred and two hundred animals scattered in eleven areas (Bergl et al. 2016).

9.1.2 *Habitat Destruction and the Bushmeat Trade*

The destruction of the rainforest has for decades been directly associated with the survival of species, especially the survival of large mammals. Brook et al. (2008) come to the conclusion in their meta-analysis of studies on species extinction over the last thirty years that the destruction and fragmentation of habitats must be the main drivers of species extinction, alongside exploitation and pollution. Furthermore, climate change, environmental variability, and invasive species are secondary accelerators of the process, and are becoming increasingly destructive. It is assumed that species are most threatened with extinction when new threats emerge that are outside their evolutionary experience or are effective at a rate that adaptation cannot keep up with. The size of a population plays an important role in that small populations would die out faster as a result of changed conditions. A high degree of species specialization reduces species' ability to adapt to the loss or change of habitat, such as shifts in available food sources. "Similarly, interactions between environmental factors and intrinsic characteristics make large-bodied, long-generation and low-fecundity species particularly predisposed to anthropogenic threats given their lower replacement rates" (Brook et al. 2008, 455). These are characteristics that apply to great apes.

The loss of biodiversity, according to Brook et al. (2008), reduces the functionality of biosystems. It is human beings who are driving this development. We are considered to be the cause of the so-called sixth extinction of species, which began about fifty thousand years ago and has intensified in the last five hundred years (Brook et al. 2008, 455, 457). Worldwide, the remaining primary forest[7] continues to be cut down, both legally and illegally. Tropical wood is either itself an internationally lucratively traded resource or it is further processed, for example, as furniture, charcoal, or paper. On the other hand, rainforest is also cut down to make room for the cultivation of crops, mostly in huge monocultures, or for the extraction of mineral resources. Of all the countries affected, Indonesia has one of the highest deforestation rates worldwide (Schuster et al. 2007, 147). Borneo's progressive deforestation since 1900, as documented by the WWF, makes the problem visible in its rapid progress.[8] Most of Indonesia's forests have now been put out to tender for production purposes, and large parts of rainforest are being illegally destroyed. Still-intact rainforest areas are mostly mountainous zones or otherwise difficult-to-access areas, such as steep slopes, where it is not economically worthwhile to use the soil (Schuster et al. 2007, 144).

This massive deforestation has dramatic ecological consequences and leads to soil depletion, loss of biodiversity, and global climate change. Indonesia's deep rainforests, in which the orangutans are native, play a special role in the global effects of the deforestation of the rainforest. Slash-and-burn and dewatering—for example, for oil palm plantations—release huge amounts of carbon dioxide, causing billions of dollars of damage when they enter the atmosphere. Even national parks, often the last retreats for great apes, are severely affected. Over 90 percent of the more than forty parks in Indonesia, for example, had already been deforested by 2007. It is not poor farmers who bear the responsibility, but "well-organized companies with heavy machinery and strong international links to the global markets" (Nellemann et al. 2007, 5). The resulting habitat fragmentation and isolation of populations are responsible for the fact that genetically healthy and permanently survival-securing reproduction cannot take place within the populations. Especially for migratory great ape species, for which it would be of central importance to be able to reach from forest island to forest island, the lack of corridors presents itself as a threat to their existence.

---

7 Primary forest is defined as "forests of native species where there are no clearly visible indications of human activities and where the ecological processes are not significantly disturbed" (FAO 2005, 4).

8 See the WWF maps in Schuster et al. (2007), 144.

With the destruction of the forest, however, vital sources of food also disappear. For example, orangutans on Borneo are increasingly invading oil palm fields and eating the young palm shoots. The plantation owners want to prevent this and have the animals hunted and killed. Even bounties are suspended, creating a practice in which the animals are regarded as a plague and killed extremely brutally. BOS reports animals buried alive or dismembered with machetes. Adult animals are killed, and young animals are sold for thousands of dollars (Schuster et al. 2007, 185, 278). Many young orangutans (as well as chimpanzees, bonobos, and gorillas affected by the animal trade in Africa) do not survive the catch and transport. At a minimum, their mothers are killed when they are caught, but most of the adults protect and defend the young. It must be assumed that several adult animals are killed for each young animal caught.

Deforestation and the unsustainable use of the rainforests through the destruction of habitats contribute both directly and indirectly to the extinction of species. Even if the massive clearing of the rainforest does not lead to the immediate extinction of species, in the long term it sets in motion a number of secondary threat factors that can cause this. Ultimately, several simultaneously effective threat factors cause synergistic processes[9] by interacting with and upon each other and reinforcing each other. They can form cascade-like extinction dynamics or chains of extinction that have not been recognized and have been underestimated for a long time (Brook et al. 2008, 453, 456). The loss of habitat is thus indirectly responsible for the extinction of species, "by facilitating invasions, improving hunter access, eliminating prey, altering biophysical conditions and increasing inbreeding depression" (Brook et al. 2008, 453). Loggers' road networks allow access to remote rainforest areas and are therefore partly responsible for the increasing hunting of wild animals, which is threatening some species to an extreme degree (Brook et al. 2008, 458). Whereas until a few years ago the destruction of the rainforest was regarded as the main threat to the great apes, the role of this destruction is now attributed to the aforementioned secondary threat factors as well, which are mainly caused by deforestation—above all the consumption of and trade of wild meat,

9 "A synergy among processes implies a positive interaction (feedback) whereby the total negative effect on biodiversity is greater than the sum of each threatening process's individual contribution. Small synergistic interactions will result in population decline…; however, only when large synergies occur that push a population below their minimum viable population (MVP) size does extinction risk become nonnegligible" (Brook et al. 2008, 454, caption to fig. 1). In the case of the great apes, of which (depending on the subspecies counted) only a few thousand or a few hundred still exist worldwide, critical population sizes have been reached.

known as "bushmeat" in Africa (Milner-Gulland et al. 2003, Ammann 2001).[10] It is now assumed that the explosion of commercial bushmeat hunting is destroying primate populations faster than deforestation would ever have been able to (Rose 2002, 208, 212; Oates 1996, 1, illustrations 31–33 on pp. 420–21). "Bushmeat frontline countries" in Africa are the DRC, Cameroon, and Gabon (Ammann 2001, 71).[11]

Wild animals have been hunted worldwide in the rainforests for more than a hundred thousand years, but hunting and consumption have increased dramatically in the last few decades (Milner-Gulland et al. 2003, 351, 352). Experts therefore speak of a "bushmeat crisis." Karl Ammann (2001) states quite clearly: "The commercialisation of bushmeat has reached crisis level as far as the off-take of most hunted species is concerned" (72). The annual hunting yield is estimated at more than 23,500 tons in Sarawak (Malaysia), 67,000 to 164,000 tons in the Brazilian Amazonas, and 1 million to 3.4 million tons in Central Africa (Milner-Gulland et al. 2003, 351; Wilkie 2001, 89). The inhabitants of the Congo basin, and especially certain ethnic groups, eat as much if not more meat than the inhabitants of Northern industrial countries (73 kg per person per year). This is worrying because the population in the region more than doubled between the 1920s and the turn of the century—and because it is expected to double again by 2025 or 2030 (Wilkie 2001, 86). Africa's population increased eightfold between 1900 and 2000 (Milner-Gulland et al. 2003, 352).

A distinction must be made between those sections of the population who use bushmeat for lack of alternatives and those consumers who are not dependent on the consumption of this meat. For many consumers, eating bushmeat is not a necessity but a privilege—a privilege for which they are prepared to pay prices far greater than those for domesticated meat (Wilkie 2001, 86–93; Milner-Gulland et al. 2003, 251, 352; Wilkie and Carpenter 1999, 927). A large portion of bushmeat is smoked, transported over long distances, and sold in the markets of the big cities, where higher profits can be achieved (Rose 2002, 212; Wilkie 2001, 93). The prices can increase tenfold on the way from the hunting camps to the markets (see Auzel and Wilkie 2000). The estimated extraction of more than one million tons of bushmeat from the Congo Basin alone adds up to an annual profit of more than one billion dollars (Rose 2002, 210). Last but not least, demand is also being driven massively by a self-proclaimed

10 "Bush" is an African term for "forest." However, since bushmeat consumption is not limited to Africa, experts have introduced the term "wildmeat" to refer to the global nature of the phenomenon (Milner-Gulland et al. 2003, 352). In the following I will discuss the bushmeat crisis in Africa, especially in the Congo Basin, and for this reason alone I will continue to use the term "bushmeat."

11 See also Wilkie (2001, 86).

growing global gourmet elite outside Africa. It is estimated that around five tonnes of bushmeat is smuggled in personal baggage through Paris Roissy-Charles de Gaulle airport per week which sums up to almost 273 tonnes per year (Chaber 2010). It can be assumed that this meat is served in all major European cities like Antwerp, London or Brussels.

Great apes are amongst the species strongly affected by the growing bushmeat trade. Their distribution density in the forests of the Congo basin is less than 0.13 animals per km^2, and hunting for them is considered extremely dangerous. As a result, they are rarely found in markets.[12] Despite this, they are often bycatch of the hunt, as they fall into traps designed for other animals. Whether caught intentionally or unintentionally, their hands, feet, heads, or skulls will be sold as delicacy or as fetish objects (Wilkie 2001, 87; Ammann 2001, 77). Because they are so rare and large, and because they only reproduce slowly, even comparably low hunting rates would endanger their survival. The number of hunted animals is simply too high in relation to the remaining number of individuals and their annual reproduction rate (Wilkie 2001, 87ff.; Milner-Gulland et al. 2003, 354; Wilkie and Carpenter 1999, 927).

Various scientific studies have shown that commercial bushmeat hunting can indeed lead to the extinction of wild animals. In Gabon, for example, it led to a decrease of 43–100 percent in animals within the area under study. As expected, primates and generally large animals were the most affected. Six out of fourteen species had been virtually extinct in hunting grounds, four of which were primates, including gorillas and chimpanzees. Trap hunting wipes out wildlife within 10 km of a village. At least 25 percent of the caught animals rot in the traps or are lost to scavengers. In remote forests, up to 39 percent of animals are left to rot in traps. The traps are left for between twenty-nine and seventy-seven days, but often go unchecked during this time.[13] Wilkie (2001) concludes that, although previous research has used different methods and investigated different areas in the Congo Basin over different periods of time, the results combined leave little doubt as to the following conclusion: "The bushmeat trade is clearly having an adverse impact on forest wildlife throughout the region, and commercial hunting of wildlife is likely to be unsustainable for primates, all large-bodied duikers, and in many areas all mammals larger

12 "Duikers, bush-pigs, primates and rodents are the most commonly hunted groups of animals. Both numerically (>75%) and in terms of biomass, duikers are the most hunted bushmeat species; primates constitute 20% of the animals sold in markets, apes fewer than 0.5%" (Wilkie 2001, 89).

13 Wilkie (2001, 90–91) and Milner-Gulland et al. (2003, 354) also point to the wasting of animals left behind and assume that this practice is also a reason why the hunting rates are strongly underestimated.

than rodents" (Wilkie 2001, 91). Rose (2002) agrees and concludes that external influences cause both economic and social pressure and consequently make hunting a modern process, which ultimately leads to a destruction of wild populations (Rose 2002, 210).

According to Ammann, this modern form of hunting includes an entire network of illegal and corrupt practices that have become a daily scenario in Central Africa (Ammann 2001, 72). From the Republic of Congo, Cameroon, and Gabon, for example, he reports that bushmeat from countless different wild animals is always offered on the observed markets, regardless of whether the hunting season is open or not. Fetish objects such as gorilla skulls, elephant tails, and chimpanzee feet, which come from animals that could not have been hunted legally under any circumstances, are also on the market as byproducts of the bushmeat trade. Politicians use television speeches to urge their country's children to use their holidays for hunting and persistently deny that there is a poaching problem. According to Ammann, trains stop in the jungle so that bushmeat hunters and traders can load and unload their goods. Both army and police officers and other officials are involved in the trade. Employees of the Cameroonian Ministry of the Environment and Forestry (MINEF), for example, lend their weapons to commercial poachers and bribes are also paid to rangers (Ammann 2001, 74ff.). Under these conditions, the establishment of national parks and protected areas does not lead to the actual protection of flora and fauna, and laws against poaching or trade in live and dead protected animals are not enforced. Ammann says: "The trade has now gone so commercial, with so many parties depending on it, that there will be no way to enforce any aspect of the law without causing some hardship" (Ammann 2001, 78). Already more than two decades ago, Steel (1994) calculated for Gabon that the annual consumption and trade in bushmeat was a $47 million industry in a country with only one million inhabitants (Ammann 2001, 79). High profits are made from the trade in bushmeat, although the weapons and ammunition involved are used illegally in 99 percent of cases. And this is the case even though the transport of bushmeat and hunters on timber company trucks is illegal, although traps from cables are prohibited, and commercial hunting is restricted by law and regulated by a designated hunting season. In addition, hunters are not allowed to settle in many areas, and hunting for endangered species and trade in products made from these endangered species is strictly prohibited (cf. Ammann 2001, 81).

In addition to the points mentioned above, some tropical timber companies and their employees have been criticized for being involved in the illegal arms trade for both bushmeat hunters and civil war parties, thus benefiting from or further promoting a country's often unstable situation. Greenpeace has researched and documented the harsh criticism of the German Danzer veneer factories and their Swiss trading department, Interholco. According to Greenpeace, the close relationship between Interholco and the Inland Logging Company (ILC) in Liberia, whose owners were close advisors to the former Liberian President Charles Taylor,[14] has proved to be particularly problematic. According to Greenpeace, the ILC was accused of maintaining a private army that harassed the local population (Greenpeace 2005). Other NGOs also point out that private armies of timber companies are actively involved in civil wars. Global Witness wrote in a report for the UN Security Council: "Liberian logging company militias are actively engaged in fighting, allegedly on both sides.... Logging company militias commit grave human rights abuses, act with virtual impunity and are used as a surrogate security force by the Liberian government" (Global Witness 2003). The Danzer/Interholco example has prompted additional discussion about the increasing threat of armed conflicts to the great apes in Africa.

9.1.3 *New Wars and Political Instability*[15]

For decades, armed conflicts have been taking place in all countries where great apes are widespread, affecting not only the human population but also nature and animals. The great apes are also either caught between the fronts of civil wars and other conflicts or are indirectly affected by them.

Worldwide, purely private, nonstate or substate conflicts and wars (e.g., between militias and warlords) are on the uptick; these are not covered by classical definitions of war and conflict. They are referred to as *New Conflicts* and *New Wars* in conflict-research theories (see Benz and Benz-Schwarzburg 2010). Only about ten years ago have conflict researchers begun to turn their attention to the specific characteristics of such warfare, which is gaining in importance, especially in Africa (see Kaldor 2007; Chojnacki 2005; Münkler 2002). The *ecological* effects, especially the effects on endangered species, are enormous, but have so far been virtually completely neglected. Benz and Benz-Schwarzburg (2010) make it clear that almost half of the countries that are home to great apes have experienced both conventional conflicts between states and nonstate conflicts. Fighting has shaken these countries for years or even decades. Most countries that are home to the great apes have hardly had a moment's rest between (often

14 Charles Ghankay Taylor was President of Liberia from 1997 to 2003. He was the first African head of state to appear before an international tribunal in the Hague for war crimes and crimes against humanity. On April 26, 2012, he was found guilty and sentenced to fifty years in prison.

15 For a more detailed discussion, see Benz and Benz-Schwarzburg (2010) and Benz-Schwarzburg and Benz (2012).

overlapping) armed conflicts in the last sixty years. This is particularly true in the Congo Basin, where the world's largest number of great apes live in the wild. One of the most important countries for the protection of the great apes is the DRC. The last free-living mountain gorillas in the world live there. Here, too, the political situation has been extremely unstable for decades: "Since independence in 1960, this country has spent 18 years in internal warfare. The DRC is the only nation in the world which experienced not only several civil wars but also more than one New War (as well as numerous less intensive New Conflicts)" (Benz and Benz-Schwarzburg 2010, 397).

The violent clashes in the DRC have increasingly taken place in the province of North Kivu. In the very center of this region of conflict is an area of about 440 km^2, which includes the Mgahinga Gorilla National Park in Uganda, the Volcanoes National Park in Rwanda, and the Virunga National Park. The latter is located in the border area between Uganda, Rwanda, and the DRC and is the crown jewel among the African national parks. The national park provides habitat for about half of the world's last endangered mountain gorillas that have survived in the wild. Until the 1990s, before armed conflicts flared up, animals were the main tourist attraction of the country (Jenkins 2008, 32).

Although the park has long been a World Heritage Site, it is also on the UN's list of the most endangered landscapes in the world. Virunga lies in the middle of a region that has experienced cruel war like no other in recent history. In 1994 more than eight hundred thousand Tutsi were killed by the Hutu people in neighboring Rwanda. More than five million people lost their lives in the Congo in two wars—in 1996–97 and from 1998–2003. This killing has claimed more victims than any other conflict since World War II (Jenkins 2008, 32).[16] Violent conflicts have been going on in Virunga, even around and directly within the southern part of the park (in the Mikeno sector), where many of the mountain gorillas live. The conflict situation in the area continues to intensify (Benz and Benz-Schwarzburg 2010, 401ff.).

The privatization of violent actors typical of New Wars can be observed in North Kivu. The main parties to the smoldering and ever-flaring conflict were and are different rebel groups and militias, like FDLR, CNDP, or the MaiMai militia (Jenkins 2008, 33). Since 2006 (and repeatedly in 2008 and 2009), disputes between rebels and counterattacks of the UN and the Congolese army

16 The Second Congo War alone turned out to be one of the most serious conflicts since World War II and led to about 5.4 million dead—at least 4.6 million of whom died in the five eastern provinces of Congo (Coghlan et al. 2007, 16). Experts describe what happened in the DRC as "wars within wars," "poly wars," "Africa's Great War," or "Africa's First World War" (Shearer 1999; Laakso and Hiekkanen 2002; Fahey 2009).

have taken place near or even within the park boundaries. Thousands of soldiers were and are still stationed in the area (BBC 2012; Jenkins 2008, 41). The gorilla sector also temporarily fell under the indirect control of rebel groups that set up army camps near the park or in the park (Jenkins 2008, 31, 38, 41, 50; Hickey et al. 2018; Gorilla Diary August 2007–March 2009, January 30, 2009, November 21, 2008; Crilly 2008). In several cases, armed combatants were responsible for the killing of gorillas (UNESCO/WHC 2009a, 2008b; Gorilla Diary August 2007–March 2009, September 14, 2007; AWF 2007). Furthermore, soldiers also kept great apes and other wild animals as domestic animals and participated in the illegal animal trade (Gorilla Diary August 2007–March 2009, November 13, 2008).

In addition to the animals themselves, those who wish to protect them are also repeatedly victims of the violent actors in the conflict. Many park rangers have been killed in service. Most of them were shot—not by poachers but by rebels. They were also forced to flee the park several times. Scientists and NGO staff also had to interrupt their work or the care of the animals (Roberts 2008; Boedhihartono and Lumpu 2008, 3; Jenkins 2008, 33, 39; Rose 2002, 215). As far as attacks on ranger stations and the control of gorilla areas are concerned, events in 2011 and 2012 have turned upside down. An attack on the rangers by the MaiMai militia in April 2018 was the deadliest attack to Virunga staff in recent years and brought the total number of rangers killed to 176 (see https://virunga.org/news/ranger-freddy-killed-in-action 2019/05/17). Virunga National Park regularly documents recent developments via the park's homepage and used to provide a blog in which the situation was continuously documented (see https://virunga.org/, 2019/05/17).

In addition to the privatization of actors, a typical New Wars establishment of war economies in the DRC can also be observed, which in turn also affects the apes in the region. The fighting groups aim to gain or maintain control over the production of so-called "conflict resources" (Benz and Benz-Schwarzburg 2010, 407ff.). All parties involved in the conflict in North Kivu are accused of plundering and continuing to plunder the rich mineral resources of the region. In addition to minerals, tropical woods, charcoal, and bushmeat, there are also conflict-relevant goods, around the mining and trade of which illegal industries have developed, which enable and promote a continuation of the conflicts (Benz and Benz-Schwarzburg 2010). The massacre of the Virunga gorillas documented by Stirton's photos was also linked to the charcoal mafia and corruption. In December 2007, the rangers of Virunga National Park estimated that illegal charcoal production had increased tenfold within three months due to the unstable political situation (Gorilla Diary August 2007–March 2009, December 14, 2007, March 3, 2008). Rebels therefore make a profit from the extraction of resources, but they have also long since realized the economic potential of the great apes themselves. Due to fighting and rebellion, tourism programs such as mountain tours to the summit of Nyiragongo, and gorilla programs in the Mikeno sector had to be discontinued (Jenkins 2008, 38, 50). However, CNDP rebels resumed gorilla tourism in Virunga National Park under their control in 2007. In September 2007 and June 2008, rangers reported that the CNDP rebels were using Bunagana on the Ugandan border to cross into the Mikeno sector of the park. The money paid for such tours not only supports the rebels and their warfare; the rangers are also concerned that the rebels bring too many tourists to the park and fail to observe any safety regulations for such a tourism program. For example, they fail to maintain the necessary minimum distance of seven meters between tourists and animals in order to avoid the transmission of diseases (Gorilla Diary August 2007–March 2009, September 21, 2007, June 27, 2008; Jenkins 2008, 38, 39, 50). In times of peace, gorilla tourism conducted and controlled by rebels as an additional source of income would be impossible. These practices are therefore also a prime example of a war economy.

As already mentioned, there are also concerns about the links between tropical timber companies and the bushmeat market. Rebels and loggers alike lend their weapons to commercial poachers. Hunting guns are produced in the garages of the concessions and with the help of the material of the companies, cables are used, for example, for snares for trapping. The hunters themselves, their prey, and their ammunition are transported on the trucks and boats of the companies. The illegal trade in ammunition is carried out by independent dealers on the boats and controlled by the boat drivers. The workers of the companies are regularly excused from their work for hunting or even instructed to hunt bushmeat. Their family members often also participate in illegal business as traders (Ammann 2001; Wilkie 2001). Commercial industries, such as the timber companies' businesses, may come to a standstill in times of fierce warfare, but this does not apply to illegal logging and the bushmeat trade. As soon as fighting subsides (at least temporarily) and a certain degree of political stability reigns, these industries also return to an environment in which the mentality of the struggle for survival has suppressed every ecological moral (Rose 2002, 208, 216).

A final characteristic of New Wars, the brutalizing nature of violence, also affects the survival of mountain gorillas in the DRC. A main method of territorial control in New Wars is the massive forced displacement of the population. A high number of internally displaced persons (IDPs) and refugees are indicators that this is taking place. However, territorial control is also achieved by force through the involvement of civilians as a buffer in fighting, ethnic cleansing, attacks against civilian targets, forced famines, and the use of systematic rape as a weapon of war (Kaldor 1999, 98ff.; Kaldor 2007, 8–9, 104ff.; Münkler 2006, 137). Precisely the same strategies can be observed in North Kivu, where fighters of all parties committed indescribable atrocities against the civilian population (Grignon 2009; Jenkins 2008, 36, 33). Expulsions in recent years in the DRC have also taken place in the region North Kivu near the Virunga National Park and thus close to the area where great apes are distributed. Hundreds of thousands of internally displaced people who are unable to return safely to their settlements remain in the area (ICG 2011). They are temporarily accommodated in permanent or spontaneous, large, and often poorly equipped camps near the park. This is of great importance for species and nature conservation because the militias, just like the population and the many refugees, need both food and charcoal (Jenkins 2008, 39–40). Bushmeat is an easily accessible source of protein for people in such situations. In this sense, says Rose (2002), slaughter and displacement of people "stimulates the destruction of wildlife".

Additionally, "emergency hunting" drives the slaughter (215–16). In their *Gorilla Diary*, the Virunga Rangers explicitly link the increasing brutality in the war and the intensification of fighting in or near the park at the end of 2007 with the increasing flow of refugees and increasing poaching as well as illegal charcoal production: "Some 4000 people are camped at Kibumba, which is right next to the gorilla sector. It is not an official camp because the UNHCR says there is no water.... These people have nothing—and nothing is being provided. So, naturally, these refugees turn to the park. Here, there are animals to be poached, and charcoal to be made" (Gorilla Diary August 2007–March 2009, December 7, 2007). In accordance with this, they report an increase in poaching during the fighting. Although traps are normally set to catch antelopes, they may lead to limb loss or even death in gorillas (Gorilla Diary August 2007–March 2009, March 23, 2008, June 13, 2008, December 12, 2008). In addition, weapons left behind by soldiers drive the bushmeat trade (Boyes 2012).[17]

17 Hart (2008) reports from the TL2 forest area: "Truly long distance hunting for big game came with weapons of war. This was no longer villagers hunting for subsistence or to sell meat for a little cash to buy the family a round of new clothes. No. This was military and ex-military seeking ivory to fill the coffers of military officers in large cities far away. Their weapons of war could destroy a herd of elephant. The origin of these weapons is a bloody footnote in the long war that has ravaged Congo and that drags on in the Virungas.... National troops and local families fled into the forest; abandoned AK-47s made their way north, down the Lomami, and are now killing the remaining elephants along the TuTu River."

Violent actors control the population not only by threatening violence but also by controlling (access to) bushmeat hunting. In view of the danger of starvation and of invading armies, refugees, and rebels, traditional taboos about eating great apes would also fall by the wayside (Rose 2002, 214).

Yet the increase in bushmeat consumption is not the only indirect result of New Wars for great apes. Soldiers, the population, and the large number of refugees and internally displaced persons also depend on charcoal (UNEP 2009, 21). Robert Muir, project manager of the conservation program of the Frankfurt Zoological Society, believes that the CNDP rebels will not leave the park until the Hutu militias of the FDLR have left; the Congolese army will not withdraw until both rebel groups have disappeared. But neither party is interested in a withdrawal, not as long as money can be made with charcoal in the sense of a war economy (Jenkins 2008, 48).

The connections drawn make it clear that rebels control the habitats of the great apes as well as the bushmeat trade and illegal charcoal production. Their brutal strategies are also responsible for driving out the population, which depends on protein, wood, and coal for their survival. IDPs and refugees camp in nature or in often poorly equipped areas. They cannot grow their own food or take part in trade. To meet their needs, they also penetrate national parks. The result of these relationships is a self-accelerating cycle, which is kept going by violence and further stabilized by lawlessness and corruption. The massive consequences of the conflicts for the protection of species pose major challenges to conservationists. Possible solutions cannot be discussed in detail here.[18] It is clear that the described characteristics of today's violent conflicts in their full significance for nature and species conservation must be recognized more clearly and be met more specifically. This applies to their direct and indirect effects and requires local, national, and international initiatives.

In light of the fatal situation of the great apes in the wild, one could think that, luckily, they are also kept in human care. However, the reasons why we humans keep animals in captivity and the husbandry conditions we impose on them are often problematic. This is clear (and will be further discussed later) with regard to zoos. But it is also the case when it comes to the lives of animals in laboratories. Great apes are still kept and used in animal experimental research, so we will investigate their situation there as a case example for how humans treat great apes in captivity.

18 For more details, see Benz and Benz-Schwarzburg (2010).

9.1.4 *Animal Experimentation: the Beginning of the End of Chimpanzee Experiments?*

Worldwide, it is estimated that between 115 and 127 million animals per year are used in animal experiments, but global figures are extremely difficult to obtain, and the assumption that the actual number is far underestimated is reasonable. Taylor et al. (2008) estimate the number of animals used annually worldwide at 115 million. Their study is the first statistical analysis of global animal testing figures published in a peer-reviewed journal, the British journal *Alternatives to Laboratory Animals* (ATLA). In a summary by the Dr. Hadwen Trusts for Humane Research, whose scientists, together with colleagues from the British Union for the Abolition of Vivisection, carried out the study, the authors write:

> However, even this massive figure could be an underestimate due to the way figures are compiled. Despite widespread public and political interest in animal research worldwide, our research reveals that a mere 21% of countries (37 out of 179 countries) actually collect data on their national animal experiments. For the rest, no official record is kept of the animals used in their laboratories or the suffering that they might have endured. We reached our total of 115 million animals by using available official data as our starting point—50 million animals from 37 countries that do publish official figures. We then developed a statistical model based on animal research publication rates for countries with no official data, which added a further 8 million animals from 142 countries. We also included those animals used in laboratories but excluded from most statistics, such as genetically modified animals bred to maintain colonies, animals killed as surplus to requirements and those killed for tissue samples. These animals are still bred for scientific purposes, live in a laboratory environment their entire lives and are killed at the end of their "use" and yet most countries don't record them at all. These "missing" animals were estimated to add a staggering 57 million to the global total, bringing it to 115 million animals. (Dr. Hadwen Trust for Human Research 2008)

Knight (2008a) uses a modified method to calculate the number of animals used in animal experiments. He gives the considerably higher number of 127 million animals per year (see Knight 2008a; Taylor and Gordon 2008).

The advocates of animal testing assume that it cannot be dispensed with at present, although more and more alternative methods are now being used. In the case of diseases, people expect effective medicines and therapies. Furthermore, society relies on the fact that new, potentially toxic substances and products offered on the market have been adequately tested so that a danger to humans, animals, and the environment can be excluded (BMELV 2009). In recent years, proponents of animal experiments and especially proponents of primate experiments have repeatedly come forward with these and other arguments. In 2005, for example, some leading scientists from American experimental laboratories with primates pleaded in the journal *Nature* for the continuation and increased funding of experiments with chimpanzees (VandeBerg et al. 2005). This, so they argue, is necessary to reduce the risk for human subjects and patients. Chimpanzees are required as experimental animals, especially for testing future therapies with monoclonal antibodies:

> The need for chimpanzees to test monoclonal antibody therapies is rapidly increasing as more antibodies are designed. Monoclonal antibodies, genetically engineered to be almost identical to human antibodies, hold great promise for treating cancer; autoimmune diseases, such as rheumatoid arthritis, lupus erythematosis, multiple sclerosis, psoriasis, and Crohn's disease; asthma, poisoning, and septicemia; and virtually any disease caused by viral infection.
>
> VANDEBERG et al. 2005, 30

The antibodies obtained from mice were targeted at biological receptors that, according to the authors, only occur in humans or chimpanzees. Monoclonal antibodies also degrade too quickly in the blood of other monkeys and mammals to be able to test their safety and efficacy in repeated administration. Hundreds of such therapies were awaiting licensing by the American authorities, including seventy therapies in the final stages of development (VandeBerg et al. 2005, 30). Some of these antibodies were tested on chimpanzees before being used in clinical trials in humans. Only data on antibodies proposed for clinical trials in humans, not on antibodies that had side effects or no effect at all on chimpanzees, had been made available to the Food and Drug Administration. However, these are not published data (VandeBerg et al. 2005, 30). In addition, pharmacokinetic tests would have to be carried out to determine the dosage of drugs, which would provide information on the periods of time during which a drug is absorbed, distributed in the body, metabolized, and excreted. VandeBerg et al. (2005) believe that "it has also recently been demonstrated that chimpanzees predict human pharmacokinetics … more accurately than other animal models, including rats, dogs and other non-human primates" (VandeBerg et al. 2005, 30). They continue: "All drugs are initially tested in non-primate animal models for toxicity and pharmacokinetic characteristics. However, sometimes the results from several species are inconsistent"—a disturbing fact that causes critics to question the dogma of animal testing, but leads the authors to the conclusion that chimpanzees must then be used as the animal model most similar to humans: "in such cases, chimpanzees are required for the final pre-clinical testing" (VandeBerg et al. 2005, 30). The number of newly developed drugs will increase in the future—and with it the need for pharmacokinetic tests. However, according to current regulations, these tests only took a few days, which means that a single chimpanzee can be used to test a variety of medications within a year. The completely decoded genome of the chimpanzee could be used to identify genetic and phenotypic specializations of human diseases in a genome comparison with humans and thus develop new therapies. For example, certain types of cancer are much more common in humans than in other animals, including other primates, and humans are unique in their susceptibility to neurodegenerative diseases such as Alzheimer's (VandeBerg et al. 2005). Critics may disagree with this logic and argue that, precisely because of the lack of comparability and because it is never possible to reproduce such diseases comprehensively in the animal model, many diseases can only be researched on human patients. However, VandeBerg et al. (2005) argue that chimpanzees are the only animal model that can be successfully infected with Hepatitis B and C. In terms of malaria, they can be infected with a parasite (*Plasmodium reichenowi*) that is very similar to the malaria parasite most dangerous for humans (*Plasmodium falciparum*). And even if the hope of creating a suitable AIDS model with chimpanzees has been dashed, since HIV-1-infected animals did not develop immune deficiency, chimpanzees are still, it is argued, important for AIDS research (VandeBerg et al. 2005, 31).

According to VandeBerg et al. (2005), our closest relatives are ultimately regarded as "biomedical resource," as a "top model," and as a bearer of hope in animal experimental research. The animal experiment dogma is adhered to without restrictions. We simply do not know which challenges for medicine might arise in the future. Therefore, chimpanzees for use in experiments should continue to be bred and kept on hand in labs (VandeBerg et al. 2005, 32).

On the other hand, there is criticism of animal experiments from the part of animal ethics and animal welfare, which is directed against very different aspects of experimental practice and which is applied in very different scopes. Animal testing in general, the right of humans to use animals for mostly exclusively human purposes, and the ethical justifiability of experiments that cause particularly severe suffering to lab animals are questioned.

Ferrari (2009) summarizes the discussion as follows: in the past two decades, many scientific studies have massively questioned the concept of indispensable animal testing, as they have shown, for example, that some animal experiments do not meet the requirements of proper scientific work, contradict the results of clinical studies, and are carried out unnecessarily. These criticisms specifically question the main argument of the advocates of animal experiments, namely the high scientific benefit of this research provided that the results can be transferred from the animal model to humans. On the basis of sixteen applications approved by the authorities in Bavaria and classified as necessary in their implementation, the German study by Lindl et al. (2005), to which Ferrari here refers, investigated, among other things, whether these studies have actually achieved their postulated experimental goal. This was assessed on the basis of the importance and scope of these studies for further scientific research. Therefore, their citation frequency, the course of citation, and the question in which research the primary citations were included (whether in further animal studies, in vitro studies, clinical studies, or review articles) were analyzed. Lindl and his colleagues speak of a "disappointing result" (Lindl et al. 2005, 143), since although ninety-seven clinically oriented publications could be determined, only in four of them was a direct connection established between the animal-experimental findings and the results found in humans. Thus, the value of these studies and the transferability of the results obtained from the animal model to humans are very significantly called into question.

According to Ferrari (2009), a further criticism of biomedical animal experiments is that many animal experiments are carried out simultaneously with or even after clinical studies. This is not in line with the idealistic notion that testing on animals is only the last (indispensable) experimental stone to be turned over.

Furthermore, critics argue that the costs of the experiments are often underestimated and experimenters also show a tendency to underestimate the suffering of the animals. Ferrari is right to say that this is particularly precarious because applicants are also regarded as "experts" in assessing the harm inflicted upon the animal models in their experiments.

The increasing existence of available and established alternative methods would, according to Ferrari, additionally give rise to criticism. Nevertheless, the fact that animal experiments are much more strongly supported politically and economically than alternative methods ensures that the talk of the indispensability of animal experiments is to be seen as a kind of "self-fulfilling prophecy" (Ferrari 2009, 287).

So, in addition to ethical questions, there is a serious need to put these practices under the scientific microscope and to review the scientific and social benefits of animal research. As early as 1959, Bill Russel and Rex Burch proposed the "3R principle" as an ethical guideline for animal experiments. The three Rs stand for *reduction*, *refinement*, and *replacement*. The 3R principle thus aims firstly to reduce the number of laboratory animals required by reducing the number of tests (for example, by doing away with superfluous regulations or by harmonizing regulations worldwide). Furthermore, by improving and further developing the experiments and test conditions,[19] the suffering of the animals is meant to be reduced and more information obtained from the individual animal experiment or the same information obtained with fewer test animals. Finally, the idea is to completely replace animal experiments by animal-free methods.

In accordance with the 3R principle, a distinction is made between reduction alternatives, refinement alternatives, and animal-free replacement methods (see also Russel and Burch 1992; BMBF 2001, 30). However, critics of animal testing do not consider it sufficient that the general reference to the 3R principle has found its way into animal testing legislation. Instead, the burden of proof of the uniqueness and fundamentality of research on animal models versus alternative methods should lie with the applicants. They should be obliged to provide an empirical study and a comparison of their approach with already recognized alternative methods. In addition, the commissions need more time and/or staff (including ethically trained staff) to carry out a careful review. This would then either lead to a slowdown in the approval process or to an increase in the number of authorities—but probably to both (Ferrari 2009, 287–88). So far, animal experiments have been considered the "gold standard" and, unlike alternative methods, are not subject to any validation process as such. In addition, the consistency of the alternative methods is measured by whether they lead to the same results as animal experiments, which appears questionable in diagnostic and therapeutic procedures. For many animal models it is uncertain whether they have indeed the predictive power for human diseases one hopes to create. Ferrari (2009) not only refers to the measurement with two different or inappropriate measures, but also identifies the epistemological problems of this basic assumption: if the predictive power of animal models is uncertain with respect to the effects on human health and the environment, but the method is still recognized as the golden standard, all nonanimal methods are by definition compared to an overvalued standard, meaning that alternative models might in fact be more valid in their predictability.

There are now a large number of promising alternative methods, ranging from mathematical-statistical methods and cell culture techniques to

19 After Young, refinement is often equated with environmental enrichment (Young 2003, 21; on enrichment, see Sections 14.2–14.8 below).

alternative materials (such as incubated chicken eggs and roe), ranging from the use of micro-organisms (such as bacteria, fungi, algae, or unicellular animals) to the use of invertebrates (such as the use of water fleas instead of fish to assess the toxicity of wastewater) (BMBF 2001, 30ff., 49–50). Much of the research on alternative methods is in the field of toxicology and will be applied there. At the moment, it still seems difficult to say how far their general applicability goes. A look at the programs of world congresses on alternative methods (such as that in 2005 in Berlin or 2009 in Rome) and at the publications of recent years in journals like ALTEX reveals, however, that there are or could be other fields of application (for example, in training and teaching) (see the ALTEX Archive, www.altex.ch). Ferrari (2008) provides an overview of alternative methods for transgenic animals, both in drug toxicology and in basic research, in the field of animal models for human diseases, in gene pharming, and in xenotransplantation (Ferrari 2008, 295–310). However, it is difficult to get such an overview of the research, application, and scope of alternative methods—even in the limited area of transgenic animals—because this ongoing and developing research is very diverse and differentiated. Alternative methods, in contrast to this, are subject to a lengthy validation process that slows down acceptance and applicability (Ferrari 2008, 296). In the case of basic research, Ferrari assumes, alternative methods are the most problematic because their goals are most widely formulated. In the research *objective* of basic research (which lies in the formulation of certain questions on biological mechanisms and the testing of hypotheses using animal experiments), the research *method,* that is, animal experiments, is already included and partly formulated. Nevertheless, the research questions could also be posed here in such a way that they could be verified by animal-free methods, for example, with the help of cell cultures and computer models (Ferrari 2008, 296). Instead of using transgenic animals or their tissues and cells, basic research could then use, for example, the production of monoclonal antibodies, stem cell research, and analytical methods with assistance of findings from robotics, bioinformatics, and microelectronics (Ferrari 2008, 296ff.).

Let's assume that there are no alternatives to some animal experiments. This would still mean that we had to make sure that the ones we conduct are indeed of real significance. This means we have to assess whether they are used to pursue such a highly important research goal that the suffering of the laboratory animals is to be accepted. This question will probably be clarified in complicated individual case studies (Borchers and Luy 2009b, 9).[20] Additionally,

20 A case-by-case analysis of animal experiments may nevertheless also address individual types of animal experiments and identify an entire group or individual subgroups of an

the current state of scientific knowledge and the availability of alternative methods would have to be clarified thoroughly and on the basis of the current research situation. Ferrari (2009) rightly stresses that the more research objectives can be achieved through alternative methods, the less important the justification for animal testing would become, simply because we don't need those experiments anymore. Providing satisfactory justifications seems particularly difficult. It may seem doubtful, furthermore, whether the competent authorities and animal testing commissions can always provide a satisfactory assessment of the justifiability.[21] As reasons, most applications for animal experiments cite the insufficient understanding of the correlations in the development of diseases or the lack of therapies in humans (Lindl et al. 2005, 143). This corresponds to widespread public opinion that animal experiments are justified, above all, and possibly exclusively, if they really serve the understanding of and fight against threatening diseases.

We will keep this idea in mind and come back now to the situation of great apes in captivity. The following section shall give some insights into the ethical challenges that we face when analyzing the use of these animals for human purposes.

9.1.4.1 Criticism of Experiments with Great Apes

Experiments with great apes are prohibited in the EU.[22] In contrast, in the past ten years, experiments with great apes, especially with chimpanzees, have continued to be carried out in the United States, mainly in the fields of hematology, immunology, pharmacology, virology, neurotoxicology, and in research into

experiment type as ethically and/or scientifically problematic. Ferrari has carried this out in exemplary fashion for animal experiments with transgenic animals, experiments that are particularly criticized because a large number of animals are needed to produce a transgenic test animal (Ferrari 2008, 2009).

21 On the difficulty of determining ethical justifiability, see Borchers and Luy (2009a); and, especially, Alzmann (2009), who also addresses the desideratum of ethical expertise (Alzmann 2009, 147).

22 On October 20, 2010, Directive 2010/63/EU of the European Parliament and of the Council of 22 September 2010 on the protection of animals used for scientific purposes was published in the Official Journal of the European Union and had to be transposed into national law by the member states. The determinations on the use of certain animals in procedures have been formulated in Chapter II, article 8, containing provisions for nonhuman primates used in experiments and also specifically for apes. The possibility of using these animals in the event of a new Ebola outbreak has been left open, for example, when it would only be possible to test vaccines necessary for human survival on apes alone. Beyond this, the use of great apes in animal experiments for the benefit of mankind is excluded (see Council Directive 2010/63/EU, 41 [art. 8, paragraph 3], and 50 [art. 55 protection clauses, paragraph 2]).

potential bioterrorist pathogens (anthrax, Ebola, and Lassa virus or plague) (Knight 2008b, 281ff.). The number of primates used in the United States was estimated at fifty-eight thousand per year, compared with eleven thousand in the EU (Knight 2007, 282). Knight assumed in his 2007 paper that 1,171 chimpanzees were still kept in US laboratories (Knight 2007, 282; with reference to VandeBerg et al. 2005).

However, more and more countries worldwide have banned or severely restricted invasive experiments on great apes, including many European countries, Japan, Australia, and New Zealand. According to Knight, the United States, in continuing to carry out such experiments, is almost completely isolated internationally:

> Although around half a dozen other countries also conducted chimpanzee experiments 15 years ago, by 2008 the US stood virtually alone. Every other country except perhaps Gabon—whose future plans were unclear—had ceased invasive chimpanzee experiments. Should the US and Gabon also terminate such research, it would effectively result in the first global moratorium on invasive research for any non-human species, unless conducted in the best interests of the individual or species.
>
> KNIGHT 2008b, 2

Finally, in 2007, the National Centre for Research Resources (NCRR) of the US National Institute of Health issued a moratorium on financing the further breeding of experimental chimpanzees. This moratorium lead to a drastic reduction in the number of chimpanzees used in experiments over the next years. It was clear back then already, that most of the animals will die or retire from research for reasons of age, anyway. In addition, the Great Ape Protection Act was introduced to Congress in 2008, proposing an end to invasive research and testing on the animals still living in US laboratories. For the approximately six hundred state-owned animals, a secure retirement home was needed, which is required by various laws in the United States (such as the Chimpanzee Health Improvement, Maintenance and Protection Act of 2000 or the Chimp Haven is Home Act of 2007) (Knight 2008b, 1, 3–4). Biomedical research on chimpanzees has been waning since 2013, when the NIH announced that it would indeed phase out most government-funded chimp research and retire the majority of its research chimps to sanctuaries. In June 2015, a new rule by the U.S. Fish and Wildlife Service (FWS) stated that all U.S. chimpanzees would be classified as endangered under the Endangered Species Act (ESA). Labs can apply for ESA permits for invasive experimentation but would only be allowed for work that enhances the survival of the species and benefits chimpanzees in the wild. Even if these developments had the expected "chilling impact" on biomedical research with great apes, the NIH also planned to maintain a colony of 50 chimps to meet future research needs (Grimm 2015).

As public concern has increased worldwide regarding experiments with great apes, in many countries, such as Great Britain, Sweden, Austria, Italy, Norway, Germany, Belgium, Japan, Australia, New Zealand, Liechtenstein, San Marino, Spain, where there are now de jure prohibitions or de facto solely non-invasive animal experiments with great apes. The EU as a community has also developed legislation on experiments on nonhuman primates (Knight 2008b, 2–3, 38). In addition, the responsibility for chimpanzees released from experiments is moving into the awareness of responsible governments and corporations worldwide. More and more facilities are being created in which the often sick animals used can spend their old age (Knight 2008b, 3–4).

Knight points to the importance of these developments as an unprecedented opportunity to end chimpanzee experiments worldwide. However, as has already been shown, scientists were (and probably still are) also committed to continuing to conduct them (VandeBerg et al. 2005; Jha 2006). They argue for the great importance of experiments for biomedical research. Knight followed this question in a similar study to Lindl et al. (2005), the results of which will be presented below.

Experiments on chimpanzees are usually justified, even more than other animal experiments, by the fact that they will lead to a better understanding of threatening human diseases as well as to the development of drugs and therapies against them (see, e.g., VandeBerg et al. 2005). Knight (2007) sums it up as follows: "Chimpanzees are postulated to predict human pharmacokinetics—the time course of absorption, distribution, metabolism, and excretion of pharmaceutical agents—more accurately than other species in the laboratory: rats, dogs, and even other NHPs (non-human primates), making them the test model of choice for human toxicity testing and drug development" (282). However, the alleged similarity and transferability, together with the extraordinarily high costs of keeping and using primates in captivity, also lead to ethical and financial concerns. Scientists such as Frans de Waal, Jane Goodall, Marc Bekoff, and others, according to Knight, would have pointed out the possibility that the genetic, cognitive and emotional similarity in particular lead to an extraordinary capacity for suffering in chimpanzees. Against this background, and in view of the rising competition for increasingly scarce research funds, an evaluation of the scientific significance of these experiments with chimpanzees becomes all the more urgent (Knight 2007, 282). Knight delivers such an investigation in his article. He searched three major databases of biomedical publications[23] for worldwide studies from 1994–2004 using chimpanzees. He focused on invasive studies with captive animals (see Knight 2007, 283–85, for his method and selection criteria). To determine the general significance of such studies for the progress of biomedical knowledge, a randomly selected subset of these studies was examined in more detail as to how often and in which contexts later publications referred to them. The central question was whether these studies actually made a substantial contribution to subsequent research. Above all, in order to be able to make a statement about their value in combating human diseases and developing therapies, for this subset of studies Knight also determined their citation frequency, specifically in subsequent publications on *prophylactic, diagnostic,* or *therapeutic methods in the fight against human diseases.* The cited studies were analyzed in more detail (Knight 2007, 283ff.).

The result of Knight's investigation was as sobering as that of Lindl et al. (2005) for the Bavarian animal experiments cited earlier in this section. Of the 749 invasive studies with captive chimpanzees he found, 48.5 percent were biological experiments (mostly cognitive, neuroanatomical, and neurological) and 41.5 percent were virological experiments (most frequently studies on hepatitis C and HIV). Therapeutic research took place in 3.5 percent of the studies and parasitological research in 3.1 percent, most frequently on the pathogens of malaria, river blindness, and schistosomiasis (Knight 2007, 285–86). Of the ninety-five randomly selected studies in the subset of Knight's study, half (49.5 percent) were not cited in any subsequent publication. Thirty-four of the ninety-five studies (35.8 percent) were mentioned in 116 subsequent publications, but they did not deal with prophylactic, diagnostic, or therapeutic questions concerning the treatment of human diseases. In just twenty-seven later publications that dealt with diagnostic, prophylactic, or therapeutic methods of combating cancer, rheumatoid arthritis, or the Epstein-Barr virus, for example, some of the chimpanzee studies were cited. However, only fourteen of the ninety-five chimpanzee studies were cited in these publications, which is only 14.7 percent (Knight 2007, 290). Knight concludes that invasive studies with captive chimpanzees have contributed insights into a number of biomedical issues. But not every bit of knowledge gained is valuable knowledge that outweighs ethical and

23 These were CAB Abstracts, the most comprehensive bibliographic database of biosciences and life sciences, with over 4.5 million entries; EMBASE, the Excerpta Medica Database, which contains more than ten million entries from the fields of biomedicine and pharmacology; and Medline, the leading database in the medical and health sector, with over twelve million entries (Knight 2007, 283).

financial concerns. It is a recognized principle of scientific progress that the expected benefit of an experiment must exceed its costs (in terms of finances, but also in terms of harm done to the animals experimented on). For this reason, ethics committees throughout the world review the scientific value and ethical acceptability of animal experiments before approving them. It would therefore have been expected that the approved experiments would have been of outstanding value and benefit (Knight 2007, 290).

Of course, as Knight himself says, the frequency of citations is not a definitive indicator of the value of a scientific study. Studies not quoted later could also contribute to scientific progress not by being published but by having their results orally passed on among colleagues, by being used without citation or only by being quoted with a time delay (see Knight 2007, 290–91). Knight thoroughly considers a number of other limitations on the significance of his scale, but concludes that the frequency of citation is a quantifiable and almost objective approximation of the value and benefit of a study: "Research that makes a significant contribution to a field—such as by confirming or refuting hypotheses—is very likely to be cited by future papers, as is research that produces interesting or controversial outcomes. On the other hand, research that is inconclusive or of little interest or significance is much less likely to be cited." He continues: "The disappointing citation rate of these chimpanzee studies is therefore cause of considerable concern" (Knight 2007, 293–94). Although it is possible that unquoted studies nevertheless made a contribution to scientific progress in one way or another, it is highly unlikely that a large number of studies do so in a significant way even though *none* of them are mentioned in any subsequent publication. The ethics committees that approved these studies and expected the applicants to adequately assess the (ethical and financial) costs and benefits of the studies fail here, according to Knight. This would not fundamentally undermine the system of ethics commissions, but clearly points to the existence of weaknesses in its implementation: "This flaw has resulted from over-reliance on the assumption, in the absence of supportive evidence, that animal experiments are likely to be of substantial use in advancing biomedical progress" (Knight 2007, 294, 302).

According to Knight's central conclusion, only 14.7 percent of the studies were cited, in twenty-seven subsequent publications. If the invasive chimpanzee experiments investigated had been important for further research, or if they had inspired further research, it follows that a larger percentage of them would have been cited in a larger number of publications. The few studies found that actually develop prophylactic, diagnostic, or therapeutic methods of combating human diseases and also refer to chimpanzee experiments for this purpose deserve attention, but they must also be examined in more detail as to their importance and the actual contribution of chimpanzee experiments (Knight 2007, 295), which is what Knight subsequently did.

Knight's more detailed analysis of these *cited* publications shows that the chimpanzee experiments themselves are usually only of peripheral importance in the publications that mention them: of the twenty-seven citing publications, 63 percent (seventeen of twenty-seven) alone are general review articles that list an average of 104 references. According to Knight, the chimpanzee experiments cited are not of central importance here and only make a small contribution. In twelve of the citing publications, the chimpanzee experiments are redundant because humans or human serum were examined simultaneously or because they only served to confirm knowledge previously gained with humans. In seven cases, the method tested in the chimpanzee experiments and later cited was not further developed, for example, because later clinical tests in humans, in contrast to the results from the chimpanzee model, raised concerns with regard to safety and effectiveness. In five cases the chimpanzee study had investigated a disease or method that was only of peripheral importance for the medical method of the publication citing it, and in three cases the chimpanzee study illustrated only a historical finding or was only used in a medical historical discussion. In another three cases, the cited result of the chimpanzee experiment contradicted other results on nonhuman primates, and in two cases there was such contradiction to data obtained on humans. In two studies, only the likewise mentioned results of studies on human patients were discussed further, not the results from the chimpanzee studies. Finally, in only one case, the chimpanzee study may have helped to highlight the desideratum of a new diagnostic method, but then no longer contributed to its development (Knight 2007, 296).

Knight continued his in-depth investigation for the five chimpanzee experiments that are quoted *by more than one* publication. But here, too, it was demonstrated that the studies were either redundant, because more meaningful results were already available from studies with humans, or that the studies were only of peripheral importance for the publications citing them. It was also the case that their results contradicted other studies on nonhuman primates or studies on humans, that they were only cited to show differences between animal models and humans, and that they did not contribute to the actual further development of therapeutic, diagnostic, and prophylactic methods. It was also noticeable that some of the chimpanzee research had been misleading and that the research it stimulated had been stopped again later (see Knight 2007, 296ff.). Several of the authors of the citing medical publications also raised the issue of the difficulty of transferring these studies to humans. Knight quotes Matthews et al. (2003): "Results to date illustrate the profound difficulties in translating animal model success to the clinical arena." Koff (2002) is also cited describing the role of the chimpanzee experiment for the study of hepatitis A and B in his publication as "potentially complicating our understanding of escape viruses" (Knight 2007, 298).

Overall, it could be observed that it was not the quoted chimpanzee experiments but rather quite different studies that had made a central contribution to the research of the publications citing the chimpanzee studies: "Detailed individual examination of these papers revealed that in vitro studies; human, clinical and epidemiological studies; molecular assays and methods; and genomic studies contributed most to the development of the medical methods described....These results suggest a lack of utility of chimpanzees as an experimental model for combating human diseases" (Knight 2007, 299).

Despite a close genetic similarity, which varies by only 4–5 percent, chimpanzees are not a suitable animal model for human diseases: "The substantial differences in protein expression ... and the further distortions of normal physiology that result from the stressful laboratory environment and procedures common to chimpanzee experimentation are the most likely causes of the observed lack of utility of chimpanzee research" (Knight 2007, 300). These animals develop neurological, endocrinological, and immunological abnormalities resulting from the stress of captivity and use in the experiment. Small, barren, and sometimes overcrowded cages along with immobility and social isolation contribute to this. The stressful stimuli the apes are exposed to include aggression caused by anxious or frustrated conspecifics, trauma, chronic boredom, and a variety of stress-inducing experimental treatments (Knight 2007, 299–300).[24] Knight's criticism that the results of the animal experiment are influenced by physical and psychological abnormalities in the animal, which in turn result from housing conditions and use in the experiment, should be further clarified. Even if one doubts that the consequently compromised informative value of the results is directly related to the weak transferability of the results to humans, but these are nevertheless serious animal welfare issues. According to other authors, too, the significance of animal test results due to such impairments can be called into question. Poole formulated the credo "Happy animals make good science" (see Poole 1997), and Würbel's

24 In fact, it is assumed that inappropriate housing conditions are generally both directly ethically problematic because they affect the welfare of the animals and indirectly relevant: poor husbandry conditions also have an effect on the validity of studies (Poole 1997; Garner 2005; Würbel 2001). The way in which laboratory mice are kept, particularly, has been shown to have effects on the brain and behavior. Better results in terms of refinement can be achieved if enrichment measures are used (see also Würbel and Garner 2007; Würbel 2001).

(2001) investigation about housing effects on rodent brain and behavior came to the conclusion that the standard housing conditions for rodents in the laboratory "could compromise the utility of rodents for research, especially in behavioral neuroscience" (207). At the very end of the discussion section of this book, I will go into such questions of well-being and animal protection in captive animals in greater detail (Sections 14.2–14.8).

Knight concludes that his analysis clearly shows that chimpanzee experiments do not make an important contribution to the development of effective methods for combating human diseases. Some exceptions are conceivable and could not be excluded by his study, which chose a random subset for further investigation, but colleagues who also examined nonrandomly selected studies nevertheless came to similar results (Knight 2007, 300–01).[25] It is highly likely that other species used in animal experiments in biomedicine and toxicology are even more unsuitable as models for humans than chimpanzees. Systematic investigations of the value of other laboratory animal models are still urgently required (Knight 2007, 302). By 2007, there had already been twenty systematic studies on the value of animal experimental research for clinical implementation in humans. The authors of these studies arrived at a similar conclusion as Knight himself, namely, that the animal models had made significant contributions in only two cases—and one of them was controversial. These experiments included both studies that ethics commissions had hoped would make an outstanding contribution to scientific knowledge and articles that were published in high-ranking journals. Knight sums the matter up as follows: "Clearly, the assumption that animal experiments are likely to be of substantial use in advancing biomedical progress is not supported by most existing evidence" (Knight 2007, 302). His systematic study has shown that this is no less the case when it comes to chimpanzee experiments. Until there is credible evidence that invasive experiments with chimpanzees in captivity can actually make a substantial contribution, and thus offset their enormous costs with an extraordinary benefit, it seems appropriate to agree with the opinion of many experts and to demand an end to such experiments in those countries that have not yet banned them (Knight 2007, 302, 303).

When considered with respect to the claimed and often uncritically assumed significant importance of animal experiments in general and chimpanzee experiments in particular, the results of the studies by Knight (2007), Lindl et al. (2005), and others are sobering, to say the least, and perhaps even

25 Knight refers here to Hackam and Redelmeier, who came to the conclusion that of the seventy-six studies they examined, only 10.5 percent yielded relevant results which ultimately actually benefited human patients (Knight 2007, 300–01).

ethically alarming. Viewed in light of the often high costs and the profound problems, these studies would not only have to reach a medium level in their importance for scientific knowledge, but even an outstanding one. This seems far from certain. Ethics commissions should not be uncritical of the gold standard, both in assessing the "outstanding scientific importance" demanded by animal protection laws and in examining the ethical justifiability bound to it. The value of the requested experiments will always be assessed prospectively and thus cannot be determined exactly as a mere assumption. However, both the ethical requirement that the desired results must be important for "essential human or animal needs" and the determination of this significance as "outstanding" also allows for the ethically justified rejection of animal experiments.

Invasive experiments on great apes have not yet been banned worldwide and clearly constitute an ethically problematic context in which humans use these animals. A second, firmly established but ethically problematic area of use of these animals remains their captivity in zoos and the entertainment industry. More such areas surely exist in addition. Both in the wild and in captivity, humans' interaction with our nearest relative is characterized by our fascination regarding the similarity between us and them and our simultaneous inability to do justice to this similarity in ethical terms. This problem is not limited to the interaction of human beings with their closest relative, the great apes. In the following two sections, this subject is examined regarding a second group of cognitively related animals, namely, dolphins.

9.2 Case Study: How We Treat Dolphins

In the field of ethics, there are increasing demands for animal rights, even personhood rights for animals. These have so far been formulated primarily for two groups of animals, namely, great apes and dolphins. For these two species, the contrast between ethical demands and the animals' actual situations shall be demonstrated. The example of the great apes served as a first case study, in which humans' problematic treatment of their cognitive relatives in the wild and in captivity became apparent. The situation of the animals in captivity was only briefly discussed, in the case of laboratory chimpanzees. Let us now turn to our second case study: the situation of dolphins. Here, we will proceed in an opposite manner. I will concentrate on the situation of these animals in human care and only briefly introduce the situation in the wild. This is a choice due to space restrictions and does not mean that dolphins and other whales in the wild are not threatened by many human-made factors. Such factors include, for example, the direct hunting of these animals or the indirect effects of the fishing industry from drift nets to by-catch problems. The increasing environmental contamination of the world's oceans with pollutants and waste or the increasing noise caused by military activities and shipping are also decimating stocks and impairing the welfare of animals in the wild. Nevertheless, all these factors will not be addressed here, but the case of the dolphins will now serve to illustrate how humans deal with their cognitive relatives in captivity.

Like great apes, dolphins are among those wild animals that are mainly kept in zoological gardens or aquariums. Therefore, one should take an initial look at the ethical criticisms of zoos before dealing specifically with the keeping of dolphins and especially their use in dolphin-assisted therapy or swimming-with-dolphins programs. The fact that the zoo is not only about the animal and also not only about the protection of species constitutes the central tension surrounding the topic to the present day. In the following, it should become clear that the problem of zoo animal husbandry is not only due to the zeitgeist or the epoch-bound progress of zoo animal husbandry—and thus also is not to be solved on all fronts by modernizing the zoos. Rather, the concept of the zoo is replete with ethical problems, which have become more acutely apparent in developments in modernization over the last few years. Of course, the problems described here also concern dolphins, but not exclusively so. Most of the following criticisms of the zoo concept concern great apes as well, or simply all animals that are kept in captivity, especially when it comes to sociological arguments based on a misguided relationship between the visitor and the zoo animal. But many of the criticisms that refer to the physical and psychological well-being of zoo animals are also relevant for very different species in captivity.

9.2.1 *Zoos and Aquariums: Noah's Ark or Disney World?*

The zoo as an institution has undergone a long historical development from a wandering menagerie of animals to a modern conservation breeding park, which cannot be discussed in detail here (see Dittrich 2007, 13ff.; Young 2003, 3ff.). As far as the design of cages and enclosures is concerned, general architectural trends have been followed for centuries (and perhaps still are). According to Precht, the animals are sorted into their cages as animated exhibits, comparable to the possession of a stamp collector in his collector books. The result was the creation of "psychotopes,[26] which reveal more about their

26 The term "psychotope" is used in zoology, according to Precht, to describe systems that are intended to enable the visitor to gain an insight into the natural habitat of the captured animal (Precht 2000, 328).

creators than about the animals shown" (Precht 2000, 329; transl. MK). Young writes about zoo development in Great Britain:

> It was not until the 1960s with the growing interest in animal welfare (spurred by Ruth Harrison's (1964) book *Animal Machines...*) and in recognition of the need for conserving species from extinction by captive breeding, that many zoos developed more animal-welfare-friendly enclosures. This is despite the fact that some zoos had for many years recognised the potential for animal suffering.
>
> YOUNG 2003, 4

In the archive of the annual reports of the Edinburgh Zoo, for example, it should be noted that the Zoological Society had already been informed in 1911 that polar bears needed toys and objects in order to be able to live out specific needs, mainly play behavior and problem-solving.

In fact, a scientific approach to zoological gardens did not emerge until after World War II. The Swiss zoo director Heini Hediger presented the scientific basis of wild animal husbandry in the zoo (see, e.g., Hediger 1942, 1965). To this day, zoological gardens justify their existence by referring to different values: through breeding and through financial support of conservation projects they ensure the survival of endangered species. In addition, they go beyond entertainment and pursue an educational mission by providing knowledge about animals—especially about exotic and endangered species. Furthermore, many provide scientists the opportunity to investigate the behavior of wild animals in captivity. The four main tasks of modern, scientifically managed zoological gardens (in their self-image) are: entertainment, education, conservation of endangered species, and research (Hutchins et al. 2003). Since the establishment of zoo biology, the zoos themselves and scientists at research institutes have published more and more scientific papers on housing problems. Together with enormous advances in veterinary medicine, such as new diagnostic methods, successful therapies, and more advanced hygiene measures, this has certainly improved the welfare of zoo animals. Just in time for CITES coming into effect, says Dittrich, due to regular successful breeding, zoos have not had to keep coming into effect in animals from the wild (see Dittrich 2007, 19). Enclosure fences have made way for optically distinctive trench barriers that reduce the subjectively perceived distance from the animal, and zoos try to simulate the natural habitats of the animals in cages and enclosures (Dittrich 2007, 20).

Nevertheless, since the 1980s there has been clear criticism of those zoos that have not (sufficiently) followed modern developments. This criticism applies particularly to certain species such as dolphins, polar bears and other bears, elephants, and great apes and other primates. Behavioral disorders such as stereotypes are especially obvious here. They arise from frustration of elementary needs and stress. If they persist, they become an expression of neurological impairment (see Chapter 14.2–14.6 below; Würbel 2010 [quoted in Schramm et al. 2007, 38]; and the contributions in Mason and Rushen 2006). In the 1980s, such behavioral disorders in zoo animals were a signal to the already sensitized zoo visitors that the animals did not feel comfortable under these circumstances (Dittrich 2007, 20). In the course of these discussions and with a view to the development of zoos, it has become clear, according to Dittrich, that modern facilities must develop on two levels: given the current state of science, they must offer their animals the best housing conditions possible and create species-appropriate housing that can be verified as such scientifically. At the same time, however, they also have to strive to design housing conditions that match the natural habitat of the animals. By that guideline, the institutional zoo conveys an educational program and satisfies the idea of visiting a near-natural environment (Dittrich 2007, 29). While it is generally difficult to reproduce the ecological conditions of the animals' natural habitats, the facilities today—and with a resolute view to visitors' sensitivities—endeavor to "look natural."

Since World War II, the number of visitors to zoos has increased enormously (Dittrich 2007, 21). Today, more than ever before, the media is interested in zoos; they have a strong public image and attract millions of visitors every year. Despite high visitor numbers, which other public institutions such as museums, operas, and theatres can only dream of, many zoos are dependent on subsidies from countries and municipalities. Private institutions, in particular, are constantly increasing admission prices. Zoos worldwide are increasingly dependent on private initiatives and private investors.

The role of zoos in the future will not only be rooted in their economic efficiency, but above all in the balancing act between their educational, research, and conservation missions, on the one hand, and their recreational or entertainment functions, on the other. In the following, the individual arguments for zoo animal husbandry and the tension between these arguments are critically examined. The conservation argument and the entertainment function are discussed in detail, the research and educational argument in combination with the first ones. The fundamental question of whether there can even be species-appropriate zoo animal husbandry, especially with regard to cognitively and emotionally highly complex species such as apes and dolphins, is pursued in the last part of this book, with a view to stereotypes and the ethical necessity of enrichment (see Sections 14.2–14.8).

9.2.1.1 The Zoo as Noah's Ark: Arguments for Conservation, Research, and Education

The cooperation of zoos in successful breeding is institutionalized and internationalized. The international non-profit organization Species360 (formerly International Species Information System, ISIS) has been collecting zoo animals worldwide since 1974: breeding books were created, and breeding programs that manage a worldwide exchange of zoo animals were developed. Since the 1960s, zoo-bred animals have also been released back into protected, renaturalized-but-abandoned former distribution areas. Dittrich speaks of approximately one hundred such reintroduction projects in 2007 (Dittrich 2007, 21). For many visitors, the zoo may seem like a "Noah's Ark" according to its species conservation task. Zoos not only provide a habitat for many extinct or endangered animals in the wild, but also the *only* survival space remaining.

Critics, however, criticize the breeding programs as ecologically questionable contributions to nature conservation and, according to Precht, they might only be the zoos' attempts to justify their morally problematic character (Precht 2000, 319). Precht summarizes various arguments in support of this negative attitude. The fundamental criticism of humans' right to keep animals in captivity at all is initially less an argument in itself than a conclusion from other arguments and preliminary considerations. This is based, for example, on the accusation that zoos would tear animals out of their original habitat, transport them over great distances, and keep them in completely unfamiliar surroundings. Wild catches for zoos are still happening and are also criticized (see Section 9.2.3). However, the Convention on International Trade in Endangered Species of Wild Fauna and Flora and other legal guidelines no longer make them a private matter for zoos, and they are now strictly regulated. Only the refreshing of genetic material and the avoidance of inbreeding in captivity remain as reasons for this. All in all, Precht says, people have become very careful when it comes to taking wild animals for zoos. While zoo directors would have gladly and optimistically captured some of the last wild Sumatran rhinos in the 1980s and 1990s[27] to develop breeding populations in English and American zoos, today no one would take an Amur leopard or a Siberian tiger out of the wild. Instead of hoping for breeding projects that have not yet been successful, the entire species conservation budget is increasingly flowing into local protection (Precht 2000, 320–21). Despite often adverse circumstances resulting from poaching or other factors such as introduced predators, attempts

27 Very few wild catches survived in zoos. The birth of a Sumatran rhino in 2001 in the Cincinnati Zoo was the first birth in captivity in 112 years (Crew 2011, 9).

are being made to establish and secure protected reserves so that remaining or resettled populations can recover and multiply.

With many zoo animals, however, the offspring and/or the release to the wild after a life in the zoo are not successful. Precht thinks this even applies to most zoo animals, and there are many reasons for this. Some of them can never be released again because their habitat has been completely destroyed, because they have lost at least one species-specific behavior in the zoo (for example, their hunting behavior), or they have lost their shyness toward humans. In some cases their genetic material has changed due to earlier thoughtless inbreeding, and they have fertility problems or an increased susceptibility to disease due to genetic impoverishment (Precht 2000, 333). In addition, successful offspring are often not bred in zoos with publicly exhibited stocks but through local, national, or international species protection and environmental initiatives. And finally, the preservation of intact habitats into which the animals can be released or in which they can survive in the wild from the outset is of central importance. The survival of the great apes in the wild certainly does not depend on breeding in zoos and the subsequent release of these animals to the wild, but rather upon the conservation of the last animals and their habitats on Borneo, Sumatra, and the African continent. At this point, it must be seriously questioned to what extent it makes sense to breed animals whose reintroduction to the wild does not seem possible, since there are no longer any suitable natural areas (Precht 2000, 333).

Even if one recognizes the species protection mission in principle, alternatives to the "zoo as Noah's Arc" still have yet to be examined from an ethical perspective. Alternatives which, inter alia, do not require captivity and display of animals. The Kakapo Recovery Program, which successfully brought the owl parrot back to the southern island of New Zealand, may serve as an example of a reintroduction initiative that does not breed in zoos today, but is isolated from the public.

Zoo advocates emphasize that conservation successes, such as the settlement of the Przewalski horse in the Mongolian steppe, in which zoos were significantly involved, must also be acknowledged.[28] To be able to assess how far the conservation argument applies, however, it must become much clearer how crucial the contribution of zoos is in such programs. According

---

28 See the homepages of the International Takhi-Group (https://www.takhi.org/en/, 2019/05/17), Kakapo Recovery Program (https://www.doc.govt.nz/kakapo-recovery, 2019/05/17), and the great apes section of the IUCN Primate Specialist Group (http://www.primate-sg.org/section_great_apes/, 2019/05/17).

to the Association of Zoological Gardens (Verband der Zoologischen Gärten, VDZ), about fifty species that went extinct in the wild survived in zoos in captivity (https://www.vdz-zoos.org/de/themen/natur-und-artenschutz/, 2019/05/17).

It remains to be seen, however, whether such numbers range significantly below the expectations of critical scholars, including ethicists. Also, they might range significantly below the expectations of the public, which is fueled by a very positive outlook on the zoo's conservation impact provided by the public relations work of the zoos themselves and their advocates. After all, 27,000 species of animals, plants and fungi are threatened with extinction in 2019, meaning that they are listed as critically endangered, endangered, or vulnerable. Their number amounts to more than 27 percent of all assessed species on the list. Among the threatened species there are 8,490 mammals, bird, amphibian, reptile, and fish species. In addition, there are over 5,099 invertebrate species (IUCN 2019). Note, however, that extinction risk has been evaluated for only a tiny percentage of the about eight million plant and animal species on earth. The current extinction rate is now, due to the influence of humans, a hundred to a thousand times higher than the assumed natural rate; pessimistic estimates postulate that up to three species die out per hour (Loreau et al. 2006, BMU 2007, 16). In view of these figures, the few species rescued so far in the zoo seem like a drop in the ocean, and even the regional and international breeding programs being conducted for around one thousand species will not be able to halt the extinction of species. Instead of tackling the core problem of eliminating the threat factors in the wild, a practice that poses problems from the perspective of animal ethics is sustained.

One might argue that the conservation argument only justifies the keeping and breeding of *endangered* species in the zoo, whose protection in the wild is not possible. This, however, only applies to very few species in zoos and often not to those that attract the public: according to Precht, the decisive factor for inclusion in a studbook of the European Conservation Breeding Programme (EEP) is not only dependent on the endangerment of an animal species in its natural habitat, but also on the sensational value, or attractiveness, of the animal (Precht 2000, 335). For instance, Colin Tudge, an English science journalist and advocate of conservation breeding programs, advocates preferring an attractive bird of prey to a beetle when in doubt (Tudge 1993, 142). If zoo visitors only want to see very specific animals, and if they are not interested in species conservation, then this clearly confirms Tudge's cynicism. It must probably be acknowledged that the zoo visitor has the strongest interest in "beautiful," "impressive," "cute," or "entertaining" animals. The level of endangerment for the species plays a subordinate role when it comes to the most popular zoo animals. In 2009, a German radio station (Rundfunk Berlin-Brandenburg) asked visitors to the Berlin Zoo and Tierpark Berlin about their favorite zoo animals. The rankings from one to five were Knut the polar bear, Knautschke the hippo, Geordi the red panda, Adolpho the anteater, and Satu the orang-utan. All these animals (two of them were hand-raised) were featured in the marketing by the zoos and through the media, especially in the TV show "Panda, Gorilla and Co," filmed by the RBB in Berlin.[29] Among the remaining twenty-five animals of the "most popular" list, there are also some more hand-reared animals, and about half of the selected animals had already been presented to zoo visitors as young animals or at the time of the vote were only a few weeks old. Almost all were mammals. In the case of chimpanzees, gorillas, lions, tigers, and polar bears, Würbel refers to them as the "blockbuster animals" in the zoo (Schramm et al. 2007). In fact, not all zoo animals for which visitors voted, such as the suricates, sun bears, or giraffes, are listed as endangered in the wild, or even threatened with extinction.

It is crucial that only a few zoos pointedly do without certain attractive animals for ethical reasons.[30] The fear that visitors will then stay away is probably unfounded. Despite clearly discernible hierarchies of popularity, so Precht, it remains a "completely unconfirmed prejudice" that zoo visitors will in fact only go to the zoo in the future to see lions and hippos (not endangered in the wild), for example. If these animals simply were not there anymore, Precht says, one would certainly get used to observing the endangered rhino instead. Zoo visitors who would normally not be able to distinguish a leopard from a jaguar probably would not be disappointed if one or the other disappeared from the zoo (Precht 2000, 344). In the following section, we will delve further into the idea of the zoo as a form of entertainment (and part of that industry), as well as the associated orientation toward the visitor and the commercialization and marketing of the animals. For the moment, with regard to the zoo's conservation and education function, it should be noted that these can only be maintained to a limited extent as arguments in favor of zoo animal husbandry.

29 See https://www.rbb-online.de/panda-rbb/, 2019/05/17. The idea behind the genre of zoo docu-soap in German television was the MDR program "Elefant, Tiger and Co," which was filmed at the Leipzig Zoo and has been broadcast since 2003. As it gained unexpectedly high ratings, a number of other television stations soon produced similar programs.

30 The Frankfurt Zoo, for example, has abandoned elephant husbandry since the early 1980s, and the Cologne Zoo has switched to an alternative form of elephant keeping in big groups with largely reduced contact between animals and caretakers (See https://www.zoo-frankfurt.de/faq/,2019/05/17; https://www.koelnerzoo.de/en/tiere-2#elefantenpark, 2019/05/17; Precht 2000, 344).

On the other hand, both arguments are strongly emphasized in the WAZA's future orientation:

> The vision for WAZA and the global community of zoological facilities is...clear. WAZA calls on all zoos and aquariums to adopt an integrated One Plan Approach to conservation. Animal welfare and conservation must be the primary purpose, and zoological institutions should be guided by this in everything they do... . With vast living populations of animals, devoted and talented staff, and extraordinarily large, diverse and engaged audiences zoos and aquariums have the power to facilitate the vital work of caring for and conserving living wild animals and ecosystems.
>
> WAZA a, 19

Precht states that there is an urgent need to consistently convert the stock in zoos according to the criterion of endangerment. This would be the only way to create space for species-appropriate accommodation for endangered animals. What species-appropriate accommodation means in each case cannot be determined by the zoo alone. Ideas of what promotes animal welfare are not arbitrary. They are, furthermore, not determined by the interests of the zoo but follow biological knowledge (see Section 8.3.1 and Würbel 2007). In general, it is more reasonable to argue in favor of social housing for social animals than saying that social isolation is appropriate here. The argument that for some species, such as predators, species-appropriate husbandry in zoos is impossible can also be justified biologically: animals such as polar bears roam an enormous natural habitat that even the largest enclosure in a zoo cannot simulate. It seems impossible to offer them substitutes for the stimuli they are exposed to on their long hikes and on their extensive search for food (Würbel 2007; quoted in Schramm et al. 2007, 38). The decisive factor is whether the animal's behavior and well-being are significantly impaired, e.g., by intrusive shouting from visitors, by lack of retreat possibilities, by the lack of appropriate enclosure equipment, or by the misjudgment (often an underestimation) of physical and psychological needs (Precht 2000, 322). The quality of husbandry conditions in zoos should therefore continue to play a central role in the criticism of zoos, measured in terms of animal welfare.

Ultimately, both the validity of the conservation argument and the validity of the education argument are closely linked to the quality of housing conditions in a zoo. As far as the education argument is concerned, if zoo animals do not show natural behavior, then no knowledge about the natural behavior of animals is imparted to visitors. Similarly, it must be said that if a zoo does not offer any information at all about the animals (for example, with appropriate signs, keeper talks, or guided tours), then in such a case, too, no pedagogically meaningful education takes place. If endangered species are not kept, bred, and released into the wild, then the zoo is not a conservation breeding park for endangered species, either. And if threats in the wild are not clarified, then no knowledge of conservation is imparted. The conservation strategy of zoos, which relies on flagship species as ambassadors of conservation and which relies on the aesthetic appeal of some species, bears a normative burden (Precht 2000, 341).

Precht concludes that for 150 years, zoos have been educating their audiences to enjoy themselves in the zoo instead of educating themselves during the visit. There is a "cuddly toy lobby," the silent majority of visitors which regularly and uncritically visit the zoo, completely independent of how species-appropriately the animals are accommodated there. A re-education of the public to care about and respect animals must, by all means, be undertaken; otherwise, the idea of ethically motivated conservation breeding seems absurd (Precht 2000, 347–48). It is the task of zoos to do this by restricting the selection of animals themselves (up to and including the renunciation of problematic species), by providing species-appropriate and welfare-oriented housing and by providing a correspondingly high-quality range of information on the life and endangerment of their animals. Where this cannot be done, for whatever reason, zoo animal husbandry must be abandoned for ethical reasons, which means that breeding should no longer be carried out and the keeping in the zoo should therefore be discontinued (see, similarly, e.g., Würbel 2007; quoted in Schramm et al. 2007, 38).

9.2.1.2 The Zoo as an Amusement Park: on the Disneyization and Commercialization of Zoos

Even if one recognizes in a limited way the arguments from conservation, research, and education for an equally limited number of modern zoos, and even if one assumes that at least some facilities reach a justifiable welfare standard in their husbandry conditions, a fundamental problem still remains. This problem is to be assigned to the entertainment claim of zoos. Every zoo, no matter how hard it tries, remains subject to the basic conceptual problem that, in addition to the arguments of species conservation, research, and education, its task as an entertainment institution is also inscribed in its idea. If the task of zoos was exclusively to protect endangered animals, then they should not display them publicly and should be financed as noncommercial institutions. So what role do zoos play as entertainment facilities?

Alan Beardsworth and Alan Bryman pursue this question from a sociological and tourist studies perspective, which has so far been perceived less prominently than, for example, the classical ethical criticism on welfare-oriented husbandry or the conservation argument (Beardsworth and Bryman 2001). They regard the zoo as a key place for animal presentation and mass tourism, where the visitor's gaze as a zoologically interested lay person and the visitor's gaze as a tourist converge. They are of the opinion that the zoo in late modernity as this very institution is undergoing significant structural and ideological changes with regard to its legitimizing narratives. They characterize these developments as a *tendency towards Disneyization* (Beardsworth and Bryman 2001, 83). Bryman (1999) coined the term "Disneyization" with reference to a well-known thesis put forward by Ritzer (1993, 1998), who claimed that our present society is passing through a process of *McDonaldization* (Scorer 1993, 1998). Both processes are based on similar mechanisms and both would overlap in certain institutional spheres.

Disneyization is an relatively new term that is used to describe the influence of the principles of Disney theme parks on a number of other organizations and institutions.[31]

The principles mentioned are not limited to Disney theme parks and existed even before the opening of Disney World in 1955. Nevertheless, they are used in an exemplary way by Disney institutions and embodied by them. These are the following principles: the sorting of the display and the construction of the institution's own self-portrayal based on themes (*theming*), the arrangement and mixing of different consumption areas (*dedifferentiation of consumption*), marketing via comprehensive advertising and sales strategies (*merchandising*), and the targeted use of positive emotions by service personnel (*emotional labor*) (Beardsworth and Bryman 2001, 91).

Alignment with themes (*theming*) refers to the main motif or leitmotif underlying an institution, which is inherent in the individual exhibitions and attractions. Such a "main narrative" is being increasingly expressed in zoos, which are concerned with the conventional presentation of their animals in categories, such as their exhibition in the "Monkey House," "Lion House," or "Elephant House," and distribute individual areas of their exhibitions according to more abstract themes. Examples here include jungle worlds, African steppes, tropical reef aquariums, nightlife, and biodiversity houses, where whole living worlds or ecosystems are constructed through effective imitation.

31 In general, Disneyization means the process by which the principles of Disney theme parks dominate more and more areas of American society and other societies in the world (Beardsworth and Bryman 2001, 90–91).

The theme principle extends to the theme restaurant with theme dishes or themed means of transport in the park. For example, the jungle lodge or the South America hacienda in Leipzig Zoo with a corresponding menu will be a suitable location for (themed) celebrations.[32] As a Disney World visitor you will be driven through the African steppe in the middle of Florida with a Kilimanjaro Safari Jeep (Beardsworth and Bryman 2001, 91–92).

It is controversial whether the inhabitants of such themed exhibitions really benefit from their changed environment. The respective topic is first and foremost told to the *visitor* by means of appropriately designed paths and signs, brochures, and resting places. Nevertheless, the "aesthetic of naturalness" has a great attraction for contemporary zoo designers and architects. Beardsworth and Bryman close by saying:

> It is plausible to argue therefore, that this form of theming, that is the production of quasifications of "natural" environments, may be primarily for the benefit of humans. It is they who can relish and enjoy the carefully crafted artificiality of such settings. For the inmate animals, such artifices, which Desmond (1999, 164) has carefully termed "faked organic realisms," are in themselves beyond comprehension.
>
> BEARDSWORTH and BRYMAN 2001, 92

In addition to the thematic orientation within the exhibition, zoos are also exploring new themes at company level in the course of the redesign of their institutional identity. They are reinventing themselves as agencies dedicated to education about exotic animals and the conservation of rare, endangered or extinct species. This has already been presented and has even become visible on the global level of the umbrella organization: WAZA sets the zoo of the future the goal of *primarily* caring for conservation (WAZA a). This form of redesigning one's own identity ranges from zoos' websites to new naming conventions. It is a main component of the self-description of many zoos and can also be observed at Disney Animal Kingdom, the one institution that is in fact not only a disneyized zoo but part of Disney World itself. Disney World consists of many different theme parks, water parks, many hotels, several golf courses and numerous restaurants (including a rainforest café and a hotel near the zoo). According to Disney Corporation, the goal of the zoo is to promote love for animals and raise awareness for animal welfare (cf. Beardsworth and Bryman 2001, 93 f.). Beardsworth and Bryman comment: "This remark nicely side-steps the fact, that one of its lands is sponsored by McDonald's and the park itself

32 See https://www.zoo-leipzig.de/en/your-event/, 2019/05/17.

sells vast quantities of steaks, burgers, hot dogs and chicken nuggets and is built on former wetlands" (Beardsworth and Bryman 2001, 94). The use of names such as "Wildlife Conservation Park" is intended above all to emphasize the image of zoos as educational and conservational institutions. According to Beardsworth and Bryman, it is striking that a new thematic arrangement within the zoos is usually combined with a general image change towards a conservation institution (cf. Beardsworth and Bryman 2001, 94).

Under the second Disney principle, *dedifferentiation of consumption* Beardsworth and Bryman (2001) state that different forms of consumption, which belong to different institutional spheres, are increasingly interwoven and thus become difficult to distinguish from one another (cf. Beardsworth and Bryman 2001, 94). Not surprisingly, such a mixture of different spheres of consumption can also be seen particularly clearly in the example of Disney's Animal Kingdom Zoo in Disney World. As in other zoos described by Beardsworth and Bryman, elements from amusement parks are integrated into the zoo. Thus, among the animals, one finds a roller coaster, which, according to the authors, is predictably called "Countdown to Extinction". The typical amusement park attractions are thus consciously integrated into the theme of the zoo as a nature and species protection park and sold under this theme. According to Beardsworth and Bryman, scientists reported about zoos all over the world that such mixtures of amusement park and zoo can be found more often. Circus elements and carnivalesque elements were also reported, which ensured that the attraction of wild and domesticated animals was combined with entertainment elements. In this way the visitor is induced to spend more money, to come more frequently and to stay longer (cf. Beardsworth and Bryman 2001, 95). In addition to blending the boundaries between individual attractions, there is an increasing integration of other consumer areas, such as shops and restaurants in the zoo. So far, large zoos in particular have followed this trend, but it can be assumed that large shops and shopping centers and a wide range of (themed) restaurants will gradually replace the usual small café and the well-known souvenir shop at the exit. The described design and mixing of different consumption areas contributes to the increasing commercialization of zoos and aquariums (cf. Beardsworth and Bryman 2001, 95).

In addition to the thematic orientation and mixing of different areas of business, according to Beardsworth and Bryman (2001), the third Disney principle has become much more prevalent in zoos, describing the increased marketing of zoos through comprehensive advertising and sales strategies (*merchandising*). Promotional products such as t-shirts, caps, mugs, ties and pens are already an established part of souvenir shop stock. Entertainment industry advertising strategies are deliberately used (see Beardsworth and

Bryman 2001, 95). With regard to the animals themselves, advertising and sales function arise through the creation, presentation and representation of species or individuals as icons. In the case of zoos, species that are particularly threatened with extinction and/or are particularly popular, often specific iconic individuals, are exploited for this purpose. If it wasn't before, this was made quite clear on the basis of the aforementioned list of the 30 most popular zoo animals in Berlin. Knut, a hand-reared polar bear at the Berlin Zoological Garden, was right at the top of the list and will forever be remembered by many people thanks to the unprecedented media hype surrounding him in 2006/07. This media presence was then only to be outdone by the comparable presentation, or rather staging, of "Flocke", a polar bear cub born a few months later. As a me-too product, she was meant to attract similar attention and similar income to the Tiergarten Nuremberg and the city of Nuremberg (cf. anonymous 2007b; a). The presentations of both animals were justified by the fact that the polar bear (one of the most attractive species to look at and at the same time one of the most challenging to keep in captivity) was hailed as an ambassador of climate change. In their advertising rhetoric, Knut and Flocke drew attention to the fact that the ice floes are melting away under their feet and that polar bears drown or starve to death in the wild (cf. Tiergarten Nürnberg 2010). This is why Knut is at the same time a registered trademark and a commemorative stamp created by the German Postal Office to raise awareness for environmental protection (cf. anonymous 2007a; 2008). Beardsworth and Bryman say about such advertising icons:

> Such representations can [...] be directly incorporated into merchandise items. Thus, the presentation of animals and animal performances by zoos can feed directly into the generation of commoditized images, which can have considerable commercial potential.
>
> BEARDSWORTH and BRYMAN 2001, 96

Here, the animal as a symbol of untouched nature is figuratively squeezed into a commodity and thus into a marketing process contrary to her aura. Something unique and subjectively and normatively valuable is turned into an advertising product. The rarer and more threatened the animal, the more inappropriate the connections between herself and economy and commerce seem to be. There are obviously breaks between normative values (the conservation of biodiversity as a goal based on environmental and animal ethics) and purely monetary values (the economic interests of zoos that use the animal as a registered trademark and box-office hit and thus as a means to an economic end). These fissures must be concealed so as not to irritate the visitor. Beardsworth and Bryman conclude their considerations with the following important statement: "surrounding the zoo with an aura of research, conservation and education helps to legitimate this commodification and helps to resolve the irony" (Beardsworth and Bryman 2001, 96). The reinvention of the zoo as an educational and conservational institution thus also serves to overcome and conceal the contradictions that arise from the collision of the zoos self-image as conservation agency and their trend towards profitable entertainment.

The fourth principle cited by Beardsworth and Bryman (2001), which becomes increasingly important in the development of zoos, uses the term *emotional labor* to describe questionable demands placed upon the "always smiling, always helpful employee" (97). The visitor should be pampered in a cheerful environment and only be offered positive experiences. Where does the rather serious transfer of knowledge about the problem of species conservation, as it could be expected from an institution interested in education, come into play? It is questionable whether politically explosive and sometimes disturbing information on the threats to animals in the wild, let alone self-critical utterances in relation to the difficulties of keeping them in captivity, could have a place here. According to the authors, the adoption of this Disney component in zoos can have an impact on environmental and animal ethics issues as well as on the zoo's request for species protection. On the one hand, the targeted use of emotions as psychological pressure could be used to cause the visitor to feel guilty about the destruction of habitats and the human-made threats caused to the animals. On the other hand, however, the targeted use of such emotions could also be used to create a "feel-good factor" for visitors, since they feel "better" when they pay the entrance fees and buy the zoo's advertising products, in the conviction that they themselves are thus participating in honorable efforts to protect the animals (see also Beardsworth and Bryman 2001, 97). Thus the use of emotional labor would be directly linked to the commercialization strategy and questionable at the latest at the moment when the money collected fails even to flow into a successful and meaningful form of species protection.

In addition to these two possible uses of emotions, however, Beardsworth and Bryman are also thinking of a third, already in itself ethically problematic effect of emotional labor: zoos sometimes tend to degrade their animals to personnel and also transfer the task and requirement of emotional labor to them. This is especially clear in shows with marine mammals in aquariums. Zoos and aquariums indeed experiment with training their animals to do certain things to arouse the interest of visitors. In fact, it can always be said that animals are performing emotional labor when they are induced to present behaviors that are to be interpreted by the audience as emotions so that, for example, an animal appears to be "friendly" by "waving" to the audience. A prime example of such a program aimed at anthropomorphic emotions is the killer whale show in Sea World (see Beardsworth and Bryman 2001, 97–98). In addition to their entertainment value, such uses would of course also serve to advertise and sell souvenirs of products that are often tailored to those animals that perform emotional labor (Beardsworth and Bryman 2001, 97–98).

Based on Beardsworth and Bryman's (2001) analysis, it becomes clear that the zoo has developed new objectives and justification strategies that are based on an increasingly stressed identity as an educational and a species-conservation institution, but which at the same time also counteract these values. It seems that these values must be communicated to an audience that is assumed to want to be entertained, first and foremost. The theme of the wilderness and the exhibition of wild animals are thus newly formulated, thematically prepared, and playfully implemented using effective quasification techniques. As soon as the wild is turned into a sales product in this way, it is then available to the masses in an accessible form for consumption. According to Beardsworth and Bryman, with reference to Bostock (1993), in this way the dangerous and threatening aspects of the wild are also mitigated and entertainingly shaped (Beardsworth and Bryman 2001, 98). Certainly there is a danger here that something similar could happen with information content on species conservation if it were actually to be communicated seriously in the zoo. Of additional importance here is the fact that the zoo visitor does not come to the zoo as a blank page but as a consumer who brings along a certain image of animals that is particularly influenced by his previous media knowledge. We all carry some *virtual capital*. Zoos are based on this presumption and also upon the visitor's ability to react playfully to what has been communicated in the media and simulated there:

> The complexes of representations and depictions contained in the individual's stock of virtual capital will enable him or her to make sense of, and appreciate the entertainment value of, the presentations, representations and quasifications contained within the themed zoo.
>
> BEARDSWORTH and BRYMAN 2001, 99

The Disney idiom is one of many idioms that have prepared the visitor for the zoo: anthropomorphic and sentimental ideas about animals have already been conveyed to the visitor from childhood in the Disney films. In Disneyized zoos, where the animals may also be encouraged to present human-like emotions in the guise of tricks and so on, these ideas are further fostered and delivered (Beardsworth and Bryman 2001, 99).

Although museums, especially natural history museums, are developing in a similar direction, the special feature of zoos is that they are not just working with inanimate exhibits but with living animals. In this instance, there should be much more demanding ethical requirements placed on the exhibition and captivity. According to Beardsworth and Bryman, this is particularly relevant against the background that the role of animals is changing. Animals in zoos develop from exhibited objects to trained staff members, a shift that should not be misinterpreted as an increase in respect for their subjectivity or agency:

> Increasingly, they will be "staged" as attractions in quasified entertainment settings. In a very real sense, they will become "workers" or "cast members" in the playgrounds created by the leisure industry. This will be particularly true for those "charismatic" animals the public is especially drawn to.
>
> BEARDSWORTH and BRYMAN 2001, 99

Beardsworth and Bryman believe that the next stage in the development of the Disneyized zoo will be the increasing use of quasified animals (such as computer-animated exhibits) with a deceptively real appearance. This would be the next step due to the complex management and the demanding requirements that exist when keeping so many zoo animals. In their opinion, the entire quasinatural setting of an artificial wilderness will gain in importance rather than the animals themselves. This will change the view of the tourist and turn the themed zoo into a place where city dwellers experience a quasified form of nature and encounter the "quasi-wilderness" with maximum comfort, maximum service, and maximum security at the same time (Beardsworth and Bryman 2001, 100). Beardsworth and Bryman believe that "in the post-tourist context the consumers of tourism experience do not necessarily demand that those experiences be 'authentic.'" Instead, experiences are required that are entertaining, captivating and effectively staged (Beardsworth and Bryman 2001, 101). If the authors are right, then in the distant future there is the possibility that at least part of the ethical problems that zoos face will simply disappear—essentially because they will be housing fewer and fewer real animals. However, since the consumer may not want to do without a direct encounter with real, wild nature, there is a risk that the ethical problems will shift to other institutions, such as nature reserves and the safari sector.

At present, it is not foreseeable whether the zoo will eventually do away with itself in this way. In any case, according to the authors, it will orient itself more and more toward entertainment. If zoos simultaneously strive for the education of their audience, then they will mask this more and more as entertainment (Beardsworth and Bryman 2001, 101). One wonders here whether the zoo of the future will really be a conservation breeding park. The problem will not be simply swept aside with a relativization along the lines that the influence of Disney principles on zoos certainly varies from institution to institution and with the subtle suggestion that a little bit of commerce never did any harm to a good cause. Beardsworth and Bryman (2001) illustrate a highly worrying trend that is concealing structural and fundamental problems. According to the authors' conclusion, the reconfiguration of zoos into entertainment enterprises based on the themes and entertainment motifs used could above all become a new model for the exercise of power of humans over nature. As such it is then uncritically accepted by the masses and "consumed along," as it were. Humans' power over nature, which was once conceived as a right to the exploitation of nature, has now been transformed into a public duty to maintain and preserve nature and species. However, since this theme is offered to visitors to modern zoos as part of a "fun day out" package, it also represents a legitimation. It seems to be a reassuring and disguising attempt to distract from ethical concerns that leave us with the impression that zoos may not only be old institutions but also antiquated (for a similar discussion, see Beardsworth and Bryman 2001, 101).

Many of the animals whose outstanding cognitive, social, and emotional abilities have been discussed in this book are classical zoo animals. Above all, however, dolphins and great apes are not only among the most popular zoo animals but are also those species that are at particular risk of being caught up in the zoo entertainment trend. As highly intelligent animals, they can be trained well for shows, successfully used in demonstrations, and profitably marketed. Anthropomorphic productions are particularly suitable due to the physical similarity with great apes and were offered, for example, in the Swabian Park (Schwaben Park) in Gmeinweiler-Kaiserbach, Germany, which advertised "Europe's only chimpanzee show" until 2017. The park has phased out the show and is phasing out the breeding of its roughly forty chimpanzees because of pressure from the public and the authorities. In addition to amusement park attractions such as the Ferris wheel, super slide, and wild water ride, you could for decades go there to see a show featuring dressed-up chimpanzees using the phone, pushing doll prams, brushing their teeth, or driving a bobby cart.[33] The park also offered a parrot show and a pet show. At least some, if not most, of the chimpanzees were raised by hand and performed their tricks

33 This information was available on the homepage of the park www.schwabenpark.com, 2012/07/03.

several times a day. An undercover-investigation of an animal rights organization brought attention to the case and helped to shut down the shows.[34] Beardsworth and Bryman's criticism seems appropriate here, and on many levels. The animal rights activists criticized the anthropomorphic and ridiculous presentation and assumed that the young animals were separated from their mothers and raised by hand, especially with a view to their better suitability for the shows (Schertler 2012). Here it becomes clear what direct ethical consequences follow if animal parks align to entertainment and commercialization strategies.

In this case, as in many others, the concept of encountering animals in terms of education and conservation seems far removed. The zoo of the future should stay away from such strategies.

9.2.2 *Keeping and Presenting Dolphins in Aquariums*

The case of the Schwabian Park brought us back for a moment to the great apes because the park embodies so nicely the characteristics of Disneyization in Europe. Let us return now to our second case-study species, the dolphins. Throughout the world, institutions in which dolphins are held in captivity are opening their doors at an ever-increasing rate. More and more of them exhibit and stage the animals (for example, in aquariums and marine theme parks), but also allow their visitors to swim with them in the pools ("swimming-with-dolphins" programs) or touch and stroke them (in "petting pools"). Moreover, facilities with connected dolphin therapy for human patients are on the rise (see WDC b; Marino and Lilienfeldt 2007).

How the animals are kept varies. Most of the pools are tiled or concreted without natural elements such as water plants. The water must be treated either chemically or biologically and purified via filter systems. A few facilities in the world keep dolphins in natural seawater basins separated from the sea, where the animals are not exposed to massive chlorine and filter cleaning of the water and swim in a more natural environment with plants, rocks, and other aquatic animals. Sometimes the dolphins can at least in theory also leave these basins, but are reliant on regular feeding and encouraged to interact with humans (see also Dolphin Research Center 2009; White 2007, 2, 4).

In over thirty countries, as of 2008 an estimated 1,200 dolphins and 50 orcas were held in captivity in approximately 100 establishments[35] (for some

34 https://www.stuttgarter-zeitung.de/inhalt.schwaben-park-kaisersbach-zauberei-statt-affenzirkus.2874f9e7-31be-41c9-88a7-5b847d8707af.html, 2019/08/25.

35 As of December 2008, only 246 of them were listed in the species360 (formerly ISIS) database, including a river dolphin, several orcas and belugas, and a pilot whale. The database is online accessible via https://www.species360.org/, 2019/05/17.

numbers see also White 2007, 198–99). However, it is difficult to give a precise number, since not all institutions are members of the zoo and aquarium umbrella associations. It seems that WAZA does not know exactly how many dolphinariums there are worldwide. Dolphinariums in the narrower sense (in which only dolphins are kept) were not among its members in 2008, but 16 of its 244 member zoos were also home to dolphins or other whale species (such as belugas, orcas, and inias). WAZA also had no information on where dolphin therapy is routinely offered.[36] The Alliance of Marine Mammal Parks and Aquariums (AMMPA) is the largest international association of marine parks, aquariums, research institutions, and organizations, which claims to be committed to the highest standards in the keeping of marine mammals and to the protection of these animals in the wild through public education, scientific research, and species protection (see AMMPA a). AMMPA had a total of 50 members in 2008. Of the 27 US members, 21 hold a total of approximately 400 dolphins (as of December 2008;[37] see AMMPA b). The European Association for Aquatic Mammals (EAAM) had 31 members at that time, at least 27 of whom were also keeping dolphins, and one porpoises, which belong to the delfinidae. Six of these facilities were also in the AMMPA (as of December 2008;[38] https://eaam.org/, 2019/05/19.). In total, about 75 facilities, most of which are home to dolphins, were listed via AMMPA and EAAM and met their standards. In addition, however, there are a growing number of institutions worldwide that have not yet been included in the relevant associations and may not be included because they do not meet their standards.

The fact that more and more institutions are opening in which dolphins are held in captivity for different purposes is diametrically opposed to ever-increasing criticism from the realms of animal protection and science (see e.g. Williamson 2019). Any advocate of dolphinariums (and the sometimes-associated dolphin therapy) must be aware that many of the animals living in such facilities have been caught in the wild and transported before they arrived at their destination. It should also be obvious to anyone that besides the hunt, the catch, and the sudden isolation from the school, also the loading from the water into a box and the following transport in the airplane mean

36 This was the result of a request made by the author to WAZA on December 1, 2008, which was answered by staff from the WAZA Executive Office in Berne.

37 This is according to a request from the author to AMMPA on December 4, 2008, which the Executive Director of AMMPA answered.

38 This is according to an evaluation of the homepages of the member institutions and individual email requests from the author (which were forwarded to the corresponding institutions via staff from the EAAM Student Committee).

extreme stress for the animal. Even if care is taken to ensure that the animals do not dry out, they still have to remain almost motionless in the transport containers in the tightest of spaces. In addition, they are accustomed to the gravitation-free habitat of the sea. Lufthansa Cargo AG carried out such transports of dolphins for commercial facilities until one of two dolphins died during a transport to Argentina in November 1999. According to critics, the transport conditions (both animals were transported in a common container, among other things) constituted a violation of international air transport regulations in addition to the ethical issues at stake. Lufthansa Cargo decided to bear the consequences and to stop transporting dolphins and whales for commercial purposes in the future (Lufthansa Cargo 2019).

However, the main arguments against keeping dolphins in captivity relate to the housing conditions at the destinations.

With a few exceptions, where seawater basins are available, the animals are kept in small, mostly tiled basins. According to the Gesellschaft zur Rettung der Delphine (GRD), one of the major German dolphin welfare organizations, the US standard is 9 meters long and 1.8 meters deep (GRD 2019). Judging by the fact that these animals swim up to 100 kilometers per day in the wild and dive up to 300 meters deep, according to the critics, these pools hardly offer sufficient space. On top of that, enclosing dolphins in such a compact environment and in small groups with few other animals means they cannot fully live out their normal social life. Sophisticated social cooperation strategies and close ties in social alliances are eliminated. Their echo-location system, which is adapted to the oceanic expansiveness, is hardly used by captive animals because it rebounds several times when reflected from the smooth walls. Their complex communication simply atrophies (GRD 2019). Here, clear impairments become evident, also with regard to those psychosocial abilities that can be and have been described as personhood qualities of dolphins. For feeding, according to the critics, dead fish are used, which must be artificially sprayed with water, since dolphins consume water exclusively through food. A dietary supplement containing anabolic and vitamin preparations is necessary (GRD 2019). With regard to the physical well-being of captive animals, it is also important that their eyes and skin are exposed to the constant chlorine cleaning of the water, especially when people go into the water, as in the case of swimming-with-dolphins programs or petting programs and in therapy facilities. In such facilities, increased disinfection due to pathogenic (fecal) germs must be conducted. In addition, according to the GRD, the constant noise that is conducted well in the steel and concrete constructions of the pools, for example, from the filter pumps, from the music during the performances, and from the audience, is another serious form of stress. The animals have no way of avoiding this, just as they have no opportunity to flee in disputes over ranking. Dolphins in captivity therefore sometimes become aggressive, especially in petting pools, where they are exposed to the immediate vicinity of visitors. If it were possible for the animals to bear offspring, according to the critics, even the F1 generation displays no normal behavior (GRD 2019). The breeding of captive dolphins proves to be extremely problematic and only succeeds in a few institutions. Reintroduction of animals that have adapted to captive conditions is practically impossible. Mortality rates in dolphinariums are very high. Since 1971, the Nuremberg Dolphinarium, one of the last two dolphinariums in Germany, was home to 53 dolphins. 35 of them have died since, at least six of them at birth or prior to birth. Five calves and one dolphin mother died between May 2006 and June 2007 alone (WDC a).

For an ethical assessment of the keeping of dolphins in captivity, it is also important to note that dolphinariums, including those that offer therapy, share responsibility for where their animals come from and how they may have been taken from the wild. In the following section, the Japanese drive hunts are dealt with separately (Section 9.2.3). Offering dolphin-assisted therapy generates an ethical justification strategy for the existence of dolphinariums that many do not question. The promised therapeutical effect seems attractive. However, an uncritical stance is shortsighted here and dolphin-assisted therapy comes with many severe ethical problems (Section 9.2.4). In addition, (alleged) therapies are often closely linked to commercially oriented swimming-with-dolphins programs. Any kind of direct interaction with the animals is highly profitable, as is the use of trained animals in shows. While in the 1960s one dolphin was sold for $300, today the price is $100,000 or more. An orca purchased from Sea World twenty years ago for $875,000 is now worth more than $5 million, and a single dolphin can bring in over $1 million per year. Everyone involved in catching, selling, and owning these animals wins, from the fisherman who catches the animals, to the governments who sell them, to the countries where the animals attract tourists and the owners who charge up to $50 per visitor for their parks and offer shows with dolphins. Kestin (2004) concludes: "Dolphins and whales have become so valuable that parks can use them as collateral for loans. They're worth money even dead, bringing their owners million-dollar life insurance payouts" (1).

In summary, from a pathocentric perspective alone, keeping dolphins in captivity is ethically unjustifiable (see also DeGrazia 1996b; White 2007). In the ethical argument against such captivity, there is a strong reference to the fact that one cannot do justice to the complex cognitive and emotional abilities of these animals in captivity. But even if keeping dolphins were implemented as well as possible (an increasing number of institutions are, according to Brando (2010), trying to implement species-appropriate housing and enrichment), nevertheless, White's arguments regarding the dignity of animals as nonhuman persons applies. The right of animals to individual freedom and their dignity are therefore violated, however appropriately they are kept. In weighing harms and benefits and in any considerations of justice, the benefit of this captivity is advantageous for none other than humans alone.

It is striking that the extremely complex cognitive and emotional abilities of animals on the one hand encourage us to deal ethically and responsibly with their subjective needs and rights, and, on the other hand, they are the reason why we misuse and abuse these animals so easily as charismatic objects of our entertainment. In many dolphinariums, which are designed accordingly, it must be questioned whether the lucrative display of the animals is not the main reason for their continued existence. Educating visitors can hardly be spoken of if it is primarily about circus-like presentation of tricks and if the animals cannot be shown in their natural behavior. Whale and Dolphin Conservation (WDC, formerly WDCS) therefore judges that such shows do not in any way encourage respect for these animals, nor do they contribute to the understanding of the dolphins and thus represent a pure demonstration of human dominance over these individuals (WDCS b). Arguments of increasing Disneyization of zoos, as found in Beardsworth and Bryman (2001) (see Section 9.2.1 above), also apply in the case of dolphinariums, perhaps there more than anywhere else. In Disneyized facilities, as we have seen, the entertainment value and the (anthropomorphic) staging of the animals take center stage. The "Shamu" shows with killer whales in the Sea World Parks, in the United States, which are also criticized by Beardsworth and Bryman (2001), were for many years indeed trumpeted with words that give food for thought: a show called "Believe" merged "new killer whale behaviors with elaborate set pieces, state-of-the-art multimedia, music and choreography." The show surprised the visitor with a series of completely new, exciting killer whale behaviors. The show "Shamu Rocks," on the other hand, was a rock "n" roll concert of unprecedented dimension, "combining improvisational movements of killer whales with music remixed with some of the hottest rock stars in the industry." A "high-energy, rock 'n' roll extravaganza and nighttime killer whale show" awaited interested audiences. Lastly, the "Blue Horizons" show had dolphins and birds performing together. The result was "a mesmerizing display of dance, flight, color and entertainment," a "theatrical attraction" in which the sky meets the sea, dolphins fly, and birds dive.[39] The focus of zoos and parks following such staging is definitely not the presentation of species-typical behavior and information about the threat to animals in the wild, but is quite clearly multimedia-based entertainment for humans and commercial marketing of animals.

9.2.3 *Catches in the Wild*

How do dolphins get into dolphinariums? In certain countries, such as the Solomon Islands, an island country in the southwest Pacific, or in Japan, dolphins are still caught for international trade with aquarium operators and parks. In recent years, in addition to the Solomon Islands, Japan has also come under considerable public scrutiny for the removal of dolphins from the wild.

Every year from September–October to March–April, dolphin hunts are carried out in Japan, on a large scale, in the coastal village of Taiji. According to animal welfare organizations, only a small group of twenty to thirty fishermen are involved in these hunts. Taiji fishermen killed about 1,200 to 1,800 dolphins

39 In the meantime, Sea World has changed the description of the shows and now avoids language on the homepages of the parks that might indicate a relentless media staging of the animals. The quoted descriptions of the shows could be found on the homepages of the parks in 2010. However, today's descriptions give the impression that the conservation theme as a new identity strategy is still very much coupled with an understanding of the orcas as a feature show element. Entertainment and playfulness, as well as the creation of an amazing, multi-sensory experience are still main elements of the descriptions. For some shows, species are mixed up that usually don't interact with each other, creating a harmonizing and colorful, though quite naïve and artificial experience of we-all-share-the-same-world. The San Antonio park advertises the "One Ocean" show with the following words: "The energy of the ocean envelops you in a multi-sensory celebration of life under the sea that both entertains and inspires with a message that we are all part of one world, one ocean" (see https://seaworld.com/san-antonio/shows/one-ocean/, 2019/04/14). The "Ocean Discovery" show is an "amazing showcase featuring graceful beluga whales, athletic Pacific white-sided dolphins, and colorful macaws interacting with each other in a playful way" (see https://seaworld.com/san-antonio/shows/ocean-discovery/, 2019/04/14). In Orlando, "One Ocean" is described the following way: "Killer whales join you on a journey into a world that drenches your senses in vivid colors, vitality, and the global rhythms of the ocean. Dancing fountains set the stage as you connect with thrilling sea creatures, feel the energy and spirit of life underneath the sea, and realize that we are all part of one world, one ocean" (see https://seaworld.com/orlando/shows/one-ocean/, 2019/04/14). Parks like Sea World undoubtedlldy have been made aware of the rhetoric they use in the media advertisement of the animals. Facing decades of heavy criticism, Sea World announced in 2016 the end of their captive breeding of orcas (SeaWorld Parks & Entertainment 2016). Turning towards conservation as a theming strategy accompanies this move which might indeed mirror an ethical decision but surely also prevents bad press and disguises remaining entertainment aspects of the shows.

per year in 2004 and the years following, but the numbers have dropped to about 600 in 2016–17 due to public pressure caused by the joint campaigns of animal welfare organizations (http://savedolphins.eii.org/campaigns/sjd, 2019/05/17). Animals that are preferred for the hunt are species that are routinely held in captivity such as bottlenose dolphins, pilot whales, and porpoises. The catch rate is approved and regulated by the Japanese government (see SJD 2019; White 2007, 197). Save Japan Dolphins, an association of the most renowned and experienced conservation organizations dedicated to the protection of marine mammals (including the Animal Welfare Institute, the Earth Island Institute, and Ocean Care), has been monitoring events in Taiji for several years. The hunt, whose scenes have meanwhile been captured on film,[40] are described as follows:

> The fishermen of Taiji have developed a highly effective method of locating, capturing and eradicating dolphins, sometimes as many as one hundred in a single day. Before sunrise, about 26 fishermen board their 13 motorized boats and head out to deep water where the dolphins migrate. The dolphins have been using these migratory paths for thousands, perhaps millions, of years, and the hunters know exactly where to find them. When a school of dolphins swims by, the fishermen position their boats one behind the other, perfectly evenly spaced. Then they lower several stainless steel poles into the water, one on each side of each boat. The poles are flared out at the bottom much like a bell, which amplifies the sound produced when the hunters repeatedly hit the poles with hammers. The noise creates a wall of sound underwater, and the dolphins suddenly find themselves trapped between this wall of sound and the shoreline. Trying to get away from the sound, the dolphins swim in the opposite direction, toward the shore. The dolphins' panic and loss of navigational sense enable the fishermen to drive them into a small, hidden cove near Taiji harbor. The fishermen seal the mouth of the cove with two nets, and the dolphins are trapped.
>
> The Animal Fund 2019, SJD 2019

Once rounded up in the bay, the animals are slaughtered the next morning. They are stabbed from the boat with long spears and hooks. The seriously

40 CNN, for example, ran a report about Japanese drive hunts in Taiji on February 11, 2008 (CNN 2008). In addition, the documentary *The Cove*, which won the renowned Sundance Film Festival Audience Award in 2009 and the Critics Choice Award in Los Angeles in 2010, brought the drive hunt in Taiji to cinemas worldwide (see the homepage of the filmmakers, the Oceanic Preservation Society (OPS) https://www.opsociety.org/our-work/films/the-cove/, 2019/05/17 and Harnell (2008)).

injured, thrashing, and screaming animals are hoisted into the boats and further processed there with knives (The Animal Fund 2019). There can be no doubt that this hunting method causes serious damage to animals in every aspect. Surviving animals are traumatized by the experience and isolation from their school.[41] The slaughtered animals are processed into animal feed and fertilizers and for human consumption. The animal rights activists who refer to the statements of the fishermen themselves suspect that the dolphins are also decimated because they are perceived as direct competitors with the fishermen. Fish stocks are shrinking. Instead of seeing the reason for this in global overfishing and environmental-ecological contexts, the animals are being pursued as an alleged food competitor. What is happening is then just "pest control" (SJD 2019). Live-caught animals are sold to dolphinariums, mainly in Japan and China, but also to institutions in the United Arab Emirates, Turkey, and Thailand (see White 2007, 197). Observers from Save Japan Dolphins describe the direct cooperation between fishermen and dolphinarium operators as follows:

> Many times, the dolphin hunters of Taiji will drive a large school of bottlenose dolphins, pilot whales, or false killer whales into the killing Cove, and dolphin trainers and marine mammal veterinarians flock to the scene to seek out the best-looking dolphins for their display facilities. By doing business with the dolphin killers, they are helping to maintain the dolphin drive hunts. .... Working side by side, dolphin trainers and fishermen force the dolphins into shallow water, haul the dolphins ashore, and line them up. The trainers then inspect the dolphins one by one, choosing only the ones that can be used in dolphin shows and captive dolphin swim programs. They are typically looking for young, unblemished female dolphins. The grueling selection process drags on for several hours, and some dolphins die either from shock, injuries or exhaustion during this time. ... We have seen members of the international aquarium and zoo industry get in the water with the dolphin killers, tying ropes around the dolphins' tail flukes so that the fishermen could tie the dolphins to their boats. Often times, the dolphins are so exhausted at this point that they can't even stay afloat. Some have large amounts of blood coming out of their blowholes. The dolphin trainers don't seem to care. The fishermen haul the dolphins to the killing Cove, with the dolphins' blowholes

41 It is highly problematic if some of the animals so traumatized later on have direct contact with humans in swimming programs or in dolphin-assisted therapies. Such traumatized animals can be a danger to humans because their behavior can be unpredictable.

> underwater. Dolphin trainers have tormented the dolphins for hours. Some are in shock. Others are seriously injured, and they can't breathe. Now, they are going to be killed and slaughtered. Not even pregnant females or young calves will be spared.
>
> SJD 2019

In addition to the active participation of some trainers in the catch and their direct, personal responsibility for what is happening, the indirect responsibility of operators and visitors to dolphinariums, who support the catch by buying such wild catches or attending the shows, must also be discussed. The hunts are not profitable due to the processing of the slaughtered animals but thanks to the sale of the living ones. Save Japan Dolphins concluded that the meat of a slaughtered dolphin is worth about $600, while the animals caught alive in Taiji were sold for up to $155,000 per animal (SJD 2019). The entire hunt would pay off much less if the international dolphinarium industry would not cofinance it. Buyers who obtain their animals from such wild catches for parks, shows, and theatre facilities are therefore accused of being directly responsible for catching the live animals and indirectly cofinancing and legitimizing the slaughter of the remaining animals: "They are fueling the dolphin hunt by making it tremendously profitable" (SJD 2019).

The practice of obtaining wild catch animals for zoos and aquariums is not only adopted by small institutions in third-world countries where animal welfare standards do not exist or are not monitored or enforced. Western trainers have also acquired such animals in recent years, apparently even from Taiji, and apparently despite the fact that they are WAZA members. However, this clearly breaks the WAZA Code of Ethics, to which all member institutions have committed (White 2007, 197). WAZA is committed to ethics, animal welfare, and species protection. It stresses that the continued existence of zoos and aquariums depends on their profession being understood as being based on respect for the dignity of the animals in their care (WAZA a, 61). Therefore, WAZA strives for the development of a strong ethical tradition and for the highest standard of ethical behavior[42] (WAZA c). In their Code of Ethics and Animal Welfare, to which all members make a pledge, it states: "Members of WAZA will ensure that all animals in their care are treated with the utmost care and their welfare should be paramount all times." Cruel and nonselective methods of taking wild animals from the wild, in which entire populations are

42 This discussion refers to an "ethical tradition which is strong" and the "highest standard of ethical conduct" (WAZA c).

threatened,[43] are strictly rejected by WAZA (see WAZA c). Similarly, the collection of animals for exhibitions, especially aquariums, where a high mortality rate is expected, is condemned. While it is recognized that from time to time the collection of wild animals may be necessary to refresh the breeding programs (which should be designed for species conservation, education, or scientific research and, it must be stressed, not only for the entertainment of visitors), in principle it applies to all members that new animals should be acquired through the exchange between zoos and through breeding programs. If it is exceptionally necessary to remove wild animals, it must be ensured that there is no damaging effect on wild populations (WAZA c). It is quite clear: "Failure to satisfy or adhere to the Bylaws, Code of Ethics and all other rules and regulations shall be sufficient cause for suspension or denial of membership" (WAZA b). WAZA therefore rejects wild catches from actions such as the Japanese hunts.

Nevertheless, dolphinariums that have been involved in such hunts remained WAZA members until very recently. In this way, according to criticism from animal protection advocates, these institutions were granted a seal of credibility in belonging to an association of institutions to which a certain quality (and ethical) standard is assigned. Furthermore, according to SJD, dolphins from drive hunts were and are still caught for many non WAZA institutions around the world:

> A number of aquariums and swim-with-dolphins programs around the world purchase live dolphins caught in the bloody drive hunts of Taiji. Japan alone has more than 100 dolphinariums and swim-with-dolphins programs, ranging from large aquarium facilities with huge tanks and dolphin shows to small tanks at motels or floating sea pens in harbors. In the last 10 years alone, dolphin exports from Taiji, Japan, have gone to China, Korea, Ukraine, Russia, Egypt, Iran, Turkey, the United Arab Emirates, Thailand, Saudi Arabia, Taiwan, and the Philippines.
>
> SJD 2019

Before a ruling by the U.S. National Marine Fisheries Service (NMFS) in 1993 "imported dolphins and small cetaceans from drive hunts in Japan, such as false killer whales, were obtained by SeaWorld, the Indianapolis Zoological Society, Miami Seaquarium, and the U.S. Navy. Dolphin brokers from the United States are often still spotted in Taiji" (SJD 2019).

---

43 WAZA speaks of "cruel and non-selective methods of taking animals from the wild" (WAZA c).

The Taiji Whale Museum, which coordinates and carries out training and sales of dolphins caught in Taiji, was a member of JAZA, which in turn belongs to WAZA. After great public pressure, WAZA issued a press release in October 2005 on the Japanese hunts in which members were asked not to buy any animals from hunts. Non-WAZA members would also be encouraged to refrain from buying animals from such hunts. At first, this statement merely had the tone of a polite reminder to its members regarding their obligations, but it soon disappeared from the WAZA archives. Later, WAZA argued that the only direct WAZA member who obtained dolphins from the hunts had left the organization (WAZA d). Save Japanese Dolphins seems to assume, however, that many of the dolphins in Japanese dolphinariums originate from such hunts (SJD 2019). WAZA clearly wanted to avoid confronting its members with a harsh condemnation of these fishing practices. They chose to sit at the negotiating table in a nonrejectionist attitude to the hunts: such hunts have a long tradition in Japan and have to be discussed in Japanese society and with Japanese partners (WAZA d). This course of taking a calm approach was not unproblematic. According to Save Japan Dolphins, more than six hundred million visitors pass through the gates of a WAZA facility every year. They could all be reached if the questionable wild dolphin catch of dolphins would be condemned, and more explicitly at that, by aquariums. This information should also be part of the visitor's education about the various threats to animals, to which WAZA members dedicate themselves, in addition to the protection of species and the dignified treatment of animals according to their own statements.

It took WAZA and JAZA until 2015, facing over ten years of pressure, to ban all members from acquiring dolphins from the Taiji dolphin drive fisheries and similar fisheries (Elsa Nature Conservancy of Japan 2015). The Taiji Whale Museum simply withdrew from JAZA and continues to operate as before.[44] The question is if and how fast the umbrella organization's ethics codes are implemented in their actual policy. Another question is whether such changes in policy find their way into the member institutions' zoopedagogy. How would the visitor of a dolphinarium react to a multimedia show on whaling or the zoo visitor to an information center on the extinction of species? Maybe the reactions would be more positive than one thinks. Visitors who are really interested in endangered species and are looking to educate themselves would stay. Visitors, on the other hand, who are merely looking for entertainment, would presumably resort to amusement parks that ideally, from an ethical point of view, only offer inanimate exhibits and attractions.

44 See https://dolphinproject.com/blog/taiji-whale-museum-withdraws-from-jaza/, 2018/07/20.

9.2.4 *Dolphins as Therapists?*

While most of the institutions that keep dolphins tend to present the animals to visitors and sometimes use them in shows, others—usually in addition—have included an offer for dolphin-assisted therapy in their program. Since in this instance the therapeutic value of the interaction between humans and dolphins is advertised, a serious ethical argument seems to exist at first glance to justify the problematic captivity of the animals. However, this must be carefully examined.

Participating in dolphin-assisted therapy (DAT) means having the opportunity to swim and interact with (usually captive) dolphins. This form of interaction has been used as therapy since the 1970s and is now offered worldwide in the United States, Mexico, Israel, Russia, Japan, China, and Turkey, among other countries (Marino and Lilienfeldt 2007, 239). In Germany, dolphin therapy is/was being carried out in the Nuremberg Dolphinarium. DAT in Nuremberg started as a cooperative project between the zoo (Tiergarten Nuremberg) and scientists from the University of Würzburg as well as the HU Berlin.[45] Dolphin therapy is used for a large number of diseases, e.g., in children and adults with autism, epilepsy, down syndrome, multiple sclerosis, depression, AIDS and cancer. Advocates of DAT base their claims on the positive benefits of the therapeutic interaction between patient and dolphin. This therapy, according to the advocates, leads to improvements in social-emotional and communicative behavior of patients, a longer attention span, a higher level of activity and motivation, improved communication skills, a better understanding of speech, and increased self-confidence. Sick children are encouraged to show their own impulses to act in interaction with the animal and have the opportunity to actively shape the situation, while the observing parents are supposedly sensitized to the competences and signals coming from their children (Nathanson et al. 1997; Nathanson 1998; Breitenbach et al. 2006; University of Würzburg, Abteilung Rehabilitationspsychologie 2010).

Opponents of DAT, on the other hand, respond with a significant number of critical arguments (see Marino and Lilienfeldt 1998, 2007; Humphries 2003; Baverstock and Finlay 2008; Brensing 2003, 2004). These are directed both at the alleged benefit of the interaction for the patients as well as the welfare of the animals and will be examined in more detail below based on the argumentation of Marino and Lilienfeldt (2007).

---

45 See https://www.reha.hu-berlin.de/de/lehrgebiete/rhp/forschung/delfintherapie, 2018/07/22. The project's first phase lasted until 2006, and the second phase lasted from 2006 to 2011. It is unclear whether the therapy is still conducted or planned to continue.

Lori Marino, one of the world's leading dolphin experts and the founder and executive director of the Kimmela Center for Animal Advocacy, has formerly been a senior lecturer in the Neuroscience and Behavioral Biology Program at Emory University, and a faculty affiliate at the university's Center for Ethics. Her colleague Scott Lilienfeldt is a professor of psychology at Emory. Together, they published their second article on scientific criticism of dolphin-assisted therapy in 2007. In the journal *Anthrozoös*, of the International Society for Anthrozoology, they came forward with an analysis of five studies on DAT, which had been previously published in peer-reviewed journals since 1999.[46] Studies that did not meet scientific standards were to be excluded from the outset. Their article "Dolphin-Assisted Therapy: More Flawed Data and More Flawed Conclusions" follows their 1998 study of two publications already available to them at the time (Marino and Lilienfeldt 1998).

Marino and Lilienfeldt (2007) tried to measure the scientific value of the studies found using standard criteria for evaluating scientific (medical) work. The decisive factor here is whether the studies were taking into account or excluding certain effects that might influence the results and had already been known for a long time (Marino and Lilienfeldt 2007, 241ff.). The central question for Marino and Lilienfeldt was as follows: Can it be proven that it is the therapy with the dolphins that produces possible improvements, or are other effects responsible? The criteria used in the evaluation were, for example, the consideration of long-term effects, the exclusion of the novelty effect, and the exclusion of the placebo effect. Only if these are in place can a study of dolphin therapy show that the therapeutic success is sustained and that the event or interaction itself brings about the improvement achieved (and not just, for example, the novelty of the event or an expectation held by the patients or their relatives).[47] With dolphin therapy, according to the authors, treatment typically

---

46 The authors searched articles in general search engines (Google), in specialized search engines (Google Scholar and Current Content Database) and in selected journals (*Anthrozoös, Society and Animals, Applied Animal Behaviour Science,* and *Zoo Biology*). They found and evaluated the studies of Antonioli and Reveley (2005), Likura et al. (2001), Lukina (1999), Servais (1999), and Webb and Drummond (2001) (see Marino and Lilienfeldt 2007, 240–41).

47 Marino and Lilienfeldt (2007) see threats to the validity of the studies they investigated especially in two areas: "Most of these threats relate to either *internal validity*, that is, the methodological soundness of the study, or to *construct validity*, that is, the soundness of the measures as indicators of the constructs examined by the investigators." The latter area includes nonspecific effects as "improvements from influences that are not specific to the intended treatment, and that are shared by a wide variety of other treatments." Two important subcategories of these effects are placebo and novelty effects ("the placebo effect is the well-documented but little understood improvement that derives from

consists of several complex components beyond the presence of the animal. Swimming in the water, being close to the water, being active outdoors, and maintaining the attention of adults (parents and therapists) are also central elements of the therapy. In addition, the dolphin itself is a complex stimulus that can be deconstructed into various possible therapeutic components. The patient may, for example, be affected by the mere size of the animal, her touch, and the opportunity to interact with her. It would be scientifically highly problematic to ignore the complexity of the effective components of the "therapy" and to attribute the effect to a single component, although its sole effect with simultaneous control of the other conditions was not tested at all (Marino and Lilienfeldt 2007, 243, 248).

Marino and Lilienfeldt conclude that the therapeutic status of dolphin-assisted therapies claimed in the analyzed publications is untenable when measured against recognized criteria for scientific validity. All the studies they examined did not control or exclude the occurrence of the effects mentioned (Marino and Lilienfeldt 2007, 242ff.). There is no evidence that it is the interaction with the dolphin that makes a difference to the patient. In their opinion, DAT is not a valid therapy. Instead, there is a "lose-lose situation" for dolphins and patients, with regard to ethical arguments and in view of the risks associated with the interaction.

For the patients, particularly the risk of infection and injury seems high (Marino and Lilienfeldt 2007, 248; see also Clark 2007). Both risks are now well documented by scientific studies and various research reports and apply to all contact between humans and dolphins, in swimming programs in the wild as well as in therapeutic and nontherapeutic programs in captivity (see Buck and Schroeder 1990; Mazet et al. 2004; Rose et al. 2009; Webb and Drummond 2001; Samuels and Spradlin 1995; Brensing et al. 2005; Frohoff 2000; Frohoff and Packard 1995). A survey of the Wildlife Health Center of the School of Veterinarian Medicine at the University of California (Davies) in 2001 and 2002 showed that 50 percent of employees (243 of 483) in institutions dealing with dolphins and their care had already suffered injuries and diseases resulting from direct or indirect contact with the animals. Obviously, doctors trained to treat human patients have difficulties in correctly diagnosing and assessing the

---

participants' expectation of improvement. Novelty effects are the general energizing and uplifting effects of a new, exciting experience"). Dolphin therapy is particularly susceptible to both effects, to placebo effects, because it is marketed to patients and their families as a highly efficient therapy and because the treatment process is transparent for the participants. Dolphin therapies are susceptible to novelty effects because it is an obviously new and exciting experience of swimming and interacting with a large, intelligent and charismatic animal (Marino and Lilienfeldt 2007, 241, 248).

zoonoses occurring here. They often are not aware of which pathogens can be transmitted from marine mammals to humans (Mazet et al. 2004, 3). The authors write: "Marine mammal work-related illnesses commonly reported by survey participants included: seal finger, conjunctivitis, viral dermatitis, bacterial dermatitis, and non-specific contact dermatitis. Severe illnesses were less commonly reported but did include tuberculosis, leptospirosis, brucellosis, and serious sequelae to seal finger (e.g., surgery and septicemia)" (Mazet et al. 2004, 3). Breathing problems and allergies were also frequent. Injuries resulting from direct interaction with the animals were caused by bites, for example. The highest risk for injuries and diseases was borne by those employees who had direct contact with live animals (not only with their excrements, blood samples, or cages) (Mazet et al. 2004, 9–12).[48] Rose et al. (2009) emphasized that the study showed that more than half of the reported injuries (ninety cases) were serious injuries—either deep wounds that had to be sutured with multiple stitches or bone fractures (Rose et al. 2009, 30, 70n230). Even the staff working with the animals who are trained in handling dolphins are therefore exposed to a high risk of injury and infection. Vacationers, visitors, or patients who come into direct contact with dolphins in swimming and therapy programs, usually have the naive belief that dolphins are particularly peaceful animals that are kind to humans. In reality, however, dolphins and other whales are predators that can react aggressively in contact with humans. If the swimmers touch the animals on sensitive parts of the body such as the eye or the blowhole, then the animals defend themselves, ramming or biting the swimmers or pulling them underwater. In Brazil, according to Rose et al. (2009) in their summary of such events, a man was even rammed into the abdomen by a dolphin with fatal consequences after apparently trying to stuff objects into the animal's blowhole. Adults holding their babies and toddlers over the edge of the pool and over the heads of the animals in parks and dolphinariums are extremely reckless. Trainers—especially those who work with orcas—should be aware of these risks. Many cases of injuries and even deaths are known worldwide (for a summary of known cases and the associated data, see Rose et al. 2009). For Marino and Lilienfeldt, the risks of serious injuries and infectious diseases should not be underestimated, even in the DAT programs.

According to the authors' evaluation, the interaction with the animals certainly pleases patients and represents an exciting new experience for them. Nevertheless, they warn, there is no scientific proof for actual, long-term therapeutic effects (Marino and Lilienfeldt 2007, 248). On the one hand, dolphin

48 The results of this survey were also published in a peer-reviewed journal in 2008 (Hunt et al. 2008).

therapy may only improve the patient's mood in the short term, while, on the other, it gives false hopes of long-term effects and is disproportionately expensive, at a cost of several thousands of dollars. Participants are financially exploited by a profitable industry closely linked to commercial swimming-with-dolphins programs. Lori Marino is particularly concerned that helpless groups of people are exploited here when these companies play with the hopes of desperate parents and with the problems of disabled people (see Clark 2007).

In general, the fact that evaluations of the success of DAT are often carried out by parents of treated children who are asked directly after the therapy should also be subject to criticism. The fact that their statements are positive could also be explained by justification strategies. Parents are never independent, objective observers of their children (the WDCS argues similarly in its critical report on the DAT project at the University of Würzburg, where this also seemed to be one of the main issues; see WDCS a).

Lori Marino refuses to keep dolphins in captivity for ethical reasons. She also believes that the dolphin-therapy industry sometimes supports the wild catch of animals with its brutal methods. To catch a single therapeutic dolphin, many of her conspecifics would have to die. Marino says: "If people knew how these animals were captured, I don't think they would want to swim with them in a tank or participate in DAT" (Clark 2007).

The criticism from the dolphin experts is supported by criticism from human physicians. Anna Baverstock and Fiona Finlay, both pediatricians at the Community Child Health Department in Bath, United Kingdom, come to the same conclusions as Marino and Lilienfeldt (2007) with regard to the effectiveness of dolphin therapy. In the best case scenario, dolphin therapy leads to health improvements comparable to those that could be achieved through other forms of therapy (such as animal-assisted therapies with domesticated animals like dogs)—in the worst case scenario, it prevents patients from receiving more effective help (see Baverstock and Finlay 2008).

Marino and Lilienfeldt's evaluation demonstrates that the criticism of DAT can be divided into two blocks of arguments: a group of counterarguments focuses on problems that are generally to be found regarding dolphins in captivity, such as wild catches, basins that are too confining, atrophy of complex communication and complex social behavior. These criticisms are directed in principle against any form of captivity of dolphins and not specifically against that for therapeutic purposes. From an ethical point of view, it is important in this context that not every criticism directed at dolphin therapies can generally be applied to all animal-assisted therapies. Alternatives to dolphin therapy (often also much cheaper), such as equine, alpaca, and dog-water therapy, do not show the aforementioned problems at all or do so in a much-reduced form—not least of all because they involve domesticated animals and not wild animals.[49] The second group of arguments that Marino and Lilienfeldt (2007) discuss is not directed against the capture and husbandry of (wild) animals but against the alleged therapeutic status of dolphin therapy. These arguments are thus specifically directed against DAT *as a therapy with dolphins* and not against dolphin therapy *as a form of keeping dolphins*. However, they focus not only on the scientific relevance and therapeutic value of the use and exploitation of the animals but also on derived problems, such as the fact that it is medically and ethically questionable for a therapy to offer patients hopes that it cannot fulfill or that DAT is disproportionately expensive in comparison with alternative therapies.[50]

Overall, the therapeutic status of DAT would have to be systematically examined to demonstrate a clear benefit for patients. However, this would probably require further intensive use and captivity of the animals, which in itself is questionable if dolphins cannot be kept in a species-appropriate manner and suffer due to the housing conditions. Among the points to be clarified would be, above all, the question of which factors ensure an (at least) short-term improvement in patient mood.[51] Is it simply the vacation atmosphere? Is it the contact with an animal that appeals to the patients' emotionally? Is it the contact with water, which in turn involves special psychomotor requirements? Is it the playful aspect of this social contact? Combinations of factors are also

49 The therapeutic value of such alternative animal-based therapies cannot be judged here (one might pose the question; yet in this case, too, whether and to what extent these interactions have an effect on the patient). From an ethical point of view, however, it seems significant that there are obviously therapies that are claimed to be comparable without involving dolphins. When it comes to dogs, horses, and alpacas, these animals have been domesticated for thousands of years, are not taken out of the wild, are accustomed to contact with humans, are bred specifically for interaction with humans, and their husbandry in captivity is not as problematic as it is in the case of dolphins.

50 From an ethical point of view, questions of distributive justice arise when treatment options for diseases are not equally available to all patients from all social classes. *Alpakas für Menschen* offers animal assisted interventions with Alpakas and other animals like mini pigs and chicken. They claim that Alpakas are often being called "Dolphins of the pasture, but the therapy can be offered much cheaper than the dolphin assisted therapy" (Alpakas für Menschen 2019). Most families often cannot afford expensive dolphin therapies.

51 Proponents of DAT have argued, for example, that dolphin therapy changes the brain-current patterns in the patient, that the ultrasound of the animals has a healing effect, and that through interaction with the animal an intensive feeling of connection with nature develops, which promotes the well-being of the patients (in accordance with the so-called "biophilia hypothesis"). None of these claims seems tenable (Brakes and Williamson 2008; Brensing 2004; Brensing et al. 2003; Marino and Lilienfeldt 2006).

possible, such as the specificity of concentrating on social interaction in connection with buoyancy, floating, and relaxation.[52] In fact, the therapy that was pursued by Eva Stumpf and Erwin Breitenbach's team in cooperation with the Dolphinarium Nuremberg also consisted of several (in this instance, three) components: the creation of a vacation and leisure atmosphere, the social-educational care of the family, and the interaction with the dolphins. Marino and Lilienfeldt (2007) were right to note that complex action mechanisms were involved in DAT. What happens during the sessions in Nuremberg is described in the project presentation: The children are highly motivated to interact with the animal. They are encouraged to take their own actions and can actively shape the situation. They can only express their needs if they use their options to communicate. These attempts at communication succeed because the therapist shapes the situation according to the child's impulses to act. Parents observe their child from a distance; they can relax and observe as they are not subjectively responsible for the situation, which enables them to perceive even small signals of their child sensitively. They can recognize new behaviors and observe old behavioral patterns from a new perspective. This results in parents reacting more sensitively to their child's signals after the therapy week, which in turn encourages the child to continue to interact actively with his or her environment. The interaction of parents and children is better coordinated (Universität Würzburg, Abteilung Rehabilitatonspsychologie 2015).

The main aim of the whole therapy regime seemed to be to sensitize the parents for the competences and signals from their children. What Breitenbach and his colleagues could actually demonstrate is that the parents perceive positive changes in the social-emotional and communicative behavior of their children. They described them as more self-confident and certified a higher social-emotional competence. They also observed an increase in the understanding and use of verbal language as well as an increased nonverbal reactivity. Even if we accept that these are substantial, lasting improvements and not just short-term improvements in mood, the question remains whether these changes were only noticed by the parents due to their increased sensitization to their children, although they already existed beforehand as part of the child's competences. The measurement of the therapeutic success here could not provide a clear answer to this, and whether the involvement of a

52 In this instance, it is certainly important that other forms of therapy, such as therapeutic riding, also induce a multi-dimensional response in the patient. Here, too, she is challenged on a physical, sensory and psychological level.

dolphin is necessary remains questionable. The researchers nevertheless concluded that the subjective assessment of parents about significant changes and therapeutic effects tends to be confirmed by objective data and that the effectiveness of DAT is particularly evident in the treatment of children aged five to ten years. In addition, the results confirmed that not all three therapy components are equally responsible for the success of the therapy because the effects in the ability to communicate, which should not be underestimated, could also be achieved without a vacation and leisure atmosphere and without social pedagogical care of the parents (Universität Würzburg, Abteilung Rehabilitatonspsychologie 2015).

Critics accuse the researchers of "unfounded overinterpretation" (WDCS a) of their findings and criticize the fact that only one of the measuring instruments, namely the parents' perception, proves the effectiveness of dolphin therapy (in contrast to the three other measuring instruments, which seemed to exist in the perception of caregivers, in interviews with parents and in video recordings). In general, only minimal effects could be measured, and a comparison group of thirteen patients in farm-animal therapy was too small to form a meaningful control group and to allow a statement to be made as to whether alternative therapies were conceivable (WDCS a).

DAT is/was offered in Nuremberg and in a growing number of other facilities around the world, despite the fact that no specific added value of DAT compared to other therapies can be reasonably seen. The evaluation in Nuremberg leading to the introduction of DAT came to very limited conclusions concerning the value of the therapy. In fact, the conclusions render it rather questionable. Still, the dolphinarium continues to offer DAT and advertises its facilities as suitable for this purpose. Via fundraising events, money seems to be raised to support DAT in Nuremberg.[53] The therapies, if they should indeed be continued, pose a major challenge for future research. Marino and Lilienfeldt (2007) come to the conclusion that

> before concluding that DAT possesses specific efficacy that could not be attained with a variety of alternative treatment conditions, authors must offer sufficient evidence that the same results would not be achieved

53 See https://tiergarten.nuernberg.de/en/discover/hoehepunkte/delphinlagune-und-manatihaus.html, 2018/07/22; and https://www.frankenfernsehen.tv/14-benefiz-gala-zu-gunsten-des-vereins-der-tiergartenfreunde-und-fuer-die-lebenshilfe-nuernberger-land-111734/, 2018/07/20.

> with another large, charismatic, attractive animal used in the same procedures. If the proper tests were conducted and the control and treatment groups were found to improve equally, one would need to conclude that there is nothing special or necessary about the dolphin per se in DAT. If the DAT group improves significantly more than the control group, then perhaps DAT could be considered a potentially therapeutically interesting intervention, although pragmatic issues of accessibility, expense, and risk would remain.
>
> MARINO and LILIENFELDT 2007, 248

Ethical concerns also remain. If dolphins really suffer from almost every conceivable form of captivity and cannot live in a species-appropriate manner, then this would have to be weighed against the benefit for the patients. It should be demonstrative that this benefit is clear, significant, and not achievable by alternative therapies in a similar way in order to justify the welfare of the animals being impinged upon. This means that even from a pathocentric-utilitarian perspective (in contrast to an animal rights perspective), advocating dolphin therapy seems problematic, as may be the case with other animal-assisted therapies.

If there is a benefit that is proven, if this benefit is significant, and if there really exist no other alternatives for critically ill patients, an ethical argument in favor of less-extensive forms of dolphin therapy could be worth considering: some problems of keeping dolphins in captivity might be solved by natural seawater basins or river basins. Instead of keeping confined animals, wild animals could be offered access to such pools, thus creating a therapy space from which the animals could withdraw at any time. Such therapies are already being conducted, for example, with Amazon river dolphins (*Inia geoffrensis*), also called "boto" in the Rio Ariaú, a branch of the Amazon (see Project Anahata 2008). The Anahata project, founded by Igor Simoes Andrade, offers children (who were, at least initially, almost exclusively from the Manaus area) the opportunity to swim and play in the water with the previously trained river dolphins. This interaction is combined with various physiotherapeutic measures. The project cooperates with the Ariaú Amazon Towers Jungle Lodge, a hotel that offers swimming with dolphins and makes its boat available to the project. Between 2006 and 2009, about seventy children with various diseases, such as cancer and autism, have been treated every year. Igor Simoes Andrade claims that he has not observed any aggression between animals and patients or dangers due to infection risks, but he stressed in 2009 that the project was only just beginning. The animals obviously enjoyed the interaction and were very careful with the children. Play and food were a reward for their cooperation, but they were completely free to come and go as they pleased.[54]

However, such forms of therapy do not eliminate the risk of infection and risk of aggression (that animals will injure patients and vice versa). In addition, the animals are usually fed and trained. So how "voluntary," in fact, is their participation? With regard to the Dolphin Research Center in Florida (one of the few facilities with seawater swimming pools that animals can leave if they wish), Thomas White also notes critically: "After years of living in captivity, forming bonds with the other DRC dolphins, and depending on humans for food, is staying put truly an expression of their freedom? Or is it the combination of 'learned dependency', loss of the social and hunting skills necessary to survive in the wild and a predictable source of food?" (White 2007, 4). Whether or not the animals are actually free to come and go as they please is therefore doubtful.

Above all, however, even if interaction were validated as therapy and without alternatives, a commercialization of exceptional therapies with wild dolphins such as those in the Rio Ariaú would not be desirable. A large influx of patients or tourists could lead to the animals being overwhelmed. In the meantime, it has been scientifically proven that commercialized swimming-with-dolphins programs and the presence of tourist boats in the animals' habitat lead to changes in behavior in wild populations. Animals spend much more time moving and much less time resting, having social interaction, and foraging. This will have long-term effects on nutrition, reproduction, and the use of habitats (Christiansen et al. 2010, 91, 96–97). The commercialization of therapies with free-ranging dolphins would most likely not only lead to negative effects on the animals themselves but also to the expansion of the infrastructure and therapy facilities. This would ultimately have a strong impact on the environment and the habitat of the animals. In principle, therefore, commercialization of such forms of therapy and interaction, which are relatively gentle on animals, could nevertheless be problematic from the point of view of ecology and animal welfare. It could lead to humans finally penetrating into the last hideaways into which these animals can still retreat. It is highly questionable if such commercialization can be avoided as long as programs are run on a commercial basis and as long as the programs are as attractive to desperate patients as they are due to subsequent marketing.

54 Igor Simoes Andrade answered an email request from the author dated March 18, 2009, and also answered critical questions. For some limited information on the Anahata project, see also http://igorsimoes.blogspot.com/, 2019/05/17.

If dolphins are even to be ethically considered as nonhuman persons because of their highly complex socio-cognitive and emotional abilities, any form of DAT would not be ethically justifiable as long as it remained unclear whether the animals participate voluntarily and benefit from the interaction. For representatives of the position that dolphins are nonhuman persons, these animals are excluded from exploitative forms of use by humans. Doing right to them is not a question of weighing their interests against ours. If their basic rights were at stake we could not get away by conducting a harm-benefit analysis, especially if the benefits were distributed unilaterally in favor of humans and the harm unilaterally at the expense of the animals (see also White 2007, 216).[55]

55 It is clear that even human persons are subject to harms, and not everything we expect of them is based on voluntariness and consent. But, for example, if children have to go to school, although they may not yet be able to understand the benefits of it directly, we still argue that it is in the interest of their long-term well-being. However, if a child actually suffers from compulsory schooling and the benefits of education are questionable, this attitude might very well come into question over the long term. The same high standard of justification would apply as well to any constraint which we impose on an animal, according to an animal rights view.

CHAPTER 10

Summary

Animals with complex emotional, social, and cognitive abilities make very high demands on a species-appropriate handling and housing, which can hardly be guaranteed in captivity. This applies equally to the keeping of great apes and dolphins as zoo animals, show animals, laboratory animals, or therapy animals. The same would apply to the performance of great apes in circuses or the use of dolphins in the military (since 1959 the U.S. Navy has trained dolphins, for example to detect mines (see Lee 2019 for more information)). It also applies to other farm animals, lab animals, and companion animals such as pigs, rats, or parrots, which have been, due to space constraints, excluded from discussion in this book. Many of the considerations in this book could be applied to these animals—with far-reaching consequences for our meat consumption, animal experimentation, and other practices.

The example of the great apes and dolphins has shown that the reasons why these animals are kept in captivity alongside millions of other animals and used for human-oriented purposes are quite different—but often questionable. In laboratory and research animals, human health interests are cited in order to justify the use of the animals. The important, even outstanding contribution of chimpanzee experiments to biomedical progress and the therapeutic value of interactions with dolphins is strongly emphasized. These initially sound like convincing ethical arguments, but they often crumble if we take a closer look at them. The question of species-appropriate or welfare-friendly husbandry and treatment of animals also arises in zoos and amusement parks, but they have to be debated against the background of the completely different objectives and functions of these facilities. In the case of zoos and amusement parks, one (if not *the*) main focus is on people's interest in entertainment and education. The conservation of species (dependent on the educated visitor) and the research necessary for this are also cited as arguments for keeping and breeding animals in captivity. We therefore face a mixture of arguments in favor of such captivity, related to human interests, and of arguments that claim to be ultimately not directed toward the interests of the kept animals but rather toward the protection of endangered species or the improvement of the human-animal relationship in general. However, none of the arguments in favor of zoos are undisputed or unproblematic. The ideal zoo would be a conservation breeding park, exclusively for threatened species and with exclusively species-appropriate husbandry conditions. However, this is an (ethical) vision

 | DOI:10.1163/9789004415072_011

from which many modern zoos are very distant. It would have to lead to a cessation in breeding for many attractive species, and even to the closure of zoos and a radical turnaround in the profile of many institutions. At the moment, however, contrary trends, such as an increased Disneyization of zoos, are being pursued. The conservation argument also seems to play a questionable double role within theming strategies as an image-creating justification strategy. Whether breeding in zoos is a rescue measure for endangered species that leaves no other alternatives is also in doubt. Many successful efforts in the wild, in nature reserves, and in breeding facilities that are not accessible to visitors, as well as the high costs and often unsuccessful breeding attempts in zoos, seem to suggest the opposite. In addition to the difficulties of species-appropriate husbandry, there is also the problem of wild catches to refresh or bolster stocks in captivity. Finally, the reintroduction of zoo-housed or zoo-bred animals to the wild is only successful in rare cases.

Also ethically problematic, as has been shown, is human contact with human beings' cognitive kin in the wild. Both the great apes and many whales and dolphins, like many other species, are endangered or threatened with extinction in their natural habitats. What is particularly troublesome is the fact that the animals are confronted with whole networks of partially interacting threatening factors, which are difficult to understand and almost impossible to fight. I could only sketch some of these in this book. The apes have become endangered by habitat loss because of large-scale agriculture or tropical timber industry, accompanied by the deadly impact of the bushmeat and animal trade and political instability in their range states. In the case of dolphins, the situation is hardly better if we think, for example, about overfishing and pollution of the oceans. Here, too, we are dealing with an industry that is becoming aware too late of the importance of sustainable management. Here, too, the animals are caught with brutal and frequently fatal trapping methods, and the products derived from them are traded (partly illegally) worldwide. However, while in the case of apes at least one of the biggest problems, bushmeat consumption, has to be partly considered from the point of view of the poverty in the countries concerned, the question of the justifiability of dolphin meat consumption is different. Countries like Japan, which are far from being able to argue with the poverty and hunger problems of civil-war-stricken Africa, insist on fostering traditions or gourmet interests that are only represented by a fraction of their population. As dolphins are also impacted by the tuna industry and by overfishing, ultimately everyone who buys fish, at least fish from certain fisheries, bares part of the responsibility.

In its ethical considerations, the last two chapters were based on the overriding question of why we should protect animals. The three central answers that were investigated were based on the argument from animal sentience and suffering, the argument from endangerment, and that from cognitive similarity. According to everything we can say today, all three arguments apply to apes and dolphins at the same time. They are conscious and possibly self-conscious creatures with complex nervous systems and complex brain structures. According to the three possible answers to the question of why they are worthy of protection, they surely have a pronounced capacity for suffering and the broadest spectrum of complex cognitive abilities known to us to date in a nonhuman species. The reasons for their Red List status, illustrating their endangerment situation, are extremely complex. This holds for great apes as well as dolphins. They thus represent the paradigmatic case for species to which all three arguments mentioned apply.

While the argument from suffering has had a strong tradition in animal ethics and animal protection since its inception, the argument from endangerment has only gained in strength in the last few decades, in view of environmental destruction and loss of species. The argument from cognitive similarity has only gained in importance in recent years against the background of increasing, above all ethological and comparative psychological research. The latter can be seen in the (philosophical, political, and legal) dissemination of the ideas of the Great Ape Project and the Hiasl case. The third answer, that animals should also be protected because they are cognitively similar to us, was given special attention in this book because it draws directly on cognitive kinship as an argument for establishing rights. Theoretically, the violation of personhood rights of nonhuman persons should lead to strong claims for protection, analogous to the violation of basic human rights of human persons. The significance and scope of this argumentation depends on two factors: on the one hand, we will have to see whether the claim for protection derived from cognitive similarity goes beyond what the other two arguments (endangerment and capacity to suffer) can afford. On the other hand, we will have to assess whether the arguments in favor of animal personhood can overcome some conceptual problems that are already incorporated in the idea of human rights. To "overcome" these conceptual problems can mean to solve them when transferring ideas of personhood and rights to nonhuman animals. But it can also mean to accept some conceptual problems here, in the case of nonhuman persons, just as we are prepared to do in the case of human persons.

In principle, animal ethics should examine, include, and use all available and valid arguments. In view of the threats that many species face, it is to be hoped that all three arguments in combination will lead to increased protection efforts. In response to the question as to why these animals are worth protecting, the complex answer emerging from all three arguments (their suffering, endangerment, and cognitive similarity) may be particularly strong because it addresses humans on different levels, most importantly on a rational and on an emotional level. Humans, who are open to rational arguments and yet tend to argue anthropocentrically, are probably best reached with the conservation or endangerment argument: if species die out, biodiversity and ecosystems—and thus also the basis of human life itself—are destroyed. The arguments surrounding animals' ability to suffer and the cognitive similarity are also rational arguments, but they are mainly based on the recognition of analogies. The morally relevant recognition of animal suffering and of complex socio-cognitive abilities can be seen as a question of logical consistency when research results indicate that animals possess both. Strengths and difficulties of the argument from cognitive similarity are discussed in the concluding discussion chapters (especially in Chapters 12–15, on cognitive similarity). First, however, the question of how a connection with our cognitive kin can also be established on a more emotional level should be investigated. The focus thereby will be on the fact that the role of the animal as the "alien," "stranger," and "other" transforms against the background of an increasingly perceived relationship between humans and animals.

PART 4

Discussion

CHAPTER 11

Cognitive Kinship and the Concept of an Evolutionary Self

11.1 Animals as Strangers and Kin at the Same Time

It has already been pointed out that conservationists place great hope in so-called flagship species, which serve as ambassadors of species protection (see, e.g., Ammann 2001, 72; UK Parliamentary Office of Science and Technology 2005). This is based on the assumption that we are particularly willing to protect those species with which we also develop a close emotional relationship.[1] Jacques-Yves Cousteau, with reference to the emotional dimension of animal protection mentioned here, formulated quite soberly that humans would *protect* what they *love*—but only *love* what they *know* (Cousteau and Schiefelbein 2008, 255). This can be connected argumentatively with reference to the cognitive relationship. We "know" those animals best with whom we share decisive similarities. The deepest form of "knowing" an animal could consist of "knowing oneself in the animal".

For centuries we have had a special relationship with animals like great apes and dolphins that resemble us in their social and cognitive abilities (see B. Ingensiep 1999; Bekoff 2007b, 491ff.). It is becoming increasingly clear that these animals resemble us in qualities that are particularly relevant to us, possibly even in those qualities that make us persons with personalities. This could not only have repercussions on our definition of such abilities and on the rights of these animals, but could also affect the human-animal relationship and our own self-image. To explain this in more detail, it is worth taking a look at the way we have perceived animals so far. Their role in the history of the human-animal relationship was that of the classical "stranger." In his lecture on philosophy of history, Hegel even called animals "the incomprehensible" and "the stranger par excellence." He addresses the dazzling character of their strangeness, that they are perceived partly in a demonized way as something we

1 This can be claimed both as a descriptive fact ("We protect animals because we feel a closeness and connection with them") as well as a normative requirement ("We *should protect* animals because we feel a closeness and connection with them"). I will follow the first one. The fact that humans appear to be particularly willing to protect animals when they feel such closeness may well result in the call for species protection to take advantage of this to motivate people to protect animals.

 | DOI:10.1163/9789004415072_012

cannot understand and grasp at all and partly as a friendly, sympathizing life. Above all, however, what strikes us is the contrast between their goal-oriented activity, restlessness, agility, and liveliness and their mute closeness and inaccessibility (Hegel [1837] 1970, 261).

Rainer Wiedemann uses Hegel's characterization to refer to today's ambiguity and ambivalence in the human-animal relationship. Although many animals, for example, as pets, would belong to our close social circle, they are "quite different" from us and "naturally" not equal to us. This still applied in principle, although we know much more about animals and their behavior today than we did in Hegel's days. Animals were still "strange creatures par excellence: trusted, but basically little explored companions in our everyday world" (Wiedemann 1999, 352; transl. MK). For the animal as the stranger, both apply: she shares our world and at the same time represents the "outside" and the "other." With reference to Georg Simmel's sociological category of the strange, Wiedemann thinks that the formal position of the animal is characterized by an ambivalent "synthesis of near and far" (Wiedemann 1999, 352; Simmel 1958; transl. MK). However, strangeness is a "relational difference" that is to be determined relative to changing perspectives and not abstractly (Wiedemann 1999, 352; transl. MK).

Strangeness, according to Ortfried Schäffter (1991), is not a characteristic of things or persons but a "mode of relationship in which we encounter external phenomena." It is not an "objective fact" but "an experience that challenges one's own identity," since novel relationships that are alien to the self-conception of the past are developed (12; transl. MK). New perspectives on the strange thus challenge one's own identity and can also change the relationship between the strange and oneself, for example, between the incomprehensible animal and the familiar human. Such a shift in relational differences and the relationships linked to them can happen through our increasing knowledge about similarities, *especially* if they target classical difference categories (like culture, language, and theory of mind). Even if something remains strange, it can change from something "incomprehensible" into something "comprehensible." Thus, the strangeness could become a respected strangeness that constitutes a new relationship between the now better understood entity and ourselves.[2]

2 Schäffter constructs an entire theory on how we experience the strange. Within his theory he classifies the perception of the animal exclusively in the category of the "the strange as the sounding board of one's own." One could argue that animals are, however, also classically anchored in the mode of the "strange as a counter-image." It would have to be examined whether, in addition, a perception of the animal would even be conceivable in the category of the "strange as an addition" or even "the strange as complementarity." In fact, what the first

It can be assumed that the animal will never abandon its role as fundamentally strange and fundamentally other, despite our progressive knowledge of cognitive kinship. But this role increasingly differentiates into a double role in which the animal is no longer *exclusively* viewed as strange *in the sense of incomprehensible*, but at the same time as *related to humans in a substantial way*. This is because evidence of similarity and kinship has penetrated into areas that were reserved for the human being in his or her previous self-image. Whether we share the structure of our skeleton or the nature of our teeth with animals as physiological similarities may not be of decisive importance for our identity, which usually overemphasizes mental capacities. In terms of socio-cognitive abilities, however, they function as a decisive x-factor in the formula that describes us as the pinnacle of creation: "Man is an animal + X" (Wild 2008, 26).[3] Only if we recognize ourselves as related to animals with regard to such abilities as well can we become aware of our own evolutionary connection to the rest of living nature, as a nature that is also shaped by mental capacities. Only then do we feel attached to it and not detached and elevated from it; only then do we feel integrated instead of decoupled. In this sense, newly pointed out kinship relationships in the field of cognition can contribute decisively to a new *consistently evolutionary self-understanding* of humans.

However, the fact that we perceive the animal as strange and different has a centuries-old tradition. Darwin's idea of a gradual difference with regard to mental abilities and the beginning of their systematic scientific research are

three modes (soundboard, counter image, and addition) have in common is that the strange "is not left standing in its peculiarity, the dislocation with it does not happen in a dialogue with partners, but all otherness is appropriated in the shortest possible way as one's own." However, a perception like this is questionable. A naive emphasis on the cognitive similarity of animals that measures animals uncritically against what humans are able to achieve can often not do the animals justice. We should not and (do not have to) go so far as to completely dissolve the otherness of the animal. This is not necessary for an effective respect of ethically meaningful similarity. Instead, one could become aware of one's own perspective and "let the strange remain to some degree strange." With the example of the animal, strangeness may become an "arduous experience of an *opposite* boundary." This would, however, be a different kind of borderline than the one we have drawn between humans and animals so far. In the sense of a changed relationship between the strange and the own, it could lead to a "new form of 'common ground,'" which may prove to be more sustainable "than the empathy for *supposed* 'universal' foundations of humanity" (Schäffter 1991, 28, italics added, transl. MK).

3 In accordance with this formula, according to Wild, humans have been characterized as rational beings who have language, form states, and use their hands (Aristotle), possess a soul (Descartes), have reason (Kant), have knowledge of death (Hölderlin), can adapt to everything (Dostoyevsky), are not determined (Nietzsche), are eccentrically positioned (Plessner), or have a world (Heidegger) (Wild 2008, 26).

only a stone's throw away in comparison to this orthodox view. There is much to suggest that the idea of only gradual differences in cognition has not yet been consistently grasped. In an essay in his 1917 work A *General Introduction to Psychoanalysis*, Sigmund Freud rightly ranked among the three great affronts of mankind[4] an indication of the *identity-threatening potential* of the idea of having to abandon the dominant position as the only rational being in an otherwise mindless nature. New insights into the cognitive relationship between humans and animals exactly hit the mark of this offense.

11.2 On the Way to a Nature Deficit of a Special Kind

In view of the importance of cognitively related animals for our evolutionary self-image and in light of the Red List status of many such animals, we should urgently ask ourselves what we will miss if, for example, the great apes disappear. They are both our closest evolutionary kin and the creatures in which we probably recognize cognitive similarities most clearly today. Will we experience a feeling of being lost, a "nature deficit" of a special kind? A nature deficit in our self-image and our own positioning? By doing nothing to protect these animals from extinction (and by in fact accelerating this process more and more) humans would decouple themselves from nature in such a way that they would eradicate their *direct evolutionary link* to the rest of the animal kingdom. This could also affect their self-image, which is to some extent evolutionarily rooted. Today these animals are still *existing references* to a last common ancestor and the probable *most direct representative* of cognitive similarity. They embody evolutionary theory, making something that we *theoretically* know *real*

4 He understands this to be the particularly resounding effect of certain scientific research on human self-love. Copernicus's "cosmological affront" cleared up the fact that the Sun, Moon, and planets do not revolve around the earth, and the central position of the Earth is thus no guarantee for its dominant role in space. It was followed by the second, "biological affront" by Darwin, which put an end to human's arrogant self-assessment as the only beings gifted with reason and soul. It was followed by the "psychological affront" of Freud's own theory, which explained that many unconscious mental processes and impulses are only accessible to and subjected to the self through an incomplete and unreliable perception. While the first affront meant that man is not ruler and center of the universe, and the second affront questioned his claim to power over the rest of living nature, the third affront stated that *the self is not master in his own house* (see Freud 1917, 4). All three affronts thus call into question, so to speak, a self-image of humankind, which is conceived anthropocentrically and is based on a claim to power over nature, and also over one's own biological nature. For his famous description here, Freud took some input from the nineteenth-century German physiologist and polymath Emil du Bois-Reymond.

and experienceable. Most people are deprived of a direct, unadulterated encounter with them (even in the zoo, they are only a tired image of themselves and their abilities). Yet we know they actually exist. If we watch documentaries about them, look at photos of them or read reports and studies about their behavior, we know that these animals still exist at this moment and are directly affected by our actions. Documentaries about them are not reports about long-extinct species whose fate is sealed and whose life and survival are therefore no longer related to our life and actions today. They are not history, but part of our own reality. If these animals die out, then this direct reference is lost, which could lead to a special form of dissociation between us and our connection to the rest of living nature.

In recent years, environmental psychology, psychology of environmental protection, and wilderness education have established themselves as new branches of science that deal with humans' relationship to nature, even to an increasingly vanishing, original nature. In his book *Last Child in the Woods: Saving our Children from Nature-Deficit Disorder*, Richard Louv, for example, explores the question of what ecological, social, and psychological effects it will have on future generations that they lack real experiences of nature and the knowledge of a personal connection to nature (see Louv 2008). In this instance, his considerations are to be linked with the idea that animals as our cognitive kin also embody such a personal connection to the animal kingdom.

According to Louv, children and young people today (and even more so those of future generations) spend a large part of their leisure time in front of the television or computer and are neither concerned with nor in nature (see, e.g., Louv 2008, 12, 63): "For a new generation, nature is more abstraction than reality. Increasingly, nature is something to watch, to consume, to wear—to ignore" (Louv 2008, 2). There is an increasing dissociation of people or children from nature, which has concrete and abstract consequences, from the atrophy of the senses and physical fitness through a one-dimensional perception and activity to the loss of knowledge about nature, and a loss of interest in nature as well as a loss of interest in nature conservation.

Many experts share Louv's ideas or have anticipated them. His approach is based on philosophical concepts of the human-nature relationship, which assume a fundamental importance of the latter for the former. Edward O. Wilson describes the topic in his biophilia hypothesis, based on Erich Fromm, according to which humans feel an urge to see themselves in connection with other animals and have an innate affinity and love for living nature. Therefore we are attracted to other living beings and need contact with nature in order to remain healthy, to find meaning in our lives, and to become real (see Wilson 2003; Louv 2008, 43–44). From a sociological point of view, Theodore Rozak argued that no separation is as incisive and as important for the experience and self-image as the separation of humans from nature. He therefore proposed a definition of mental health in relation to nature and the environment (see Rozak 1992; Louv 2008, 44). Finally, from an educational point of view, Howard Gardner (1999) developed an understanding of intelligence in which knowledge of animals and plants also plays a decisive role. According to Louv, interests that focus on nature and animals are competences that are important for the development of one's own identity, for an identity that is based on a connection with nature and a sense of responsibility for it. Children who have such a connection to nature are the animal ethicists and conservationists of tomorrow. Louv impressively shows, however, that scientists all over the world observe that this area of competence and this self-image could actually be lost by future generations who are less concerned with nature and animals.

Louv and Wilson, Rozak and Gardner are all concerned with the psychological and anthropological significance of nature as a whole, not specifically with animals and certainly not specific animals, such as our cognitive or evolutionary kin. However, the importance of these animals can be easily connected to such concepts. The above-mentioned authors and many other scientists who conduct research in this field are concerned with intimate, close contact between humans and nature and with a self-image that becomes aware of one's own attachment to nature and its significance for one's own health and development. An extension of the presented theories by the dimension of evolutionary and cognitive kinship would be conceivable. Our awareness of the importance of our connection with nature for our identity and mental health can be strengthened and filled with meaning by the knowledge of such kinship relationships. In addition, it plays a major role in the presented considerations on the nature deficit that with the decoupling of humans from nature the sense of responsibility for its preservation is lost. But this is the motivational cement of the animal and environmental protection movement. Louv believes that its future success depends not only on the strength of the organizations involved, but also on the *quality of the relationship* of upcoming generations to nature and whether they are emotionally connected to nature (Louv 2008, 159). Some nature and species conservation organizations are now pursuing the educational idea of bringing children closer to nature (such as the Jane Goodall Institute, with its "Roots and Shoots" program).[5] Successful nature and species protection is dependent on a personal connection with nature in general and on the feeling of connection with nonhuman animals in particular. Hearing about new findings on the amazing abilities of animals (Louv refers

5 See https://www.janegoodall.org/our-work/our-approach/roots-shoots/, 2019/05/17.

here, as an example, to studies on whale songs) cannot replace direct contact with nature, but they are astonishing: "My hope is that such research will cause children to be more inclined to cultivate a deeper understanding of their fellow creatures" (Louv 2008, 24). Here, too, it can be assumed that an understanding of other animals as "fellow creatures" can be deepened through the dimension of cognitive kinship.

In summary, it can be said that an emotional closeness to nature is neither exhaustively nor exclusively producible over the knowledge of the cognitive kinship. But it is surely impressively producible that way. Animals with which we have a lot in common can be a link between us and the nonhuman nature to which we then also feel connected in relation to central characteristics for our self-image.

In a global effort, species conservation experts, primatologists, and philosophers have stressed the similarity of humans to their closest relatives in order to argue that their better protection is our moral duty. Under the umbrella of the United Nations Great Ape Survival Project, they launched the World Heritage of Species Initiative (see Wrangham et al. 2008). It stipulates that we not only have a world heritage concerning sites of outstanding cultural or natural importance like special cities, buildings, or national parks, but also need a *world heritage with regard to other species*. The great apes are declared a family of particular importance to be protected for future generations. This political initiative is also based on the desire to draw attention to the importance of these animals as an indispensable element of our own identity and to place them above short-sighted interests. Anthony Rose aptly describes this underlying concern as follows: "A perpetually rich and thriving African rain with its apes and other ancestors alive and well is worth far more now and in the future than bundles of wood and bushmeat. Beyond the oxygen and medicine that the forests produce, and the lush beauty and mystery they provide, they give us *profound insight into our identity*. It is, after all, out of Africa that we hominids came. It is in Africa that we discover who we are and thus face our potential for being more than selfish humans ruling and consuming a vanishing natural world" (Rose 2002, 235; italics added).

The success of such global efforts, which even include the impact of the survival of related species on our identity, will have to be proven in the coming years. Skeptics may not be able to shake off the impression that despite the promotion of ethical and environmental educational concepts, we are heading for the extermination of our closest kin. The discrepancy between the ethical claims raised and our actual treatment of these and other animals can hardly be resolved as such. It can be seen as the essential feature of animal ethics and animal welfare in the twenty-first century. However, the global animal welfare and animal rights movement as well as the conservation movement are relatively recent phenomena that have achieved much in the short history of their existence. They must react to a rapidly increasing network of threatening factors and develop new strategies and powers of persuasion. In addition to the classical arguments to avoid animal suffering, animal ethics and animal protection can now also fall back on conservation arguments along with the as yet little developed but important argument from cognitive kinship. It is to be hoped that a combination of all three arguments will have an increasingly significant effect.

CHAPTER 12

A Comparison of Arguments

After considering the importance of the argument from cognitive similarity for our evolutionary self-image and the motivation for conservation, in summing up we should explain open questions regarding this argument in relation to other animal ethical arguments. It has already been pointed out that the strength and scope of this argument depends on whether the right to protection derived from cognitive similarity goes beyond what other arguments can provide.

12.1 Animal Welfare Arguments vs. Species Conservation Arguments

Clearly, the argument from cognitive similarity, much like the argument from the ability to suffer, is an argument that is oriented toward *animal* protection, i.e., ultimately toward the protection of individual animals and their needs and not a classical *species protection* argument (see Chapter 8). However, the considerations of the Great Ape Project and Thomas White give the impression that, although they work with the argument from cognitive similarity, they are aimed at protecting entire families or species (such as great apes and dolphins). This may seem confusing at first glance. However, these approaches are compatible with species conservation considerations because the demand for the moral consideration of each individual perspective of each individual ape and dolphin (the demand for the consideration of their abilities, needs and their rights derived from them) naturally means that the individual perspective of *all* apes and dolphins must be morally considered. Finally, it is assumed that the above-mentioned morally relevant cognitive abilities do not only belong to single individuals of this species, but to all their representatives. At the level of justification of rights, such abilities are considered as characteristics of these animals and thus as characteristics of their species (this latter assumption runs us into problems when not all those representatives do anymore, yet, or ever possess the respective abilities. To still include them, it seems that we need additional arguments, see my discussion about "marginal" cases and the concept of personhood in Section 8.4–8.6).

What is of importance here, is that, logically speaking, the individual is included in the perspective of the community, but the community is not

 | DOI:10.1163/9789004415072_013

included in the perspective of the individual. While animal welfare arguments, which begin with the individual, her abilities, and needs, can be continued argumentatively up to the protection of whole species, it is not necessarily possible to protect individuals by starting off with a focus on species conservation.[1]

Often, a violation of the individual needs of individual animals is even deliberately accepted and considered morally correct in species conservation if an endangered species benefits from it. For example, traps are laid out to protect endangered, flightless, ground-breeding birds such as the New Zealand kiwi or kakapo in order to decimate predators. The survival of the endangered species has a higher value here than the survival of other individual animals. Some animal rights activists criticize this and instead do not want the right to life of different individuals or their needs to be played off against each other. Consequently, they defend other methods of protecting endangered species, in particular with regard to controlling overpopulation, by nonlethal means. Even at higher financial costs, birth control methods, for example, would have to be found. Hormonal contraception or relocation options do not require killing.[2] This should make it clear that species protection can also be pursued using an animal welfare approach. Conversely, it seems to me to be more difficult to also argue for the protection of the individual within a pure species conservation approach. Those who start from the individual can consider the species as an accumulation of individuals. However, those who only attribute value to the entire species reach their limits when it comes to the value and protection of the individual.

The argument from suffering (or pathocentric arguments) and the argument from cognitive similarity therefore differ next to classical species conservation arguments because they include the perspective of the individual. On the other hand, an argumentative permeability of the first two toward species conservation seems possible, while this seems difficult in the opposite direction.

1 The Great Ape Project and Thomas White do not do this either. Nevertheless, the endangerment of apes and dolphins are their main motivation for thinking about increased protection rights for these animals.

2 Discussions of this kind are illustrated, for example, by the culling of South African elephants to regulate the population. While its supporters want to control overpopulation by shooting the animals, opponents demand nonlethal methods for ethical reasons (such as administering hormonal contraceptive drugs or resettling parts of the herds).

12.2 Utilitarian Pathocentrism vs. Personhood Rights

What is the concrete difference between different animal welfare arguments as they are used in pathocentric-utilitarian approaches and animal rights approaches? And what added value does the argument from cognitive similarity in the sense of the human rights-analogous animal rights approach represent? I would like to compare two different forms of utilitarian-pathocentric approaches (a weaker, pragmatic one and a stronger one)[3] with the argument from cognitive similarity as it can be used in an animal rights approach based on human rights.

Representatives of easy-to-implement, biologically based animal protection who are not primarily interested in questions of ethical justification often use a pragmatic, utilitarian-pathocentric approach. The central question within this approach is not whether we are morally allowed to keep animals in zoos or to use them in experiments. Instead, the focus is on how to deal with these animals *presupposing that* we actually already keep them in captivity and use them for our purposes and interests. Under these conditions, *how can we best guarantee animal welfare*? This question is then being answered from a biological-veterinary point of view, e.g., with regard to measurable suffering, the evaluable preferences of animals, and the possibility for animals to cope with their environment by exercising species-typical behavior (we revisit this biological animal protection approach in Chapter 14).

Even within such approaches, the cognitive, social, or psychological abilities of animals can and should be more strongly emphasized. However, they only represent *another (previously mostly neglected) factor*, which pertains to the well-being of animals: the fact that some animals have such abilities results in certain needs; if these abilities are not taken into account, then animals can develop stereotypies. While this approach includes the assumption that animals that possess the ability to suffer deserve direct moral consideration, nothing is said about what status the avoidance of suffering should have. Thus, there is a call to avoid unnecessary suffering in animals, but it remains open whether this also means that animals have something like a "right" not to suffer.

---

3 "Weak" and "strong" are to be understood here in the figurative sense. The second approach is more potent than the first, because it promotes the avoidance of suffering as a more assertive norm and takes an explicit position on the value of this norm. Both approaches, however, serve above all illustrative purposes, and are of a constructed nature.

A strong utilitarian-pathocentric approach, on the other hand, could also be based on the obligation to avoid suffering, but then reformulate this norm itself into a right. From the human point of view, we then face the *obligation to avoid (unnecessary) suffering*.

Since the prevention of suffering remains the central norm, this only concerns the question of the right of animals not to suffer, not of other rights (we are quite close to a rights approach then, but not fully there, so to speak). But in contrast to the weaker pathocentric approach, it takes a stand on the moral force that the norm of avoiding suffering should have for animals. Because it is possible to reformulate the norm of avoiding suffering into a right of animals, and because pathocentrism is a crucial element of utilitarian and rights approaches, a strict separation of pathocentric approaches and animal rights approaches is impossible. Ethicists often use mixed arguments. The right of animals not to suffer, or their right to good or even the best possible well-being (measured naturally by the needs of animals and not by the interests of their keepers) is conceived here as an *animal right* and not specifically analogous to human rights (this clearly distinguishes this second approach from the argument from cognitive similarity that underlies human rights-analogous animal rights).

Cognitive abilities are part of the strong utilitarian-pathocentric approach in the same way as they were part of the weaker approach; they also represent one of several factors constituting well-being. Although the right of animals not to suffer is already mentioned here, it is still relatively unclear *to what extent* the claim or the right of the animal to the best possible well-being or avoidance of suffering counts. This is because in all pathocentric *utilitarian* approaches, no matter how weakly or strongly they are formulated, all interests and rights are weighed against each other in principle (this distinguishes these approaches from a real animal rights approach). Even if the strong utilitarian-pathocentric approach incorporates something like a "right" not to suffer in animals, in the end this can potentially be overridden by other interests. Thus, the weighing procedure raises questions such as the following: Is the humans' right to entertainment, education, or research significant or strong enough to trump the right of animals not to suffer? Can it justify keeping animals in zoos, even if the best possible or even simply adequate well-being cannot be guaranteed? Is the humans' right to develop drugs and therapies to combat diseases strong enough to undermine the "right" of laboratory animals to good or best welfare and to say that their use is morally warranted? In a typically utilitarian manner, the "right" to animal welfare is still weighed against the interests and rights of others, especially against the rights and interests of human beings. Especially if one does not distinguish between *arbitrary* and *vital* interests, then the interests of an animal are often subordinated to interests of humans. This is the case, for example, when animal experiments are conducted that do not serve the survival of human beings, but do so for the approval of a new drug variant without added value, for which there are already sufficient alternatives on the market. But even if vital human interests are exclusively weighted against "equally" vital interests of animals, that is, if animal experiments are only considered ethically justified if the survival of a person really depends on them, the equivalence of vital interests on both sides remains questionable. Since utilitarianism as a "method" focuses on a harm-benefit assessment and not on the inherent value of an individual, the benefit of the experiments (for humans) can always be superior to the vital interests (of the animals). There are no rights in a strict sense that are attributed to vital interests and that are *in principle not weighable.*

Every approach that envisages such a balancing of goods has to struggle with what this book has shown in detail: even arguments that can bring supposedly high benefits such as human health interests into balance (for example, by justifying animal experiments and animal-assisted therapies) are often vulnerable and fail to show the one clear and demonstrable asset that is claimed. Often, there would be other, less ethically problematic alternatives available. The hedonistic calculation of utilitarianism also confronts us with the mathematical problem as to *how much* suffering *from how many* animals can be accepted for just *how great* of a benefit *for how many* people. Is this more a question of the number of animals suffering, the severity of suffering, or both? A pathocentric-utilitarian argument does not necessarily lead to a moral prohibition of harmful practices as long as no *limits* are established, from which great pain and damage are no longer tolerated, no matter how high the benefit for humans may be. However, it seems very difficult to determine the point at which the suffering of the animals weighs heavily enough to tip the balance in their favor. This may be for two reasons.

First, intolerable suffering in the sense of an absolute limit for each individual species has to be defined species-specifically. In addition, it must be taken into account in all species that physical and mental abilities lead to physical as well as psychological needs, which in turn lead to physical and psychological suffering. Here the task is certainly to better recognize the cognitive-emotional needs of all animal species and in fact various individual animals and to pay more appropriate attention to them.

Second, and this seems to me to be the bigger problem, such a definition of limits beyond which the suffering of animals in relation to a possible benefit is no longer justifiable is always an *ethical* question, not just a biological one. The question is not only when, in what way, and to what extent serious suffering actually exists, but also what the relationship between suffering and benefits is and how profound the suffering must be in order to reach the limit of what is no longer ethically justifiable. This not only raises the problem of (biological) understanding of suffering, but also that of the *normative evaluation* of the same. What are we (still) allowed to do to an animal from a moral point of view, and what shouldn't we do anymore? This absolute limit, above which it is no longer ethically justifiable to accept the suffering of animals, results either from the fact that the benefit in comparison to suffering is too small or that the suffering is too high in comparison to the benefit. In this instance, it is necessary to make fundamental decisions. One can either withdraw to the position that the suffering of animals and their right to well-being always, and under all circumstances, is less important than human suffering and well-being (and should then, of course, be able to justify this). Even if the lives of all the millions of lab animals were to save just one human life, the use of the animals would still be justified, according to this argument. Alternatively, introducing this absolute limit for the harm-benefit analysis might also be necessary. It could mean that research should always cease if it leads to great suffering for animals—even if significant research results were to be expected. This, too, is a normative decision. And it is the decision that comes first: the empirical determination of suffering is, so to speak, only of subordinate relevance when it examines in the respective context whether the norm is applied (because there is severe suffering) or not (because there is no severe suffering). Catalogs of criteria for the ethical justifiability of animal experiments can follow this line of argument. If one asks about the reason for the norm, why even very great benefits run up against a limit in case of great suffering, then, it seems, interests or rights come into play (in this case those linked to suffering) that go beyond utilitarian balancing decisions.

The argument from cognitive similarity, which leads to some basic, inalienable personhood rights for (some) animals, addresses precisely this problem, which remains partly open in the other two approaches. In assessing the right to physical and mental well-being, it uses a measure of value that is clearly defined and politically implemented, on the one hand, when dealing with human persons and, on the other hand, addresses the strongest, and therefore most fundamental, human rights. It thus comes down to a clear evaluation of the strength of these rights in relation to other rights, and also in relation to other human rights, and can build on the expressive and appellative power of the standard of value of human rights. At the same time, however, the conceptual contradictions and problems of a secular justification of human rights linked to cognitive abilities must also be accepted, in this case. The main problem with the idea of human rights is that its strict interpretation and application is difficult to deal with in "borderline" cases who do not (fully) possess the required abilities. This applies to *human persons who are "borderline" cases,* such as very young, old, sick, or severely mentally disabled people, as well as to animals as *nonhuman persons who are "borderline" cases*. If the disadvantages of the concept of personhood seem too great, then a justification of human rights based on cognitive capacities, which, however, was and is the central one in the history of human rights, must be abandoned—for human beings and animals. If the disadvantages of this concept fail to seem significant enough to outweigh its advantages, especially the significant assertiveness of derived rights, then modifications of the approach—again, for humans and animals—can be considered. Personally, I am not prepared to abandon the human rights approach in general. I am convinced that the required socio-cognitive abilities are being overemphasized both in the establishment of human rights and in ideas of human identity, but not that they play no role at all there. I doubt that with pathocentric approaches that follow the utilitarian calculation, equally strong rights can be enforced for human and nonhuman animals. The opportunities posed by the personhood idea and the possibilities for their modification should be investigated further.

12.3 Rights that Go Beyond the Weighing of Goods

The personhood approach seeks to take into account that among the abilities that we are increasingly demonstrating in animals (at least in some species), there are those that are of particular importance when it comes to human rights. If it is assumed that these abilities play a special role in the self-conception of human beings, then it is logical at the outset to consider the implications of this. Let us make this very clear: if we make the ability to suffer a central criterion of moral consideration and determine that animals are also able to suffer, then this has consequences. If we (additionally) adopt other criteria of direct moral consideration in humans and find these fulfilled in animals, then this also has consequences.

The animal rights approach that is construed analogously to human rights assumes that such abilities constitute, among other factors, the well-being of human and nonhuman animals, but that their significance goes beyond that. The socio-cognitive abilities at stake here are considered to be those characteristics that give a person the status of a person.[4] In this sense they are morally relevant qualities. Because people possess these abilities, they are persons, and

4 Much like White (2007), I understand "human" as a biological entity and "person" as a philosophical category. At the outset, in applying the categorical separation, nothing has been said as to whether all humans are persons or not. This depends on the criteria of personhood. It is

because they are persons, they are granted special, inalienable rights that go beyond the right not to suffer.

These rights serve a very special purpose in the case of humans: they put an end to every utilitarian weighing process. To torture, murder, or deprive an innocent human person of her liberty is not a question of weighing up interests or rights. It is simply morally wrong in and of itself. We do not kill people to eat them and do not lock them up for our entertainment—and we have good reasons for not doing so.[5] It violates their human rights and contradicts any talk of human dignity. Human beings as persons are an end in themselves, who must never be used only as a means to an end. This imperative is based on the Kantian image of human beings. Humans are gifted with reason, and they are autonomous beings capable of self-governance. The fact that each person is an end in and of herself can lend her a special positioning outside a utilitarian cost-benefit consideration. No interest other than that of respecting one person's human rights is strong enough to undermine another person's human rights. Within a human rights approach, however, ethical dilemmas always arise when human rights conflict with one another. Consciously giving priority to a right is only possible if a hierarchy of human rights is assumed. But if, under the same conditions, two equally fundamental basic rights are involved (such as the right to life of one person against the right to life of another person), a targeted preference of one person is simply morally wrong. Only in very limited, exceptional situations can it be ethically *excusable* (though never morally correct) to do so.[6]

The argument from cognitive similarity implies precisely such strong rights, analogous to human rights, for animals that demonstrate characteristics or levels of the characteristics of a person. This avoids questions of weighing and proportionality in the sense of offsetting harms and benefits. It is accepted on

---

also possible that (deliberately anthropocentric) additional criteria lead to a special protection of humans.

5 Human rights are normally respected in today's constitutional states. Of course, human rights violations do occur on a daily basis, but violations are then perceived as such. The fact that a right *is sometimes violated* cannot in principle invalidate its *normative status* or its *international recognition* (in this case by the UN member states), but certainly its *legal and political implementation* can be rendered uncertain.

6 If the fundamental rights of two people are stacked against each other, and a decision must be made in favor of one of the two, there can be no morally correct decision, strictly speaking, because either way, the rights of one of the two persons would be violated. However, here no moral reproach can be made against those who still have to make a decision if compulsion to act and lack of alternatives apply. A similar situation applies to self-defense when one's own fundamental rights are threatened by another person. It would not be more morally better to die yourself than to kill the other person in self-defense.

this basis, then, that some practices that could benefit humans while violating the fundamental personhood rights of the animals used may be completely abandoned.

The strength of the argument lies in the fact that we always make a clear and nonhesitant decision along these lines when it comes to people. With respect to human dignity and human rights, it is thus quite practicable and possible to reject utilitarian weighing decisions with reference to human rights. Human trials would also be useful for the development of drugs and therapies. Nevertheless, we renounce them. Under the same conditions, the same clear decisions may be taken with respect to animals. With the call to exclude animals from some harm-benefit considerations in principle if their rights are affected, the criticism of animal rights activists is vehemently directed against *utilitarian* pathocentrism—not primarily against the pathocentric norm to avoid suffering, but against the utilitarian approach.[7] The central criticism is that the calculation of the greatest happiness of the greatest number cannot always guarantee the happiness of the individual, not can it guarantee for goods like freedom, continued life, or absence of suffering. It is also considered problematic that the greatest happiness of the greatest number is difficult to determine and that there is a risk that this will be determined, structured, and used by those who have the upper hand politically.

The real question, then, is what we are prepared to accept, either by following a pathocentric-utilitarian approach or a personhood approach. Should we continue to adhere to utilitarian balancing scenarios and concentrate on the ability to suffer as the only relevant criterion for moral consideration? In doing so, we would ignore the fact that we also accept other criteria of moral consideration when dealing with human persons and that, so far, we are light-years removed from considering equal suffering (no matter who experiences it, dog or human) as morally equal. We also ignore the fact that there could be such a thing as individual, enforceable rights for animals—rights that clearly

7 Regan, for example, makes it quite clear that he is upset by the weighing, "aggregative" aspect of utilitarianism. Utilitarianism merely amounts to the satisfaction or frustration of the preferences of different individuals without attributing an inherent value to the individual herself. Thus, weighing goods can never suffice. Instead, Regan binds the inherent value of each animal to her consciousness and individual well-being. As sentient and experiencing subjects-of-a-life, our individual well-being is of importance to us, regardless of how useful we may be for others (see Regan 2004, 243, see also xvii). In Regan's theory, too, the ability to suffer remains—as one of several important criteria—in the discussion of the *experiencing subject*, which strives for *individual well-being* and possesses a *quality of life*. However, the inherent value of an animal is then no longer something that has to be weighed up against other interests in a utilitarian manner and, if necessary, abandoned. White (2007) argues similarly here.

demarcate the limits of what is morally permissible, and say that there are acts we should in fact refrain from completely, regardless of whether anyone would benefit from them. We may have found a *practical* way of handling animals, with pathocentric-utilitarian or simply speciesistic methods. However, such approaches often accept, without any questions, the status quo of current animal use and the normative presumptions that constitute it. It is not the task of ethics to describe what is practicable, but to identify action as morally right or morally wrong with regard to defined norms. Anyone who rejects the personhood approach must at least also be aware of the weaknesses of alternative approaches and their preconditions requiring justification.

12.4 Criteria that Go Beyond the Ability to Consciously Suffer

One strength of the personhood approach lies in the fact that it is based on a *complex picture of the individual* instead of the *lowest common denominator*. Those who only consider special forms of consciousness, such as phenomenal consciousness, as the most important aspect that animals possess have thus attained a very general criterion for ethics, a criterion that can be applied to all sentient animals—but they run the risk of not meeting the needs and dignity of all animals equally. We are also called upon to do justice to the diversity of animals with a view to possible different needs. The personhood approach, therefore, includes the criterion of conscious sentience and suffering, but does not take it as the only standard. The observance of (other) socio-cognitive characteristics and abilities, i.e., culture, language, and theory of mind, become further norms.

In this book it was basically assumed that all animals with such complex socio-cognitive abilities as culture, language, and theory of mind also have consciousness or even forms of self-consciousness. For example, the passing on of traditions through social learning requires some form of consciousness. It is important that the assertion that culture, language, and theory of mind are conscious processes is not at the same time an inverse conclusion. It is theoretically possible that there are (vertebrate) animals with consciousness that possess neither culture, nor language, nor theory of mind (or other similarly complex socio-cognitive abilities). However, the more basically we define such abilities, the smaller such a group of animals will be.

Connected with the discussion about consciousness is the legitimate question as to whether consciousness (as connected with the ability to suffer, i.e., the classical pathocentric criterion of conscious suffering) is not already a good candidate to represent the threshold criterion of every form of personhood. Personhood criteria would then have been considerably de-intellectualized and certainly apply to a large circle of animals. I leave it open whether this would be a better alternative, but I am skeptical whether this is helpful in describing what animals can do and know. As was already outlined earlier, (see Chapter 3), there are different and also varyingly complex forms of consciousness. From a philosophical point of view, therefore, the task of differentiation arises, for example, between phenomenal consciousness and semantic or propositional knowledge. Even if one wants to concentrate on phenomenal consciousness and thus on conscious suffering, the question of possible other gradations and building blocks automatically arises. This would also raise questions here, which I have explored within each individual competence—and which we can now apply within the framework of conscious suffering. Regardless of all the difficulties in determining the content of terms such as suffering, cognition, consciousness, and various individual competencies, we must sometimes draw clear lines in animal ethics and animal welfare where none seem to exist. The problem of setting borders ultimately arises in all approaches in which biological criteria are given normative relevance or in which standards are not only set metaphysically but are to be bound back to the real world, and thus ultimately to what can be found biologically. At the latest, the problem of line-drawing arises in every approach when it is applied to specific cases. This is certainly a problem, but it is not a problem that exists only in an animal rights approach that argues for an analogy to human rights.

Animal ethics, which for a long time concentrated on the phenomenal consciousness of animals in the context of their well-established capacity for conscious suffering, should not ignore other forms of consciousness, such as an awareness of states of perception, in which a simple form of self-consciousness could already manifest itself. This is also in line with the assumption of a graduality of abilities. On the one hand, consciousness should not be reduced to conscious suffering. On the other hand, strict attention must also be paid to the fact that a more complex concept of consciousness is not too intellectually demanding. This danger certainly exists when one sees consciousness exclusively in its most complex forms and bound to traditionally exclusive and monolithically defined competencies. But one can deliberately counteract this by coming up with new concept. My discussion of theory of mind abilities in children and animals has shown that there are different ways of understanding theory of mind. We don't have to ask and search for false belief understanding. An understanding of perceptual states or of desires might already suffice to speak about theory of mind. Meta-cognition does not have to be so narrow that one can say that only beliefs about *beliefs*—or, still more narrowly defined, beliefs about *false beliefs*—fall under meta-cognition. Basal meta-cognitive abilities can furthermore already be seen in a reflection on *one's own* states of perception and on *one's own* desires. It is therefore important to return to the *diversity* of mental states and processes whose conscious existence is also evident in the actions of animals. Similarly, one can argue about other competencies and the forms of consciousness involved there. Thus, although the exact connection between consciousness and socio-cognitive abilities is not entirely clear, it can be said that having culture, language, and theory of mind means less than an exclusive and monolithic understanding of such competences often claims. But it also means more than being able to suffer consciously. Therefore, consciousness and suffering do not become less relevant criteria of moral consideration. They are just not the only ones.

Those who still prefer to limit themselves to the standard of avoidance of suffering alone must bear in mind that this does not cover everything that can be covered by an animal rights approach, let alone a rights approach that argues for an analogy to human rights. Not all inalienable, basic personhood rights required by this approach can be derived from the right not to suffer in the strict sense. The right to life, the right not to be tortured, and the right to freedom cannot be obtained solely through the norm of preventing suffering. For example, a right to life is not the same as a right to be killed painlessly. I will go into this in more detail with the help of an example below.

The prohibition of torture is, to a large extent, in fact based on the avoidance of harm, pain, and suffering. However, it certainly refers not only to the experience of pain as physical pain, but also to the violation of the dignity of the tortured person (for the tight link between personhood and dignity see Williams and Bengtsson 2018). Whether the violation of dignity can be reduced in all points to the infliction of suffering is doubtful. In many pathocentric approaches, the infliction of harm, pain, and suffering is understood in a narrower and more concrete sense (primarily related to physical damage). Moreover, the right to personal freedom does not refer solely to suffering in this narrower sense and cannot be reduced to it. Rather, the need to be able to move freely and to feel free (also free from control or paternalism) is accepted in humans, because it is presupposed that we are independent, autonomous beings that can act according to our free will and are determined by personal intentions, goals, and preferences. Here, too, self-determination and the concept of dignity are closely intertwined. All these assumptions seem to be closer to the concept of human dignity and human rights than directly based on the ability to suffer. It cannot therefore be said that some animals have similar or even the same complex social and cognitive abilities as humans, but that in relation to them it is nevertheless sufficient to adopt the norm of preventing suffering as the only relevant norm. Similarly, this norm is also insufficient when it comes to people. The human rights that are universally attributed to all humans, including "borderline" cases, are simply richer and stronger in their assertiveness than anything else that we have applied to animals so far.

12.5 Painless Killing and Production of Insensitive Animals

The fact that we do not argue purely pathocentrically when it comes to human beings is clearly demonstrated by the fact that the human *right to life* is not equal to the *right to be killed painlessly*. Imagine a nurse in a hospital or retirement home killing patients painlessly at night when they sleep with a lethal injection. We would call this individual's behavior murder and condemn him or her as a murderer, even if the murdered persons had not suffered any pain as a result of the killing, and even if they had no relatives or loved ones who would suffer due to their death. In particular, as long as these persons have not given their free and informed consent to their own killing[8] and as long as there was no medical reason (such as an incurable, serious disease), the killing is morally unacceptable. And even if there were a medical reason, a number of ethical concerns would arise, as can be seen from the extensive, complex bioethical debates about euthanasia and assisted suicide. If the only relevant norm in dealing with people would be the avoidance of suffering, then a painless killing of human persons should not be such a complex ethical problem. The reasons why it is still morally wrong to kill a person painlessly are traceable to what characterizes a self-determined person and her life in dignity—and this obviously goes far beyond her ability to suffer.

A similar argument can be made about the case of the possible future production of disenhanced, insensitive animals, meaning animals that cannot feel pain. Some scholars argue that eliminating the problems associated with the suffering of animals could be a beneficial solution for animals in intensive farming, at least a solution that is better than the status quo (Varner 2012, Shriver 2008; Shriver and McConnachie 2018). Critics might argue, however, that the way would then be clear for ethically justifiable intensive farming or invasive animal testing with hardly any limits. Here too, however, not all the problems of these practices can be solved by solving the questions linked to suffering.

8 With the principle of "free and informed consent," however, only a right of defense is formulated in the sense that we should refrain from subjecting someone to something without her free and informed consent. It does not mean that everything that anyone wants is also ethically justifiable.

Instead, new concerns are emerging: Is it ethically justifiable to rob animals of such an essential ability? If pain perception fulfills an adaptive function and warns a living being of dangerous actions, how would animals behave without this warning system? Furthermore, a number of ethical problems in animal testing (such as the question of the transferability of results and the necessity of some animal tests) would, of course, not be solved, even if the animals were no longer capable of suffering from the painful treatments themselves.[9]

9 For two detailed discussions of such questions, see Arianna Ferrari's (2009) discussion of the case of genetically engineered, insentient laboratory animals and Kirsten Schmidt's (2008) article on blind chickens as a test case for animal ethical theories.

CHAPTER 13

Possibilities of Modifying Personhood Rights for Animals

13.1 Toward a Consistently Gradual Understanding

The fact that the Great Ape Project and Thomas White concentrate on certain species, namely great apes and dolphins, can give the impression that their argument is only about the rights of these particular species. One has to look closely at their argument from cognitive similarity in order to find the few passages in which elephants and some birds, for example, are also addressed. I have therefore pointed out—more explicitly than the authors of these approaches—that the arguments used relate in principle to all individuals and all species suspected of possessing the relevant capabilities. Advocates of the approach should emphasize the structural openness of their argument to other animals with personhood qualities in their external communication. Although research on animal cognition has so far been strongly primate-centric, ethics should not be allowed to follow this one-sidedness. The impression that some of the problems associated with this approach have not been discussed much so far is certainly correct. Above all, there is indeed—at least at first glance—a tension between the graduality of the above-mentioned abilities and the absoluteness of personhood status, which may also have been deliberately evaded.

While the justification of human rights is based on the ideal of the intellectually gifted human person, this ideal person may not exist in reality with a clarity that would allow a clear demarcation. To a certain extent, the personhood approach buys into this problem with its reference to human rights. It is argued that the complex cognitive abilities of apes or dolphins differ only gradually from those of humans and that they therefore deserve some basic rights. But what about all the other animals, which in turn differ only gradually from apes and dolphins (and thus also only gradually from humans) in terms of their complex abilities? The fundamental problem arises that exclusive, monolithic criteria must be de-intellectualized and broken down into their individual components for the argument. This in turn, however, seems to leave us with no rigid demarcation line. It is not expected that these abilities or parts of these abilities will only occur in species that are genetically close to humans, but also in many (social) species that adapt flexibly and intentionally to

 | DOI:10.1163/9789004415072_014

problems in their niches with the help of (socio-)cognitive abilities. Rogers and Kaplan (2004) rightly note that "only a handful of species have been researched in depth, and current findings would suggest that many more species might be found to have exceptional cognitive abilities, if we only looked" (193). It has already been pointed out in this book that the moral community must therefore be an open community that is constantly open to welcoming new members. This was identified as an important component of a "new" self-image of human beings and their position in nature, which is becoming increasingly important due to the increasing knowledge about animal cognition and behavior.

The graduality of socio-cognitive abilities can be seen in a comparison between humans and other animals as well as in comparisons of different humans. We do not have the same unambiguous or full personhood at all stages of our lives. We all go through a process of acquiring certain skills in our own ontogeny. As small children we are not yet fully linguistically talented; we only slowly learn the traditions of our cultural community and gradually acquire psychological knowledge as to what other people believe and think. The same graduality is apparent in the evolutionary path from the last common ape ancestor to homo sapiens. Early hominids already possessed parts or variants of complex cognitive and social abilities. Their bone flutes or stone choppers may not be a modern concert flute or a modern axe, but in this instance we are certainly talking about tools, the first musical instruments, and early cultures. Anyone who argues for recognition of complex behavior or capacities and the products resulting from both must emphasize their preliminary stages and the evolutionary-gradual character of these abilities. If one values such abilities in normative arguments, then one must also integrate them into one's moral theory while recognizing this gradual character.

One might doubt if this is even possible. It could be criticized that an argument such as the human rights approach, which is conceptualized with a focus on a certain ideal of a "paradigm" human being, cannot easily be applied to a much broader circle of individuals who no longer fully correspond to this ideal. Obviously, however, we are nonetheless prepared to apply the framework to human beings who do not correspond to the ideal fully. Critics of the human rights approach could interpret this as an indication that we have long since bid farewell to the Kantian image of humans. As to why we grant human rights to all people, including "marginal" cases, they might assert that this has nothing to do with them *being* persons thanks to certain Kantian abilities. We do so rather for social-political reasons. Human rights simply regulate our dealings with one another. However, this again means that we *want* to maintain a certain social-political order and have formulated laws and rights for this very purpose. We obviously want to live in a society in which the dignity of the human individual is respected, in which no member of our society is deprived of life for no reason, no one is tortured, and no one is unjustly imprisoned. If you ask people why this is so, then you will probably get answers that refer to anthropological constants as values whose recognition we have hitherto regarded as central—indeed, in every culture, in every place, and, interestingly independent of whether a human person is a person in the full Kantian sense. Sick people, elderly people, demented people, young people, and disabled people are also persons with full personhood status. The gradual nature of being a person among human beings is therefore no reason to abandon the concept of personhood status or human rights. Quite the opposite: the fact that these people have been repeatedly discriminated against has become an opportunity to press all the more clearly for the international recognition and enforcement of human rights. The human rights violations committed by the Nazis during Second World War have led to international consensus that these rights must not be invalidated. The graduality and diversity of personhood qualities among different people does not normally cause us to pause when it comes to granting human rights. This is especially the case when it comes to granting the three most basic human rights (life, the absence of torture, and freedom).

Of course, "borderline" cases are controversially discussed in bioethics, for example, when, with regard to the right to life of embryos, the question is posed whether abortion could be justifiable in the first months of pregnancy, or when, with regard to coma patients, it can be debated whether life-supporting devices may be switched off in cases of irreversible unconsciousness and irreversible brain damage. However, cases that are considered controversial here are often very far removed from the example of animals: while an unconscious and irreversibly brain damaged coma patient or an embryo might be rightfully considered as questionable cases for personhood, a conscious, suffering animal with some cognitive and social abilities (and in this instance I am not only thinking of an ape or dolphin) might be still uncontroversial. The latter clearly has more in common with a baby or toddler than with a coma patient who has fallen into irreversible unconsciousness or an embryo in the eight-cell stage. I would thus argue that the bioethical talk about "marginal" cases and the extension of this label across different cases has blurred important differences. The label "marginal cases" is ableist and discriminatory from the outset. It invites us to treat (argumentatively) all sorts of varying cases, like irreversibly comatose patients, disabled people, children, or animals, as if they were inferior to healthy, adult human beings. On top of that, it also suggests that it provides us with a fine-grained terminology to set aside "paradigm cases" from "marginal cases." This is an illusion. Summarizing all "non-paradigm cases" under the label "marginal" invites us to talk about these cases as if they were the same, but they are clearly not. We owe a disabled woman a considerably different treatment than a coma patient.

13.2 Personhood Status Despite Being a "Gradual" Person?

How can an animal rights approach that argues for an analogy to human rights deal with the graduality of the existence of socio-cognitive abilities, which in turn are relevant as personhood qualities?

On the one hand, the scope of those to be granted basic personhood rights could be very broad. If it is not possible to determine where these abilities begin, then one could revert to consciousness as the threshold criterion and also grant all animals with consciousness some inalienable rights, such as the right to life, the right not to be tortured, and the right to freedom. This approach could easily be combined with *conscious suffering* as a criterion for moral consideration and combined with the demand for some inalienable rights. However, the abilities that play a role in the human rights approach, as has already been said, also go beyond the capacity for suffering and awareness.

The second alternative is that an animal rights approach that argues for an analogy to human rights could deal with the problem in the same way as the human rights argument deals with the graduality of being a person among people. Here, the scope of *addressees of the derived rights* (all human persons and all "borderline" cases of personhood) goes beyond the scope of *addressees in the classical justification of the same rights* (only fully rational, healthy, and adult human persons). There may be good reasons why we consider such a transgression to be justified: for example, because these people were once persons or because they will be persons in the future. However, some severely mentally disabled people will never actually be persons *in the Kantian sense.* And yet it is morally right to grant them personhood status. Even if they will never fully develop cultural or linguistic abilities, a complete theory of mind, self-consciousness, or complete autonomy in every respect, they have human dignity and human rights.

Several reasons support this claim. First, White rightly pointed out that renouncing their personhood status would put too much at stake. When in doubt, the mere suspicion that whom we face could be a person must be sufficient grounds for recognizing him or her as a person. If the person has such absolute value and human rights are as important as the human rights approach says, then from an ethical perspective one cannot afford not to treat beings suspected of being persons in such a way.

Second, and this seems to me to be much more important, we also appreciate and respect those people who do not completely possess these abilities and have never done or will never be anything other than what they already are. We regard a child that we observe or that interacts with us as a small person—completely independently of what she will be like as an adult. And we value a mentally disabled person as the person she is, with all the abilities she possesses. Again, these children and adults that are sick, old, and disabled—also severely mentally disabled—still differ fundamentally and quite clearly from the controversial personhood case of coma patients with irreversible unconsciousness or brain damage mentioned earlier. They differ just as fundamentally and clearly from the embryo in the eight-cell stage. Those in the bioethical debate who often call these groups of "borderline" cases in the same breath can only be accused of being imprecise and perhaps even discriminatory at best. I would argue in favor of a stricter distinction between such cases, because otherwise the needs and rights of children, sick, disabled, and elderly people will not be met. Whoever treats them like coma patients with irreversible brain damage (be it in real interaction or in moral theoretical considerations) is doing something fundamentally wrong. Accordingly, although the relatives of coma patients with irreversible brain damage who are highly unlikely to wake up from their coma will be faced with the question of morally permitted elimination of life-saving measures, the relatives of disabled people are not confronted with such decisions. Because disabled people, including severely mentally disabled people, have physical capacities and consciousness and because they possess some social and cognitive abilities, they need special *protection* and special *support* for these abilities. This is exactly what happens in special education. If these people did not possess such skills or abilities, to whatever limited extent, then any form of early support or general support would be pointless. I firmly believe that this is not the case.

So we can very well grant full personhood status to someone who does not have full self-consciousness; who communicates with us without syntactically-complex language, even without verbal language; or who has only partial autonomy or partial moral autonomy. And we may not be able to do this exclusively, but at least with reference to the existing gradual levels of social and cognitive abilities of such people. The animals who are candidates for personhood also either do not correspond to the Kantian ideal of the person, correspond to it only partially, or are simply different persons than we are. In principle, however, the argument also seems applicable to them by further exceeding the circle of addressees. Thus, the animal rights approach that argues for an analogy to human rights contains the same inconsistency between its level of justification, on which personhood is conceptualized with regard to a Kantian ideal of a person, and its level of application, upon which current abilities and personhood status can diverge. However, this contradiction only allows for two consequences: either human rights are simply not justified by personhood, or the fact that today we also grant personhood to human "borderline" cases, although the classical justification does not include them, has justification and meaning. I think the latter is true. It cannot be claimed that people answer the question "What makes a person a person?" exclusively and in unison: "That she is capable of suffering." Replies will also claim that she has cultural abilities, uses language, can think and plan, is self-conscious and capable of morality, and much more. All these characteristics must be questioned in the way we understand them, but it seems certain to me that they characterize a person in our understanding, in addition to the ability to suffer.

So what does it tell us when we realize that today we also grant personhood to people who do not yet possess these abilities completely, or no longer completely do so, or only do so partially or almost not at all? Frankly, it says something about the possible moral progress of societies. This can be expressed by the fact that we continue to expand the circle of those we morally consider. This idea of the moral community as a circle that has expanded over the course of civilization brings together such diverse authors as Peter Singer, Jeremy Bentham, David Hume, Charles Darwin, and certainly many others. In *Animal Liberation*, Singer described the idea of animal rights and the overcoming of speciesism in line with overcoming other forms of discrimination, such as discrimination based on race or gender (see Singer 2009a, 1ff.). In his essay *The Drowning Child and the Expanding Circle*, Singer refers to William E. H. Lecky, who in his 1862 *History of European Morals* already wrote that

> the question of morals must always be a question of proportion or of degree. At one time the benevolent affections embrace merely the family, soon the circle expanding includes first a class, then a nation, then a coalition of nations, then all humanity, and finally, its influence is felt in the dealings of man with the animal world.
>
> SINGER 1997; with reference to LECKY [1862] 1869, 103; see also SINGER 1981

This seems very similar to Jeremy Bentham's already quoted statements (see Chapter 8.5), who introduced pathocentric animal ethics as a continuation of other legal and equality movements, especially slave liberation. David Hume, too, discusses an increase in moral consciousness, when sympathy is extended past those close to us in an attitude of "friendship" and "regard", which ultimately takes the interests and positions of strangers into account. Morality in the end goes in wide circles and "has a reference to mankind and our fellow-creatures" (Hume (1751, 1777) *Section v. Why Utility Pleases. Part II.* M 5.43, SBN 230 and M 5.44 SBN 230–1). He also condemns the alleged superiority of "civilized Europeans" over "barbarians," of men over women and of humans over other animals. In the course of civilization, justice and humanity are being extended to all these living beings. This he considers moral progress: increasing humanity means increasing consideration for those that have not yet been considered (*Section III. Of Justice. Part I.* M 3.19, SBN 191 and M 3.21, SBN 192). Similarly, Charles Darwin sees the progress of civilizations as an extension of social instincts in the course of human evolution and cultural history (see Engels 2007, 185; with reference to Darwin [1879] 2004). Eve-Marie Engels comments on Darwin's position: "Because…the instinctive focus of compassion is extended beyond the immediate family and tribal community, human beings can include the members of other nations and races, even the remote ones…, unknown members of the world community, and finally all sentient beings in their benevolence. Darwin considers this expansion as *moral progress* in human history." According to Darwin, this moral progress is evident in the willingness to support the weak, sick, and helpless (Engels 2007, 185–86; transl. MK)—and probably also in animal welfare. Engels comes to the conclusion that Darwin sees the climax of humanity reached when it extends not only to all humans but also to nonhuman animals: "The fact that the human being is the only morally capable living being who is an individual 'moral agent,' as we say today, does not mean for Darwin that he is also the only 'moral patient,' i.e., the only being that deserves moral consideration." We could be "citizens of the world" and also, "by transcending the human species, show consideration for other, nonhuman beings" (Engels 2007, 186–87, transl. MK). Although this would express a "very optimistic attitude toward moral progress" in view of continuing experiences of injustice, "one could certainly speak of moral progress, which is expressed in the recognition of universal human rights, even if these are not yet implemented everywhere" (187).

It is a conscious decision of democracies to include these groups and "borderline" persons in the circle of those to whom human dignity and human rights belong. Because they are defenseless, they require particularly strong protection (see Wimmer 1999, 339). The same historical violations that have proved this very fact warn us against suspending human rights or restricting them. This is not only about ensuring a pragmatic social order through them but is also about wanting to live in a world in which personhood rights are valued and indeed have a value.

13.3 Species-Specific Inalienable Rights?

The problem of the gradual nature of cognitive and social abilities can possibly be alleviated by conceiving inalienable personhood rights that animals are to be granted specifically for their species. It is clear that despite the attribution of human dignity and personhood, some human groups do not have all the derived rights and duties that belong to the healthy adult human being. This applies, for example, to people who are too ill or too young to make any kind of decision on their own responsibility. In this instance, for example, a guardian is appointed. In the case of Hiasl, exactly the same was strived for when it came to a chimpanzee. Thus, it was not claimed that Hiasl was fully autonomous, that he had the right to manage the money donated to him, or that he could represent himself in court. Instead, it was demanded that he be treated in a legally similar way as a small child might be.

It could indeed make sense and be important to grant different species different rights, especially in view of their gradual abilities and the special needs of "borderline" cases. For example, while a healthy adult has a human right to work in addition to basic human rights, the UN Convention on the Rights of the Child, CRC, adopted by the United Nations General Assembly in 1989 and ratified by most countries of the world, provides for the protection of children from child labor. This means that rights derived from general human rights, such as the CRC, are based on the specific needs of different addressees and individual groups have their own specific legal catalogs. The relationship between children's rights and human rights would have to be clarified here and then a formulation of analogous, inalienable, species-specific animal rights would have to be discussed. Analogous to the CRC, a charter of the rights of great apes, dolphins, etc. could possibly also be drawn up. To this end, it would have to be decided for which species which type of basic rights are important. The discussion about the Great Ape Project has silently made a selection of three basic rights that are important for apes, but so far this selection or the nonselection of other human rights has not really been addressed and justified.

It is also not clear whether all three rights chosen *refer equally* to complex social and cognitive abilities in animals, or whether some of them relate primarily to the ability to suffer. The latter could be the case, as already stated, especially in the right not to be tortured. This right could then apply to all animals that can consciously suffer as an inalienable right not to be tortured.

13.4 Differentiation between Clear and Less Clear Cases?

Anyone who doubts the gradual existence of social and cognitive abilities in animals, or at least does not yet consider it proven in many species that they possess sufficient parts of such abilities, should at least consider that their skepticism applies to different species to varying degrees. In some cases, at least, there may already be sufficient knowledge with respect to clear research results that could have consequences. In the case of great apes, and perhaps also dolphins, it seems more likely that they could be considered nonhuman persons than other species, given the data currently available. Individual results are also available for a large number of other species, but probably not in abundance and related to such a large number of social and cognitive abilities as in the species of these two families.

However, skepticism about the lack of research results in relation to cases that are not clear cut does not justify skepticism with respect to cases that are much clearer. It may be ethically necessary to consider clearer cases now, even if the social and cognitive abilities of other species cannot yet be assessed.

CHAPTER 14

Alternative: Turn the Focus Back to the Suffering of Animals?

If complex cognitive and social abilities exist in many more species than previously assumed, and if demands for a personhood status are problematic or can only be enforced slowly and only for few, exemplary species, then one must think about where such abilities can be more strongly integrated into already established approaches in animal ethics. Therefore, as indicated earlier, we will now examine whether cognitive and social abilities can play another role in animal ethics apart from the direct justification of rights. For this purpose, the book addresses considerations that aim at the avoidance of unnecessary suffering of animals as the central norm. Considerations on the ability to suffer could, on the one hand, tie in with animal rights approaches, also when it comes to the animal rights approach that argues analogously to human rights, but, on the other hand, could also form the basis for pathocentric-utilitarian approaches. In the following, we refer to authors who follow the second path, and are often strongly inclined to follow a biologically and veterinarily oriented animal welfare approach.[1] This places the connection between the ability to suffer and socio-cognitive abilities, which has already been suggested several times, at the forefront of our considerations.

1 Animal welfare approaches have a long tradition, especially in England, where the world's first chair for animal welfare was established at Cambridge University in 1986. In contrast to representatives of a rights approach, their advocates assume that there are no fundamental animal rights, such as the rights to freedom or life, as demanded by animal rights activists (this naturally includes inalienable rights analogous to human rights). They also presuppose a logic of consideration, which is criticized by animal rights activists: "The adoption of an animal welfare stance means that we accept that society's use of animals is justified. This justification is normally based on two key principles: (1) the cost in terms of animal welfare to a few animals is outweighed by the benefits obtained by the many, be they humans or animals of the same species; (2) everything possible is done to ensure that no animals suffer unnecessarily" (Young 2003, 15). These conditions are background assumptions of the discussions described below. The following should be clarified with regard to the use of terms: the separation of animal rights and animal welfare approaches does not always provide clarity, because the term "animal welfare" is also used more generally, and not only in contrast to "animal rights." It is also sometimes used as a collective term for all efforts to improve the situation of animals (regardless of how they are justified). In this general sense, animal rights activists are therefore also animal welfare activists, but animal welfare activists are not always animal rights activists.

 | DOI:10.1163/9789004415072_015

14.1 The Relationship between the Ability to Suffer and Socio-Cognitive Abilities

Rogers and Kaplan (2004) see the goal of animal ethics and animal protection as follows: "We do not just want animals to survive but want them to have a quality of life commensurate with their needs and dignity: physical, psychological, social, and cultural" (196). The psychological, social, and cultural needs of animals are increasingly being recognized in their importance for the quality of life of animals and for a treatment of animals by humans that is guided by values like respect and dignity. The task then arises of recognizing suffering that stems from the disregard of such needs or is intensified by such disregard and to avoid it in animal husbandry. However, much is still unclear about the connection between suffering, abilities and needs. Thus, it is generally disputed whether complex social and cognitive abilities, such as those expressed in a comprehensive understanding of correlations and causalities or an anticipatory idea of the future, ensure an increased or a diminished ability to suffer. It is conceivable, for example, that those animals that do not possess such abilities *suffer less* because they do not anticipate pain from cues in their environment or because they memorize painful situations differently and do not remember to the same extent. But it is also possible, that those animals actually *suffer more* than animals who possess such abilities, or even more than humans. Not having a comprehensive understanding of the context and the interrelationships shaping it or not having an idea of the future can mean being more directly at the mercy of one's own pain. Think about a painful medical procedure that might ultimately help to cure a disease. At the time of the experience of pain, a living being without the described understanding can take no comfort from the foreseeable end to the pain or the possible "good" that might come from it. In contrast, beings with a greater understanding of the correlations, causalities, or reasons and beings with an idea of their future may be better able to bear pain because they know what the pain's reason, meaning or cause is. They also know that the pain will soon be over. It is possible that animals, precisely because they cannot integrate their suffering into the bigger picture and thus "master" it, are very much at the mercy of suffering. In pain, they are, so to speak, only in pain—above all if they are kept under inappropriate conditions where they cannot react to pain, for example by escaping it in an attitude of flight or by showing aggression (see Spaemann 1984, 78; and, in a similar vein, Schicktanz 2002, 225).

So then, do complex social and cognitive abilities relieve or intensify suffering? Presumably, this question can only be decided on a specific case-by-case basis—if at all—from animal to animal and from situation to situation. Thus posed, it may also fail to make the connection between socio-cognitive abilities and the ability to suffer. I think it is obvious that such abilities are not connected in an easy way with "more" or "less" suffering. We cannot simply ask if two animals with different cognitive ability levels suffer from the *same* pain *in varying* intensities. After all: first, every sentient animal that suffers when harmed does indeed suffer. A cat that might do worse in classical intelligence tests than a dog will suffer at the same degree as the dog from the same damage done.[2] Whether animals with more complex socio-cognitive abilities suffer less because, for example, they know about the imminent end of the pain inflicted on them, or whether their suffering, on the contrary, has an additional level of intensity due to their emotional and cognitive capabilities, I cannot and do not want to answer. Even a child who, at the age of one or two, does not yet have pronounced socio-cognitive abilities, suffers, pure and simple, when she is in pain. In some cases, her suffering may be less if she is exposed to the same pain years later with mature cognitive abilities; in some cases, the opposite may be the case; in other cases, nothing will change.

But the question can be asked differently: Is it possible that an animal with outstanding social and cognitive abilities and an animal that does not possess these abilities may not *suffer differently from the same damage* inflicted, but that the animal with more complex social and cognitive abilities may in addition suffer from *other situations and from other things* than an animal that does not possess these abilities to such an extent?[3] Imagine, for example, a chimpanzee with complex cognitive, conceptual, and logical abilities, a theory of mind, and cultural capacities, as well as multilayered social interactions with its own kind. In comparison, we can imagine an animal, for example, a rabbit, which also has abilities in these areas, but as far as we know not in the same distinctive way as the chimpanzee. If we fail to recognize that the abilities of chimpanzees gradually exceed those of rabbits, then we will not be able to deal with them in a species-appropriate way. In order to do justice to them as living beings with a species-specific dignity, these abilities must be observed. This is of course a particularly big challenge for an animal whose cognitive and social abilities can be compared to those of a small child. Of course, the rabbit's abilities must not be underestimated either. Except that no physical pain should be inflicted on her and that she needs enough space, food, and water not to suffer,

2 This applies if the structures, such as the nervous system, necessary for pain sensation, are assumed to be similar and if the situation is assumed to be similar.

3 This again applies if the structures necessary for the perception of pain are assumed to be similar and if the situation is assumed to be similar.

in this case, a living situation is required in which her cognitive abilities do not atrophy and social structures are maintained. However, the measures an animal keeper has to take to provide all of this are different from the case of the chimpanzee and can probably be implemented appropriately by informed and conscientious pet owners. This does not apply to great apes. What I try to illustrate is that rabbits can be kept in private households as pets, but apes should not be kept in such a way. It is simply very demanding if not outright impossible to keep them as pets, not least because of their psychological needs. Even experts in zoos can only really master this task with great difficulty and effort, if at all. This means that in an ethical consideration the reasons for keeping these animals in captivity must be very severe so that they can outweigh the arguments that speak against it (see Section 9.2.1). Therefore, socio-cognitive abilities and the ability to suffer are related in that animals with higher socio-cognitive abilities can, for example, suffer *additional psychological harm* due to circumstances that would cause no harm or less harm to other animals that do not possess these abilities. A rabbit suffers when she is not offered attractive surroundings and is kept in a solitary environment. A chimpanzee suffers equally from this. But she also suffers intensively and for a long time if, for example, group members die or if she is not given the possibility to exercise her complex abilities through intensive enrichment.

In an ethical evaluation of zoos, a few animal ethicists try to do more explicit justice to the topic of animal welfare with regard to the psychological needs of animals. David DeGrazia, for example, does this in his book *Taking Animals Seriously: Mental Life and Moral Status* (see DeGrazia 1996b, 1996a). He considers animals with sentience as morally relevant beings that can suffer and, like all suffering beings, have an interest in the avoidance of suffering. This is based on the very general ethical principle of nonmaleficence, which is set out in fifteen different principles or rules (for a summary of the principles, see DeGrazia 1996a, 279–80). From principle 3, for example ("Don't cause significant suffering for the sake of your or others' enjoyment"), it follows that ethically permissible zoos would have to present much heavier arguments in justifying their existence than the goal of entertaining visitors (292–93). Principle 14 is based on positive obligations we have towards animals who have special relationships with us; pet owners and zoos have equal responsibility for the welfare of the animals they own. It reads: "Provide for the basic physical and psychological needs of the zoo animal, and ensure that she has a comparably good life to what she would likely have if in the wild" (274ff., 294). The principle designates physical and psychological needs and a quality of life that is comparable to that in the wild. Both will be examined in more detail in the following section.

DeGrazia points out that the essential psychological needs of animals are often neglected (274) and that the keeping of some animals in captivity according to such ethical principles is hardly possible, or is not possible at all. In his opinion, it is only ethically justifiable to keep great apes in captivity if family structures are preserved and the animals are offered sufficient space and a strongly enriched environment with many different possibilities for climbing, discovering, problem-solving, and playing.[4] When it comes to dolphins, it is practically impossible to create housing conditions that could guarantee them a life that is of as high a quality as they have in the wild. Therefore, we shouldn't keep them in captivity (see DeGrazia 1996a, 297). DeGrazia focuses on these particular animals due to their special abilities and needs. He tries to consider findings on cognition, especially in apes and dolphins, and to combine them with an ethical evaluation of zoos. Note, that his account starts with a very basic, welfare-oriented principle, the principle of nonmaleficence. Still he arrives at conclusions regarding an ethical treatment of animals that come very close to animal rights. Cognitive abilities are not used in the direct justification of (personhood) rights, but come into play as important factors for the welfare of animals. Thus, they are indirectly relevant but still crucial for the ethical claims he derives.

14.2 Socio-Cognitive Abilities, Animal Welfare, and Species Conservation

Animal rights activists assume that zoo animals can suffer from boredom and lack of movement, due to cages that are too small and barren and other restrictions. They then develop behavioral disorders because they are unable to move in a species-typical manner, are unable to exercise and practice species-typical behaviors, and remain mentally unchallenged. Zoo critics use the term "zoochosis" to refer to behavioral disorders that typically occur in zoo animals. They reject zoos without exception, and point to such behavioral disorders in support of their arguments.[5] No matter if we follow this criticism with regard to

4 Without a doubt, this list is not exhaustive. For example, it should also include opportunities for animals to forage.

5 The term is composed of the words "zoo" and "psychosis" and was introduced into the animal welfare debate by Bill Travers (see, e.g., https://www.worldatlas.com/articles/what-is-zoochosis.html, 2019/04/24; and https://www.peta.org/features/animals-driven-insane-zoochosis/, 2019/04/24).

all, some, or only a few species, this is a problem that veterinary medicine and animal ethics must address while zoo animals are still in captivity.

Generally speaking, according to animal welfarists, animals do not suffer from being in the zoo but rather from being poorly kept there (see Precht 2000, 323). In terms of content, this argument is based on the fact that for many zoo animals, we must assume a highly complex social behavior and/or movement behavior. This also explicitly refers to complex social, cognitive, and emotional abilities. For example, this can be said in reference to extensive cooperative hunting strategies in some predatory species, with reference to the cultural behavior (like tool use) of some highly social animals or with reference to the movement radius of migrating, deep diving, or flying species. In the zoo, active search for food is practically and inevitably replaced by animals simply being fed, and natural opportunities for development of such skills are reduced to a minimum. Even if improvements to barren cages can often be successfully implemented, and some zoo animals may grow old in captivity or reproduce successfully, this is not necessarily a sign of well-being or species-appropriate keeping (see Precht 2000, 323).

If species-typical behaviors cannot be lived out, then this is not appropriate for the species. However, different definitions of "species-appropriate" are possible: species-appropriate could refer to that which is found in the natural habitat; if this standard were enforced, then almost no larger vertebrates could be kept in the zoo in a species-appropriate manner. It could also mean "that to which a species can adapt"—but almost all animals in the zoo "somehow figure out a way to get by" as they can adapt to new environments. Or it could mean that they can live out their essential, species-typical behavior (see Precht 2000, 324). To a certain extent, the latter definition represents a middle way, since the exercise of "natural," species-typical behaviors could also be made possible with artificial means in the zoo.[6] Nevertheless, according to this definition, a justified rejection of keeping those species in captivity for which an exercise of essential, species-typical behavior cannot be guaranteed would be conceivable.

In DeGrazia's words, the central ethical requirement is that of a quality of life in captivity, *comparable with the quality of life in the wild*, as it meets the

6 The environment in the zoo can be artificially enriched (see Section 14.4). Such measures strive to create an artificially natural environment: "Often, environmental enrichment devices appear to be highly artificial; however, the appearance (i.e. physical form) of the behaviour expressed may be the same as if the behaviour were being naturally stimulated in the wild" (Young 2003, 10).

essential, species-typical, physical and psychological needs of the animals. This does not mean that captivity must simulate the *factual natural environment* in a one-to-one relationship, but that the housing conditions must simulate the quality of life in the natural environment. This also does not necessarily mean that *all wildlife behavior* must be replicated. Research rightly warns against uncritically using all behavior in the wild as the standard for welfare-guaranteeing behavior in captivity. First, very little is known about the natural behavior of some species in the wild; second, behavior in the wild is highly variable and depends on local environmental conditions; third, some behavior in the zoo could simply represent an unproblematic adaptation to new environmental conditions; and fourth, behavior that contradicts the idea of well-being in captivity (for example, behavior that is associated with health impairments) is shown in the wild. Referring to "natural" behavior is therefore not easy. It cannot always and under all circumstances be said that an animal suffers if it does not have the ability to exert behavior it would display in the wild (see Shepherdson 1998, 6). Instead of referring to the category of "natural" it is therefore important to focus on the following two factors. First, what is decisive for the assessment of the quality of life of animals in captivity—despite all due caution—is what falls under species-typical behaviors *that are relevant for well-being*. Second, an assessment must be made of when this maxim will be fulfilled or violated *for each individual species*.[7]

It would certainly be possible to identify complex cognitive and social abilities as well as the species-typical behaviors determined by them that are relevant to such considerations on the quality of life. But it is possible that this argument will then reach the limit it sets. For some animals we will surely not be able to guarantee a satisfying standard of quality of life in captivity. Precht, like De Grazia, regards this as achieved in animals with particularly complex social behavior, such as elephants, dolphins, and apes. Due to physical needs, the keeping of birds of prey is also particularly problematic: animals that sail in thermal lift for several hours a day should not be locked up in small wire aviaries, he says. Their natural behavior is thus undoubtedly drastically impaired (Precht 2000, 331).

7 Antelopes, deer, buffalo, and ibex, for example, can be kept in captivity in social groups. Even according to Precht's fairly critical analysis, the active foraging of such herbivores also seems possible in the zoo. If, in addition, retreat areas were created in large enclosures, it would be possible to speak of species-appropriate husbandry in this case, despite some reductions (cf. Precht 2000, 330).

14.3 Indicators of Well-being in Captivity

How can the effects of captivity on the physical and psychological well-being of animals be defined more precisely?

The British Farm Animal Welfare Council (FAWC) provided an influential answer with its five freedoms. Animal welfare is defined as (1) "Freedom from Hunger and Thirst—by ready access to fresh water and a diet to maintain full health and vigour," (2) "Freedom from Discomfort—by providing an appropriate environment including shelter and a comfortable resting area," and (3) "Freedom from Pain, Injury or Disease—by provision of rapid diagnosis and treatment." Added to these are (4) "Freedom to Express Normal Behaviour—by providing sufficient space, proper facilities and company of the animal's own kind" and (5) "Freedom from Fear and Distress—by ensuring conditions and treatment which avoid mental suffering" (see https://webarchive.nationalarchives.gov.uk/20121010012427/ http://www.fawc.org.uk/freedoms.htm3; and Young 2003, 16; with reference to Farm Animal Welfare Council, UK 1992). These are negative rights (that oblige inaction), not positive rights (that oblige action), which are widely recognized and form the basis of many animal welfare approaches. However, it is pointed out that freedom (4), which considers the importance of the exercise of normal behavior for animals as an indicator of welfare, is the most difficult to justify. If this freedom is implemented, it does not necessarily result in physically measurable improvements in well-being (cf. Young 2003, 16), an indication that, just as with freedom (5), it is at least partly related to indicators of psychological well-being and suffering that are difficult to determine (cf. Young 2003, 11, 31–53).

Of course, indicators of poor animal welfare can also be set up in line with what has been said so far and with the five freedoms. Broom (1999) has done this with a list that includes the following possible indicators: shortened life span, limited reproduction and breeding, physical damage such as injuries, diseases, immunosuppression, attempts of physical coping,[8] coping attempts through behavior, as well as behavioral pathologies (such as stereotypies, apathy, self-mutilation, and learned helplessness) and self-narcotization. Other indicators of poor welfare are aversive behavior, suppression of normal behavior, and restriction of normal physiological processes and anatomical development (see

8 "Coping" generally means coping with negative influences. In its broad, scientific understanding, the term covers a variety of stress-responsive behaviors. This includes both learned behavior, such as active avoidance behavior, and unlearned behavior, such as hiding or fighting behavior. Coping is also mentioned in connection with stereotypes (this will be discussed later) (see Würbel et al. 2006, 14).

Young 2003, 18–19; with reference to Broom 1999). Here, too, suffering is addressed that does not primarily or not only refer to physical damage and that manifests itself in abnormal behavior, such as in coping strategies, stereotypies, apathies, or aversions.

14.4 Enrichment as Occupational Therapy

As described in Section 9.2.1, it has long been known that zoo animals can suffer from being in captivity. Systematically and from a scientific point of view, however, this topic has only been dealt with for a few decades. The solutions proposed to remedy this situation also illustrate how closely the problems relate to the cognitive and social abilities of animals. The American primatologist Robert Yerkes (1925) and the aforementioned zoo biologist Heini Hediger (1942, 1965) were among the first scientists who pointed to the necessity and to different possibilities for improvement of housing conditions. Today, this is conceptualized under the term *enrichment*:

> Environmental Enrichment is an animal husbandry principle that seeks to enhance the quality of captive animal care by identifying and providing the environmental stimuli necessary for optimal *psychological and physiological wellbeing*. In practice, this covers a multitude of innovative, imaginative, and ingenious techniques, devices, and practices aimed at *keeping captive animals occupied*, increasing the range and diversity of behavioral opportunities, and providing more stimulating and responsive environments. The artificial termite mound is a good example of environmental enrichment in practice.
>
> SHEPHERDSON 1998, 1; italics added

Animal training in the zoo, such as feeding and medical examinations, is also a form of enrichment, "because it *engages animals on a cognitive level*, allows positive interactions with caretakers, and facilitates routine husbandry activities" (Shepherdson 1998, 1; italics added). According to the above definition, it is assumed that it is a qualitative improvement of animal husbandry (it "enhances" the "quality of ...care") if animals are cognitively employed and can exercise these behaviors. According to Young (2003), too, environmental enrichment is "a concept which describes how the environment of captive animals can be changed *for the benefit of the inhabitants*" (Young 2003, 1; italics added). According to both authors, enrichment targets both optimal physical health and mental health of animals (Young 2003, 11; Shepherdson 1998, 1).

According to Young, the term enrichment applies to a multitude of methods for improving well-being. There are forms of "social enrichment" (which can mean both the production and the blocking of contact with other species or humans), "occupational enrichment" (which occupies the animals mentally or physically), "physical enrichment" (which refers to the size, complexity, and equipment of the enclosure), "sensory enrichment" (which provides visual, auditory, olfactory, tactile, or other sensory stimuli), and "nutritional enrichment" (by means of a change in the type or the administration form of diets) (Young 2003, 2–3; with reference to the type classification of Bloomsmith et al. 1991).[9]

While improving the physical and psychological welfare of animals is the main goal of enrichment, other (derived) goals are also pursued. Enrichment can provide for a more successful breeding by creating an environment that allows for reproductive and rearing behavior in the first place and where animals behave in a species-typical manner, are healthy, and are capable of reproduction.

The reintroduction of zoo animals to the wild also depends on animals that behave as complexly and naturally as possible and can therefore later survive in the wild. Seidensticker and Doherty (1998) and Seidensticker and Fortman (1998) have argued, for example, that zoos should not only preserve the animals as genetic material for future generations of these species, but also their behavioral competencies through their breeding programs. Key skills for survival in the wild should not be lost.[10]

Also, according to Shepherdson, enriched cages are finally a more informative and interesting form of keeping for zoo visitors.[11] For him, the goal of enrichment is to guarantee a form of animal keeping that is "natural" for zoo

9 Often, the term "behavioral enrichment" is used synonymously with "environmental enrichment," but actually a change in behavior is only one of many possible changes (see Shepherdson 1998, 2). The connection between the two terms can be established in such a way that behavioral enrichment means possibilities of behavior and activity that *arise* or are *strengthened* by environmental enrichment. Environmental enrichment triggers, among other things, behavioral enrichment (see Young 2003, 1).

10 Note that this addresses a wide range of skills with key qualifications that are essential for survival and for which it is rather unclear whether, how, and to what extent enrichment can serve to maintain them (see Young 2003, 25–26). So far, not many studies have investigated the exact relationship between captive housing conditions and the chance of survival after release (see Shepherdson et al. 1998, 2, 97–112, 113–128).

11 For the latter to apply, however, enrichment measures must also be connected with explanations and information for the visitors. Only those who get an explanation why exactly this measure was taken with regard to what specific animal abilities and needs will learn something about the natural behavior of the animal. This, however, would also mean that zoos more openly and honestly admit that captivity comes with serious shortcomings regarding animal welfare.

animals and visitors alike (see Shepherdson 1998, 2–3). However, the zoo visitor's idea of what kind of animal behavior is "natural" may sometimes differ significantly from scientific understanding and from what an animal actually needs for her well-being. There is a danger that within the trend towards the Disneyization of zoos, more and more attention will be paid to the zoo visitor's feelings. It would be ethically highly questionable if jungle worlds were created where visitors see their clichés of "naturalness" fulfilled, while the needs of the animals come to be of secondary importance. In a themed world that is perceived by visitors as perfect nature, this illusion can give the impression that the animal feels comfortable. Signs to the contrary are easily hidden behind this façade.

In addition to increasing stimulus complexity and providing unforeseen new stimuli in other targeted definitions of enrichment, other authors have pointed out that abnormal behavior should be reduced, the behavioral repertoire of animals expanded, or time spans in which a particular repetitive behavior is shown should be reduced or normalized. Furthermore, enrichment enables animals to master challenges in a normal way in the sense that they are provided with tools and means to deal with challenges (see Shepherdson 1998, 31; with reference to Chamove and Anderson 1989). Others have emphasized that enrichment serves to establish stability and safety for animals and to enable them to pursue and achieve goals (see Shepherdson 1998, 31; with reference to Poole 1992). It can be easily seen how this last aim, for example, corresponds with the ability of conscious, intentional, goal-directed, and at the same time flexible behavior, which I have pointed out to be a signpost for complex cognition.

The cited definitions overlap, but they are also very different and raise new questions on their own. For example, to what extent do animals have goals? And what does security mean to them? The common denominator of all these definitions seems to be that there are indeed deficiencies in the keeping of animals, which have long been attributed primarily to the fact that zoo animals are not sufficiently engaged according to their physical, cognitive, and mental abilities or needs (see Hediger 1942).

14.5 Cognition—Motivation—Frustration: the Need to Gain Information and Engage in Exploratory Behavior

An awareness of stereotypies and the necessity of enrichment could not have been created without a new understanding of animals. According to Shepherdson, in recent years research has clearly moved on from behaviorism, and there has been a paradigm shift toward a mentalistic understanding of animals:

> Something of a paradigm shift has taken place within the field of animal behavior. It is now accepted by many behavioral scientists that models of animal behavior acknowledging and discussing the existence of mental states are not only acceptable but are necessary to explain many of their observations.... These findings have clearly influenced the way we think about animal well-being and the care of animals in captivity. If animals do have complex mental states, analogous in principle to our own, then they also may have similarly complex psychological needs.
>
> SHEPHERDSON 1998, 5

With this so-called paradigm shift toward paying greater attention to mental abilities, further concepts emerged, above all a concept of needs in animals that is directly linked to the exercise of species-typical abilities.[12] To what extent it matters for the well-being of an animal if certain behaviors cannot be practiced depends on whether the animal has an internal or external motivation to practice this behavior and how strong this motivation is. The need to avoid a predator only arises when a predator is approaching—this is usually not the case in the zoo. However, other behaviors are also internally motivated, such as courtship behavior, which is partly controlled by hormones. Food-search behavior is also triggered by an internal motivation, namely the feeling of hunger. In principle, even in the total absence of external stimuli, animals will display behavior that has a vital function and is strongly internally motivated (see Young 2003, 8–9). According to Shepherdson, since Konrad Lorenz and the behavioral research based on his work, it has been known that animals have motivations to practice certain behaviors, sometimes even when there is no external compelling reason for exercising them. This includes, for example, the playing behavior of cats, which still demonstrate prey-catching behavior even when they are full. It has also been observed that animals in behavioral experiments (rats, pigeons, and many other species) continue to perform the experimental task (such as selecting a light to move a lever and release food) even when food is freely available (Shepherdson 1998, 6). The feeling of hunger alone cannot explain the choice of more complex behavior. There seem to be other internal motivations here. If the motivation to perform such behaviors

12 In this instance, Shepherdson (1998) speaks of the "concept of behavioral needs" (6–7). Young also strongly emphasizes this concept, which Hughes and Duncan (1988) introduced into research as early as the late 1980s.

cannot be followed up, stress and frustration can arise. The emergence of behavioral disturbances is closely related to the fact that animals are unable to follow their behavioral tendencies and achieve their goals (see Shepherdson 1998, 6; Clubb et al. 2006, 12).[13] In fact, enrichment can also be defined in that it primarily serves to offer the animal possibilities to live out behaviors for whose execution she is highly motivated (Young 2003, 45).

Choice tests or behavioral-economic approaches can be used to measure animals' preferences and how much effort or time they put up with in order to achieve what goals. The basis of these approaches is that animals are able to make decisions and, in the case of a limited time or energy budget, behave like logical, human consumers. If there is little time or energy available, it is not used for the procurement of luxury items but for essential ones. The less flexible a demand is, the more important it may be for well-being (this has already been mentioned in Section 8.3.1). Mason et al. (2001) could show that minks are willing to push a weight that was more than their own body weight to get access to a room with a water basin. A measurement of the stress hormones in the urine of the animals showed that the withdrawal of the water basin meant comparable high stress for the minks as the withdrawal of food (see Mason et al. 2001; and the discussions of this study in Young 2003, 47; and Würbel 2007, 25).

The execution of behaviors that are clearly related to cognitive and social abilities can also be strong motivations for animals. For example, it is known from experiments with laboratory and farm animals that it is particularly relevant for their well-being to be able to assess, anticipate and also control aspects and changes in their environment. As early as the 1970s, rats were found

13 Club et al. 2006 explain: "The failure of…negative feedback loops is often thought to underlie stereotypies. For example, in captivity, some highly motivated consummatory behaviours (e.g. mating) may be impossible, regardless of the degree of appetitive behaviour (e.g. mate search) performed. In other cases, consummatory behaviour (e.g. feeding) may occur without the normal appetitive behaviours (e.g. grazing…) being possible. If such constraints leave animals in states of high motivation, this can result in frustration-related stress…, and a number of behavioural phenomena…including 'intention movements' (e.g. the restless escape attempts of a migratory bird confined to a cage …), 'redirected movements' (e.g. the sucking that calves direct to pen-mates and pen furnishings in the absence of a teat…),'vacuum activities' (e.g. the 'mimed' dustbathing movements of hens kept on bare wire floors…), and/or 'displacement activities' (e.g. 'out of context' ground-pecking or preening by birds during conflict situations…). If sustained, these activities can then develop into stereotypies…" Clubb goes on to say that there is a common factor of all such conditions under which stereotypies arise, namely frustration: "In all situations some tendency is being thwarted, some goal cannot be reached, some homoeostasis is disrupted," In the meantime, many studies have gone on to study frustration in animals (Clubb et al. 2006, 12).

to display fewer stress symptoms when they could predict the occurrence of damage and control it through appropriate behavior (see Shepherdson 1998, 6–7; with reference to White 1972).[14] The need to exercise exploratory behavior and to gain new information about the environment also seems to exist (see also Shepherdson 1998, 7; with reference to the studies of Carlstead et al. 1993; Inglis and Fergusson 1986; and Mench 1998). Animals are curious, pursue new stimuli in their area, and occupy themselves mentally. Shepherdson believes that the studies of Inglis and Fergusson have suggested that gathering information and acquiring new knowledge is the primary activity of all animals that would only be abandoned if very specific motivations, such as breastfeeding hunger in cubs, reached a very high level (Shepherdson 1998, 7).

Thus, if the conditions in captivity prevent animals from exercising important needs (e.g., the need to exercise control over the environment or the need to explore their surroundings and gain information), this can lead to stress and frustration and thus to suffering. At a minimum, we would at least speak of *boredom* for a person in a highly irritable environment who cannot meet such needs and cannot live out her physical and psychological needs. We would assume that a chronic lack of exercising physical and mental abilities would ultimately lead to frustration and behavioral disorders. Wemelsfelder (1994, 2005) suggests also discussing boredom with respect to animals when their ability to actively pay attention to and interact with their environment is impaired. The disappearance of interactions would be reflected in a decrease in behavioral variability and an increase in self-directed abnormal behavior. This problem can be seen in all animals that interact flexibly with their environment, show exploratory behavior, play behavior or innovative learning. They could be understood as subjects for whom meaningful interaction with the environment is important (Wemelsfelder 1994, 1–4, similarly Wemelsfelder 2005; Špinka & Wemelsfelder 2011). If this interaction breaks down, it leads to boredom and depression:

14 Young emphasizes that not only the control of the occurrence of negative events is important for animals, but also the control of positive events, such as access to an enrichment object (cf. Young 2003, 38). Control over the environment should be understood in general terms: "By control over the environment we mean that when an animal directs a motivated behaviour towards the environment, the environment provides opportunity for this behavioural expression and ultimately negative feedback that terminates this behaviour. Wild animals control their environment by expressing behaviour, the expression of foraging behaviour by a canary increases the probability that the animal will find food…". Later Young says: "It is thought that this control may help prevent the onset of behavioural conditions such as learned helplessness, i.e. animals become withdrawn and inactive" (Young 2003, 37 f.).

> The initial stage of attention impairment, I submit, may be characterized as boredom. As the animal is deprived of means for behavioural interaction, his or her attention becomes increasingly dispersed towards inappropriate stimuli, such as another animal's tail, or his or her own limbs. One could say at that stage that the animal 'does not know what to do'. The animal suffers from the general lack of meaningful behavioural goals, and becomes increasingly listless and withdrawn. In the final stage of impairment, attention disintegrates. The animal may respond chaotically to his or her environment, or become apathetic. In both cases, he or she becomes virtually helpless to cope with and/or change the situation. This stage may be regarded as evidence of depression and/or anxiety. (4)

Wemelsfelder's mention of boredom and depression in animals was controversial when she first came forward with it. Still, she pointed to a lack of concepts, especially with regard to mental needs and psychological suffering in animals. It is important to note that many of the needs mentioned so far very much revert to cognitive and social abilities. Mench (1998) also pursues the hypothesis with reference to the studies of Inglis and other authors that animals exert an intrinsically motivated exploratory behavior that primarily serves to gain new information about their environment. He believes that this thesis has become increasingly important in recent years. It is closely related to animal welfare and it clearly relates to the increasing evidence of cognitive abilities in animals and the idea of cognitive continuity (Mench 1998, 32–33). If the thesis is correct, then on the one hand there is the necessity to provide animals, which show a strong need for strongly exploratory behavior, with complex, constantly new and highly variable stimuli. On the other hand, one must ask oneself which animals are most likely to have such exploratory needs (cf. Mench 1998, 32 f., 35). Mench suggests that the determination of which animals have psychological needs should not be primarily based on the size of the brain[15] but on the complexity and variability of the natural environmental conditions of the respective species (Mench 1998, 35).[16] Although he generally believes that all animals need alternation and enrichment to keep their abilities stimulated, he suggests that three types of animals in particular should be taken into special consideration: first, species that are generalists or that are adapted to highly

15 Poole (1992) would have come to the conclusion that only mammals have psychological needs, which Mench obviously rejects as too limited.

16 Mench (1998) argues: "I contend that the complexity and variability of the animal's natural environment are more important predictors of cognitive needs than brain size, at least among homeotherms" (35).

variable environments that, for example, only provide them with very scattered or highly seasonal food. This means that one is dealing with animals that are able to generate, store, retrieve, and apply predictions about the location and relative quality of food. They also need to develop effective strategies to gain access to these resources. Second among the species that particularly need enrichment are those that have developed complex behavior against predators, learning and using several escape routes and strategies. Third, Mench points to species that exhibit complex social behavior. He justifies the latter as follows:

> It has been suggested that social behavior is a primary factor driving increases in cognitive ability because a social animal needs to communicate information accurately, predict the behavior of other individuals by assessing their mental states, and detect deception during social interactions.
>
> MENCH 1998, 36

It is clear that, among others, the species of apes and dolphins are being addressed here. Animals equipped with complex cognitive and social abilities especially suffer from the impoverishment of mental stimuli in captivity and need enrichment. Of course, here, too, it is clearer which animals are to be taken into account for such measures, while it is less clear for which animals enrichment measures might be less important. According to only gradual differences in complex socio-cognitive abilities, even Mench's list can again only provide us with a prioritization, which one can be forced to use in view of scarce resources for enrichment measures. It is not claimed that some animals definitely need *no* enrichment at all, but that some species definitely need *a lot of* enrichment, *particularly* due to the characteristics mentioned. In this instance, the *quality* of the enrichment is addressed, which must be bound to the complexity of behavior.

For animals who naturally engage in complex tool behavior, it is not sufficient to distribute food in a large enclosure as a way of enrichment: "Scattering food in a large enclosure can create a foraging time budget for captive chimpanzees that is identical to their conspecifics. However, wild conspecifics do not simply walk from place to place picking up food, they often use complex food processing skills, e.g., removing the skin of fruits or cracking nuts open with stones." The quality[17] of the behavior shown is described as an often

17 Young actually speaks of "quality of the behaviour expressed." This is certainly not meant in a judgmental way, but in terms of the complexity of behavior. According to my previous

ignored or neglected part of the behavior and needs of animals (Young 2003, 27–28)—and it depends strongly on the cognitive and social abilities involved. In a restrictive way it must be said that Mench is particularly concerned with an impoverishment in cognitive activity, and especially with regard to the need for exploratory behavior and the need to gain information. Animals that do not possess these abilities and needs to the extent that it can be assumed for the types of species mentioned by Mench may nevertheless have these needs in a weaker form. Their other, no less important physical needs (such as their natural climbing or resting behavior) can and must be addressed by enrichment measures.

It could be problematic in general that exploratory behavior, and probably other behavior aimed at intellectual activity, is often regarded as luxury behavior, which is put aside when higher-ranking motivations are effective. Accordingly, Mench does not want to claim that enabling such behavior is a main goal of enrichment: "Intrinsic exploration still appears to be a comparative luxury. Searching behaviors decrease significantly under conditions of food deprivation…as does another form of exploration—play behavior." Satisfying needs that have a higher priority and which are based on a stronger internal motivation should be given higher priority. This refers above all to behaviors related to food acquisition (Mench 1998, 40). But just because nutritional and reproductive needs usually outdo other interests, this does not mean that the well-being of zoo animals is sufficiently ensured if many young animals are born and the animals are provided with sufficient food and water. This would be a very abbreviated concept of welfare. It is likely that with a gradual increase in social and cognitive abilities, the need for mental activity and social interaction also increases. These may then be more appropriately understood as *key components of well-being* than as "luxury" behavior—even if there may be even more basic needs. Young warns, that when focusing on behavioral demand curves, one should not forget that the problem with captive animals is that they have a lot of time and nothing to do. It has therefore been pointed out in research that luxury behavior can play a *vital* role there: "Normally, we provide captive animals with all of their requirements for physical well-being, and so it is the luxuries that are usually missing from their existence." Inglis et al. (2001) would

argumentation, I understand Young here as meaning that more complex behavior, which requires socio-cognitive abilities, is not (normatively) more valuable or ethically more remarkable than less complex behavior. However, noncompliance with this greater complexity can become ethically problematic. If a rabbit and a chimpanzee are given the same quality of enrichment, for example, by distributing food in the cage as in the example described, then this may lead to typical foraging behavior for the rabbit, but not for the chimpanzee. The latter's mental abilities would not be fully utilized.

have followed this and stressed that especially when animals are equipped with everything necessary for their physical well-being (and this is ideally the case in the zoo), they increasingly change to a behavior that serves psychological needs, such as gathering information about their environment or exploration (Young 2003, 47–48). These needs can therefore be reinforced by captivity and should be given special consideration.

14.6 The Difficulty of Assessing Behavioral Disorders, Using Stereotypies as an Example

Stereotypies are among the most discussed disorders that animals develop when they are not provided with an environment in which they can exercise important, species-typical behavior. They are a particular form of behavioral disorder, which is understood as a reaction to stress and consists of repeated action sequences (McFarland 2006, 187). Stereotypies are constant in form and do not follow any purpose (Mason 1991).[18] Estimates of the number of animals developing stereotypies in captivity are still fragmentary and far from covering the world's livestock, laboratory animals, and zoo animals, but they are still alarming. It is estimated that about 91.5 percent of pigs, 82.6 percent of poultry, and 80 percent of caged minks are affected. Fifty percent of mice in experimental laboratories or breeding stations are affected, and at least 18.4 percent of horses in stables (Mason and Rushen 2006; with reference to Mason and Latham 2004).

One of the main problems with stereotypies is that they become increasingly independent of the factors that triggered them and thus increasingly resistant to therapy (Mason 1991; Young 2003, 52). This is mainly associated with the fact that, triggered by stress, significant changes with long-term effects can take place in the brain.[19] Cabib (2006) has therefore proposed following a new, biologically richer understanding of stress and its neurobiological connection to stereotypes. Some animals react to certain types of stress with a series of neurobiological changes (Cabib (2006) speaks of "stress sensitization"), which have different effects on behavior and include stereotypies.[20] Stereotypies have

18 While stereotypies in human psychiatry represent the repetition of a certain, apparently driven, nonfunctional motor reaction, the abnormal repetition of a certain behavior is understood as *compulsory behavior*. Both are considered under abnormal-repetitive behavior (see American Psychiatric Association 2003, 169, 514–15).

19 "Repeated exposure to uncontrollable aversive situations can cause profound, long-term changes in brain organization" (introduction to Cabib (2006), by G. Mason, 227).

20 Cabib (2006) investigated this for two different mouse strains (DBA/2 and C57BL/6).

been associated to specific activities in specific brain regions (for example, they can be the product of increased behavioral activity by the basal ganglia and explained with reference to input and output of dorsal striatum (Garner 2006; Lewis et al. 2006)). In addition, there are not only stress-related changes here; there are also changes in the function of the cortex. Neural changes, including neurochemical changes, may occur, mainly in a disturbed cortical/subcortical function ratio (Cabib 2006, 227, 243, 250). The reason why an animal shows stereotypic behavior at the time of observation no longer necessarily has to be the boredom consciously felt in this moment but can already be a reaction caused by such neural damage. Although it is assumed that stress-inducing environmental conditions do not immediately cause stereotypical behavioral responses in animals, they can arouse an underlying predisposition to stereotypical behavior (Cabib 2006, 246). All in all, according to Cabib (2006), neurobiological hypotheses interpret stereotypies as symptoms of pathological adjustments of the central nervous system. This includes the assumption that some genotypes are sensitive to pathological stress experiences or stress-inducing environmental conditions. At the same time, the absence of stereotypies does not necessarily imply either normality or low stress (Cabib 2006, 250).

Stereotypies are thus coping strategies, that can be explained by the animal's motivations, are triggered by stress and partly triggered by a number of pathological, neurobiological changes. As such, they are used to respond to stress and frustration. This characterization of stereotypies means that treatability is difficult. Among other things, the welfare of an animal can in fact also be negatively affected by stereotypies when the developed stereotypical behavior routines in turn cannot be performed. To illustrate: if one intervenes with a predator that always runs the same route in her cage, for example, by blocking her way with a boulder, then one has at best eliminated the phenomenon, but not the reason for the behavioral disorder. Young refers to studies that have even shown that bank voles that had developed stereotypies, for example, actually preferred their old, nonenriched environment to an enriched one: in an almost empty cage, literally nothing was standing in the way of their abnormal, repetitive movement behavior (see Cooper and Nicol 1991).

The coping explanation for stereotypies concentrates on learned behaviors, which actually bring an advantage for the acting animal. However, not every stereotypy helps an animal to deal with a stress situation, and not always only one single mechanism such as stress reduction is involved. Würbel et al. (2006) state that "increasing evidence indicates that at least some stereotypies, in some species, do have some beneficial effects.…However, whether any coping effects are truly *causal* in the development and continued performance of stereotypies, or merely beneficial side effects, remains unknown" (14–15). Studies of whether stereotypies occur together with other indicators of poor well-being have shown, first, that this is the case and, second, that animals with highly stereotypical behavior often appear to be in better condition than animals with less severe stereotypies (Mason and Latham 2004). Mason and Latham (2004) suggest: "Beneficial consequences from performing the specific source-behavior of the stereotypy ('do-it-yourself enrichment'), or arising from repetition per se ('mantra effects'), may ameliorate welfare in poor environments" (57). Stereotypies are therefore only seemingly "pointless" in several respects: they are not senseless in the sense of "groundless" and can be understood as coping strategies with regard to typical species-specific motivations, but also with regard to other explanations (such as neurobiological reasons). They are also not pointless in the sense that they don't fulfill any function for the animal. Accordingly, it can be very difficult to treat such a behavioral disorder in a zoo animal. Often it cannot be reversed at all, not even by moving the animal to another enclosure or zoo and also not with targeted enrichment (Young 2003, 52). The only thing that seems certain is that the first causes of the development must be avoided and that one must try to counteract the behavioral disorder as long as its initial cause is still the cause of abnormal behavior (Young 2003, 52). In addition, the exact relationships between husbandry conditions and behavioral disorders and physiological and psychological problems are often still unclear. Solution strategies have yet to be found and tested.

14.7 The Effects of Enrichment

The same uncertainty applies to the understanding of the relationships between physiological, neurological and behavioral improvements that can actually be achieved by enrichment (Young 2003, 43). In the overwhelming number of cases of enrichment measures, success on different levels can be measured and verified.[21] They can be seen in the improved learning ability of animals, in the quantitative and qualitative increase in exploratory behavior, in weight

21 Some measures have led to negative or neutral results, but these can often be explained. For example, sometimes an enrichment object can become a territorial marker or a general good of high value, but not available for all animals in the enclosure. In such cases the object is then competed for and the enrichment measure might have a negative, aggression-intensifying effect. Why some enrichment measures had neither a positive nor a negative effect on behavior has not been clarified, and Young criticizes the fact that, generally, studies that found no positive effect tend not to be published (Young 2003, 33).

gain, in increase in size, in number and complexity of synapses, and in many other indicators (Shepherdson 1998, 7; for further discussion, see Young 2003, 31–44).[22]

Young mentions a range of improvements that enrichment demonstrably brings about, especially with regard to the behavior of animals. Initially, a reduction in the exercise of *abnormal behavior*, including stereotypies, has been observed (enrichment has, for example, revealed a demonstrative reduction in abnormal behavior in seals, horses, hares, bush rats, mice, baboons, chimpanzees, leopards, rhesus monkeys, and bears (Young 2003, 34, table 3.1)). Furthermore, enrichment is used to reduce *aggressive behavior*,[23] on the one hand, and encourage *desired behavior*,[24] on the other (32ff.). In addition, enrichment can result in other behavioral improvements. For example, daily contact between keepers and animals can cause stress for the latter, which can impair cognitive functions such as memory. However, animals depend on their cognitive abilities to remember, for example, where food is located: "Thus, *protecting cognitive functioning* is an important aspect of environmental enrichment" (36; italics added). The same applies to the maintenance or training of other cognitive abilities. Animals with intact or trained learning abilities can learn in the zoo, for example, that rattling is the key to switching to the indoor enclosure. Thus, stress-related captures, to move the animals to other enclosures,

---

22 I will only go into improvements in greater depth regarding *behavior*. The *physiological* improvements include, for example, a reduction in stress hormones, an improved immune response, and an increase in weight with the same food consumption. With regard to *neurological* improvements found, Young (2003) writes: "Psychologists have long been interested in the effects of environmental enrichment because it appears to speed-up or facilitate the repair of brain damage and often results in the improvement of cognitive or motor functions.... It has also been found that environmental enrichment can protect against cognitive deterioration that results from stressors or simply from ageing. Furthermore, there is evidence that it can improve *cognitive functioning* in animals, especially in the area of spatial information processing and memory" (Young 2003, 39–40, table 3.2, 41–42, table 3.3; italics added).

23 In livestock farming, this includes, for example, feather picking in chickens and tail biting in pigs, or the frequent deaths of piglets in pig farming who are crushed to death by the sows (Young 2003, 33–34, table 3.1.). With regard to aggressive behavior, it is important that this also plays a significant role in social hierarchies and is part of natural animal behavior. A reduction in aggressive behavior *per se* is therefore not equivalent to an increase in well-being. However, whenever captive animals show more aggressive behavior than normal, going far beyond the aggressive behavior of their species in the wild, a reduction can provide better well-being (see Young 2003, 36).

24 This might include foraging behavior, exploratory behavior, general activity, and play behavior. Studies on the positive effects of enrichment on these behaviors are available, for example, for fruit bats, chickens, rhesus monkeys, wild dogs, cats, boas, rats, pigs, bears, and dogs (see Young 2003, 35, table 3.1.).

can be avoided. This avoids fear, which, according to Young, represents an improvement in well-being (38–39, 44). Despite all limitations and uncertainties, the data clearly suggest that enrichment improves the welfare of captive animals (39). It is worth noting, although little research has yet been done, that enrichment in itself is rewarding: "Enrichment is intrinsically rewarding to the animal, i.e. the animal welfare benefits are accrued from the 'enjoyment' the animal receives" (44).

14.8 Enrichment in Captivity as an Ethical Necessity

Meanwhile, many zoos are trying to create more species-appropriate housing conditions that can allow animals to exercise and develop their cognitive and social skills, such as cultural activity in chimpanzees, as demonstrated in termite fishing. They present zoo animals with the challenge of removing their food from blocks of ice, or let them play with cardboard boxes or rubber balls. All this adds stimuli to the environment of the animals that serve to evoke mental and physiological events that improve the quality of life of the animals and should prevent or counteract stereotypical behavior. Unknown objects are made available to the animals; olfactory stimuli are also used. The desired behavior that should be triggered by this is curiosity or the joy of discovery, which results in mental and physical stimulation and in the training of mental and physical abilities (see Hutchins et al. 2003, 209–10). Unfortunately, not all zoos are transparent with what is being done to keep the animals busy. But some of them document in detail what they offer and work with behavioral biologists from cooperating institutions. The Wolfgang Köhler Primate Research Center of the Max Planck Institute for Evolutionary Anthropology, in Leipzig, which is integrated into the Leipzig Zoo, even provides instructions for enrichment ideas to other facilities.[25]

All in all, enrichment measures have become increasingly important in recent years and have almost become fashionable in practical application (see Jantschke 1997, 406; and, for a similar discussion, Young 2003, 5). Mason and Rushen (2006, 3, 58ff., 256ff.) have observed an enormous increase in scientific interest in stereotypies in zoo animals since the mid-1990s. This interest should theoretically be part of everyone's thinking when speaking of "zoo philosophy" and a central point within every management guideline for the health and well-being of animals in captivity. However, this should not hide the fact that in many zoos enrichment is still hardly or not at all used. It is also an open

25 http://wkprc.eva.mpg.de/english/files/enrichment.htm, 2019/05/17.

question in many cases whether what the animal encounters and what the visitor sees is really effective enrichment.

Shepherdson warns against seeing the implementation of enrichment as an easy task. Since the research on subjective mental processes and states, including subjective well-being, is still in its infancy, and many key concepts in this research area are not yet clearly defined and completely researched, it could still be difficult to evaluate what constitutes a successful enrichment measure (see Shepherdson 1998, 9). Young (2003) writes: "Environmental enrichment is perhaps unique in animal welfare in that there has been more application than scientific evaluation of its effects on animal welfare, especially in zoos and laboratories" (22). The individual measures must be further systematized and also critically assessed as to whether they actually always serve the needs of the animals or rather the entertainment value for the audience (see also Robinson 1998).

Another problem is that so little is known about the cognitive abilities of many animals and their needs in captivity that any necessary enrichment measures are not yet taken for them. Research to date has focused on the study of cognitive abilities, mental states, and processes of a handful of species and environmental impacts on their development and health. This, combined with a taxonomic preference for our closest kin and a sometimes anthropocentric attitude, has led to an asymmetrical orientation of research on enrichment measures. According to Shepherdson, consideration of the needs of other species in their diversity in much more detail is long overdue. The fact that their sensory and communicative possibilities are more difficult to understand should not be a reason to continue ignoring them (Shepherdson 1998, 9).

However, as has already become apparent with Mench (1998), it can be important to determine which species need enrichment most urgently for their well-being, given the problem for which species enrichment is now being used. This may correlate with the fundamental ethical decision we found in DeGrazia (1996b): there may be species that can only be kept in captivity in an ethically justifiable way due to their complex cognitive and social abilities if an unusually high amount of enrichment is carried out. According to DeGrazia, this is clearly the case with great apes. In view of the increasing knowledge about the graduality of such abilities and their existence in so-far neglected species, this question will gain in importance. Young (2003) also suggests a prioritization of enrichment. However, he criticizes the approach of binding the ability to suffer to special cognitive abilities, especially to the ability to have an understanding of future events: This proposal "produces only a two-point scale, humans/chimpanzees and the rest of the Animal Kingdom, which is not very useful for our purposes" (Young 2003, 48).

The problem here, however, is not really the fact that complex socio-cognitive abilities are only being observed in a few species, but rather that they are defined too restrictively and oriented toward people. Therefore, Young's solution to enrichment is not to leave cognitive abilities aside, but to find out whether a *universal* definition of cognition exists. Young asks: "Does any kind of universal measure of animal cognitive abilities exist?" (48). Young rejects frequently quoted standards. The mere size of the brain simply correlates with the size of the body and is therefore not a good measure. The encephalization quotient, which determines the ratio of brain to body mass, is also insufficient, since the brains of animals do not have a uniform structure.[26] The ratio of the brain areas that control bodily functions to those that are associated with cognitive processes is central, i.e., the ratio of the neocortex to the remaining brain areas. Species with a high neocortex ratio such as great apes and dolphins would also typically master cognitive tests particularly well and often live in complex social groups. Following this, Young also proposes prioritizing such species should we, due to scarce resources, be compelled to align the implementation of the measures to a nonarbitrary yardstick. He does not think that animals with a lower neocortex ratio could not suffer at all, only that they might suffer less (Young 2003, 49). In addition, a scale could also be formed from studies of comparative psychology, a proposal that goes back to Maple and Perkin (1998). They propose using the inherent level of curiosity and exploratory behavior to prioritize what species needs enrichment most urgently. In addition, they point to the importance of ecological data, since omnivores and opportunists, i.e., species that can adapt to many different habitats and lifestyles, demonstrate a special degree of such abilities (Maple and Perkin 1998).[27] Flexibility and problem-solving competence are clearly relevant here, as they are expressions of complex cognitive and social skills (if social learning is involved).

Young is also aware of the limitations of such standards and points out that although the neocortex ratio is applicable to mammals, for example, for many birds, reptiles, fish, and amphibians the importance of the neocortex and the

26 "They do not operate like a central processing unit (CPU), different areas of the brain control different abilities or functions" (Young 2003, 49).

27 Other authors have pointed to the range of motion of species as an indicator of the likelihood of abnormal behavior in captivity. The more an animal moves each day in her search for food, the more likely she is to develop behavioral disorders when kept in a confined space, and the more important it could be that she able to pursue this motivation. These are also ecological data that can be used to prioritize the implementation of enrichment (see also Clubb and Mason 2003; Young 2003, 50; with reference to Clubb and Mason 1998 and Williams et al. 1996).

fact that they do not have a comparable neocortex to humans has not yet been fully clarified. In fact, recent research points to the possibility of complex brain processes despite the lack of a human-like neocortex (Güntürkün and Bugnyar 2016). Here, for example, we are dependent on the correlate of social complexity. Among birds, for example, parrots are known to live in large flocks, form lifelong partnerships, and also perform very well in cognitive tests. The decisive factor is not only cognitive complexity tied to certain areas of the brain and to the relative relationships of certain areas of the brain. What is decisive is what animals do with the help of their brain, what they can know and what they reveal in their behavior about their cognitive and social abilities. Even if we still know little about the neocortex ratio and about the behavior, abilities, motivations, and needs of reptiles, amphibians, and fish in particular, the principle of "benefit of the doubt" still applies to them (see also Young 2003, 49–50).

Young comes to the conclusion that—provided that behavioral disorders continue to prove to be recognized indicators of bad welfare—such considerations have at least helped us to know which species need more thought before we keep them in captivity. Similar to Mench (1998), he derives a list of criteria for prioritizing the implementation of enrichment that clearly builds on the importance of complex social and cognitive abilities. So the following questions should be asked: Do animals of the species in question have an understanding of future events? Do cognitive experiments suggest that they act at a high cognitive level? Do they have a high neocortex ratio? Do they live in a large social association with complex, lifelong partnerships? Do they show a high level of curiosity and exploratory behavior? Do we know that they develop abnormal behavior in captivity? Is it an omnivorous or opportunistic species? Only those who ask themselves questions such as these *before* beginning to keep an animal in captivity do justice to the ethical claim of animal welfare as it is expressed in the principle of nonmaleficence or the five freedoms. Ideally, behavioral disorders must be avoided from the outset. If they already exist, then experience shows they must be counteracted as early as possible.

As already explained, zoos justify their existence apart from their entertainment value by two arguments in particular: they secure the survival of some endangered animal species through conservation breeding programs as well as nature and species protection projects, and they pursue an educational mission through knowledge transfer (see Hutchins et al. 2003). Also the German Animal Welfare Act refers to these ideals:

> Despite the important purpose of the zoological gardens as a recreation area for people, which at the same time serves educational purposes as

> well as science and species protection…the standards of animal protection there should be set higher rather than lower.
>
> KLUGE 2002, 250, transl. MK

According to § 2 of the German Animal Welfare Act, zoos must keep their animals in such a way that they are fed and cared for according to their species and needs and are accommodated in a way that is appropriate to their behavior. For example, the animal's ability to move in a manner appropriate to her species must not be so restricted as to cause pain, avoidable suffering or damage (Kluge 2002, 101). The "intrinsic value," the "dignity of the creature," and the "basic needs" of zoo animals must never be disregarded (Kluge 2002, 123; transl. MK). Article 3 of the EC Directive on the keeping of wild animals in zoos also speaks of a "species-appropriate design of enclosures" (European Union Council 1999). It is not clearly stated that a species-appropriate design of the enclosures is necessary not only because of physical needs (for example, because of the need for exercise mentioned in the Animal Protection Act), but also in order to do justice to cognitive and social abilities. More specialized regulations are needed and being developed. After all, Eisenhart von Loeper states in his explanations to § 2 of the German Animal Protection Act that neurotic, often aggressive or apathetic zoo animals that became that way because of inappropriate housing conditions "don't serve an educational purpose" (Kluge 2002, 123). Others agree. Young states: "The educational value of animals within zoos is directly related to them expressing normal behaviour patterns" (Young 2003, 29). Zoos are dependent on good and healthy specimens of those species that are the focus of their environmental education. It is difficult to explain to children about the adaptability and natural behavior of a tiger if the exhibited animal is merely walking up and down in front of them. Kreger et al. (1998) had shown that the aesthetics of the presentation not only affected the appearance of the walkways and cages but also the behavior of the presented animal. If a natural design of the cages occurs with an unnaturally behaving animal, then this is viewed as contradictory and confusing by the visitor (Young 2003, 29).

Against the background of the educational aspirations of zoos and the fact that zoos have a decisive influence on the relationship between humans and animals, it seems particularly urgent to take the ethical implications of animal cognition research more seriously and to emphasize them more clearly in animal protection legislation. However, it must be noted that the argument "If there have to be zoos, then they must include enrichment" naturally presupposes that the educational and species protection claim of zoos is accepted as a valid argument in favor of their maintenance. Section 9.2.1 has shown that this is only possible to a limited extent. Critics may therefore rightly point out that an ethical evaluation of zoos is quite limited and actually insufficient if it only or predominantly concentrates on husbandry and enrichment questions.[28]

Again, not only is a statement necessary *that* enrichment is required, but a statement of *what normative weight* this requirement has is also required. If animals have a *right* for all their physical and mental needs to be taken into account, then they also have a *right* to enrichment. If the degree of enrichment required is determined by the complexity of cognitive and social abilities, then there may be animals whose well-being in captivity can no longer be guaranteed, despite enrichment. Thus, an argumentation that starts with the general capacity for suffering and the needs of all animals for well-being and the exercise of species-specific behavior leads us to ethical limits in dealing with some species that have particularly complex socio-cognitive abilities. And although these are to be understood as gradual, they nevertheless result in absolute limits for what is ethically justifiable. What began as alternative reasoning in the tradition of the criterion of the ability to suffer instead of the tradition of animal rights leads in the end to points of contact of both levels of reasoning. No matter from which side one looks at the increasing evidence for complex socio-cognitive abilities in more and more species, it presents science with fundamental challenges—both in terms of the moral status of animals and in terms of the practical interaction with animals.

28 Young (2003) quite correctly recognizes that "animal rights groups believe that the environmental enrichment work is increasing the justification for keeping animals in captivity" (15).

CHAPTER 15

Final Evaluation of Personhood Rights for Animals

Overall, the demand for personhood rights for animals is proving to be an acceptable and justifiable position. However, it is also fraught with difficulties and tensions. A positive aspect is that it follows the initially honorable concern to strengthen the individual rights of animals. More and more scientists are following this path and are doing so against the background of the increasingly hopeless threat to apes, dolphins, and other animals in the wild and the problem of keeping these animals in captivity. In most cases, a correspondingly stronger consideration of these animals actually seems possible without fundamental human interests and rights being at stake. It is not a matter of human survival to stop dolphin drive hunts, nor does it mean the end of biomedical research to refrain from chimpanzee experiments. Species conservation and conservation education do not have to take place in the zoo. And behavioral research can be conducted outside from laboratories.

Furthermore, scholars who argue in favor of human rights–like rights for (some) animals take into account the emotional significance of our cognitive kin for our self-image, and a possibly special significance of great apes as our closest kin. The obvious similarity in essential characteristics furthers empathy and motivates animal protection, and these animals certainly play an important role as flagship species. As long as no direct suffering is associated with their role as ambassadors for their last living conspecifics in the wild and for other threatened species, their public visibility should be used. This would of course have to be critically examined on a case-by-case basis. It is conceivable that even then only a representation of these animals via different media (e.g., in documentary films as opposed to a live-presentation) would be ethically justifiable to increase public knowledge. It could be possible that their observation in the wild in the context of controlled ecotourism is possible, while keeping them in captivity is to be ruled out as long as highest welfare standards cannot be guaranteed.

It seems that the personhood approach is a serious proposal, but one that few have really thought through so far. Problems based on the concept of human rights have not been sufficiently taken into account and have hardly been discussed. However, these problems can be addressed—at the cost of the same inherent tensions that we have seen in the discussions about human "marginal" cases, human dignity, human rights, and personhood status. The central point here is that on the level of the philosophical foundation or justification

 | DOI:10.1163/9789004415072_016

of human rights it is referred to an ideal image of a human being, gifted with reason, who possesses consciousness, self-consciousness, sentience, the ability to suffer, culture, language, a theory of mind, autonomy, and many other qualities and abilities. In reality, however, neither all people possess these abilities, nor do they possess them to the same extent at any time in their lives. We are confronted with "borderline" cases, such as infants and disabled people, and are prepared to respect them as people with inalienable rights. We strongly point out that these people in particular need the protection of human rights because they are very vulnerable. In order not to fall behind achievements in terms of morality and civilization, we are prepared to accept the tension between the ideal of personhood at the level of justification of human rights and the diversity of gradations of personal characteristics at the level of application. The indeterminacy, generality, and graduality of being a person is met here by the fact that we know our peers well enough to understand all people as belonging to the moral community. In the case of fellow human beings, "we (anthropologically) have always been familiar with being a person or with becoming a person of everything that has a human face" (Wimmer 1999, 339).

We are faced with the challenge that we now increasingly come to know other animals better—especially with regard to their complex social cognition. Theoretically, it is possible to consider them as nonhuman persons and to grant them at least some basic human rights—despite the fact that they do not meet all criteria for personhood. This would then have to be done in full awareness that animals only have (some of) the criteria for personhood to a certain degree. This, however corresponds nicely with a concept of personhood that can be characterized as a cluster concept: in order to be a person someone need not show all characteristics and not all characteristics to the full extent (Andrews et al. 2019). A cluster concept can deal with "borderline" cases without thinking about them as inferior. Applying a cluster concept of personhood and human rights can furthermore make it possible to understand the required abilities for what they are: they consist of individual building blocks and are present in degrees—de phenomena. The challenge, then, is not to elevate the importance of "human likeness" nor to use the abilities of healthy human adults as the yardstick. Instead, in an interdisciplinary effort, nonanthropocentric minimum definitions for the required socio-cognitive competencies (preferably arrived at with substantially de-intellectualized presuppositions) must be found, which nevertheless describe the core of these abilities. Just what this can look like in some aspects has been illustrated in this book, using the examples of culture, language, and theory of mind. A definition of culture as applicable to animals does not need to presuppose a high degree of human technology; language definition does not require a phrase structure grammar; and the definition of theory of mind can go without an understanding of false belief. Complex competences can be reduced to their core, which is as free of prerequisites as possible, or they can be reduced to their core "lowest common denominator," which could include the abilities of animals.

Just which and how many animal species ultimately fall under a broad definition of "nonhuman person" depends (a) on the state of research on animal cognition, and (b) on how basic and gradual personal qualities can actually be defined. Of course, the criteria for personhood and thus for some basic, absolute rights, could even be considered more broadly. Awareness combined with the ability to suffer could, for example, be a candidate for the initial stage of any form of personhood. Both certainly apply to a large circle of animals. Just what a broader idea of human and nonhuman persons leads to is an open question. One may want to stick to clearer categorizations that can also be easily implemented legally and politically, but then one should be careful not to preserve the status quo out of pure pragmatism. It, too, should be in some way ethically justifiable.

It is certainly a justified criticism of the Great Ape Project that the concept of human rights has so far been put forward too unilaterally for human beings alone. But anyone who wants to cite complex socio-cognitive abilities as criteria for personhood per se, and not only as criteria for human resemblance, cannot only refer to chimpanzees, either. Other species, some of which have been covered in this book (such as dogs, ravens, parrots, whales, dolphins, other great apes, and other primates in general), need increasing consideration. Many species that have not been discussed here are also added, from elephants to pigs, from chickens to rats. This means that not only classic zoo animals, but also classic pets, farm animals, and laboratory animals need to be included in the ethical arena. Frans de Waal and Pier Francesco Ferrari (2010) rightly arrive at the following impression:

> Over the past few decades, comparative cognitive research has focused on the pinnacles of mental evolution, asking all-or-nothing questions such as which animals (if any) possess a theory of mind, culture, linguistic abilities, future planning, and so on. Research programs adopting this top-down perspective have often pitted one taxon against another, resulting in sharp dividing lines....A dramatic change in focus now seems to be under way, however, with increased appreciation that the basic building blocks of cognition might be shared across a wide range of species. (201)

The personhood approach to animal rights must, if it is to provide a modern picture of social and cognitive competences in their graduality, examine whether it is necessary and viable to specify the concept of basic, inalienable rights for very different species. Work must therefore continue on an appropriate concept of personhood rights for animals. One focus will be on the specific abilities of different species in order to meet their specific needs.

Another unsolved problem in the personhood approach is the relationship between cognitive abilities and the ability to suffer. Both seem to be taken into account in the Great Ape Project and in Thomas White's writings. However, how both criteria are related in biology and ethics is still largely an open question. Some basic human rights, such as the right not to be tortured, are certainly largely due to the ability to suffer. While all animals capable of suffering, also in other pathocentric approaches and other animal rights approaches, have a right not to suffer (unnecessarily), representatives of the personhood approach must respond to which species capable of suffering should have an *inalienable right* not to be tortured. That the right not to be tortured is a human right, and that it is based so fundamentally on the ability to suffer and not solely on cognitive abilities, should give cause for thought—both to the representatives of the personhood approach and to those who take a pathocentric position but avoid defining the normative weight of the suffering of animals in terms of quality.

Finally, the concept of species-appropriateness (as in species-specific welfare needs) seems a suitable concept to emphasize the importance of sociocognitive abilities in the practice of dealing with animals. Needs that are particularly linked to social and cognitive abilities can certainly be identified in many animal species. Enrichment in captivity can clearly be identified as an ethical necessity and should be applied in all types of systems of captivity. When it comes to pets, the private owner is responsible; in research facilities and laboratories, it is corporations and universities; in livestock husbandry, it is the farmer; and in zoos, it is the staff and public or private sponsors. In general, many of the ethically problematic ways of dealing with animals discussed in this book are determined by all of us in our role as consumers. This also means that each and every one of us can demand a more species-appropriate husbandry of animals through a variety of means, from boycotts, to targeted purchasing behavior, right up to the renunciation of animal products altogether.

However, the concept of species-appropriateness alone may not be sufficient to establish and enforce ethically sound handling and husbandry conditions for different animal species. This is due to the fact that the concept itself does not necessarily contain a normative statement. The idea widely accepted in veterinary medicine and biology that animals should be treated in a species-appropriate manner (provided one knows what this means for the respective animal species) is not necessarily reformulated into a right to species-appropriate treatment—and certainly not into an inalienable right similar to human rights. In other words, species-appropriateness is an element that can be incorporated into various animal-ethical approaches, but its normative weight has yet to be argued for and defined. With respect to ethics, it is not sufficient to withdraw to this concept without clarifying its normative consequences. Theoretically, however, it would be possible to demand a right to a species-appropriate treatment and a species-appropriate life for animals. This could then apply to all living beings, from humans to laboratory mice, from fish to chimpanzees. For some animals, this right would mean that they have to live in a socially and cognitively demanding environment. This reveals interfaces at which animal welfare approaches would reach their limits and would require rights elements for their continuation. These rights elements would identify the limits of what is permissible and reasonable with a view to species-appropriateness, for example. It could be said that some animals cannot be kept in captivity in a species-appropriate manner and may therefore not be kept in captivity at all. Under two conditions, the right to species-appropriateness could then be of equal importance to what was discussed in this book as personhood rights for animals that are analogous to human rights: first, if the right to a species-appropriate life would take into account the personhood qualities of animals as fundamentally equivalent to comparable personhood qualities of humans, and second, if the right to a species-appropriate life were an absolute right that could only be undermined if the species-appropriate life of humans would otherwise be endangered. Under these two conditions, the keeping of dolphins in aquariums, for example, could, in principle, also be prohibited by recourse to this right, specifically with a view to their complex social and cognitive abilities and needs. Such a strong demand for species-appropriateness would then have to be reflected in a very strong animal protection law. If, on the other hand, species-appropriateness is cited without a normative designation of its significance, then it is initially only a biological concept that has yet to be developed into an ethical concept.

The presented possibilities and opportunities, ambiguities and problems of the personhood approach are to be considered in further debates. The fact that there are ambiguities and problems does not mean that they are inexplicable and unsolvable. Nevertheless, the call for personhood rights could still be more far-reaching and promising than other alternatives. Compared to pathocentric-utilitarian approaches, for example, it can better protect the individual—whether human or animal—and sometimes even exclude it completely from weighting scenarios. Purely pathocentric arguments are preceded by the fact that the personhood approach can establish a right to life. Animal rights approaches clearly have the advantage of not starting from square one. A personhood approach can find support in other arguments and debates. This is true with regard to the *content of required rights* (life, freedom, and the absence of torture), and with regard to the *foundation of these rights* (with recourse to certain personhood qualities of an individual). Finally, by relying on the existing concept of human rights, *the strength of the rights demanded* is made clear (we are concerned with inalienable, absolute rights).

The recourse to the tradition of justification of human rights thus has a specific function in contrast to other animal rights approaches: it puts forth, above all, the *strength* of the required rights, a strength that is not achieved by other approaches. It is derived from the expressive and appellative power of philosophical terms such as *human dignity* and *human rights*, which are originally thought of as absolute (Rippe 2011). However, personhood rights for animals are up to now often considered an addition to other approaches. In the way they have been put forward so far, they are currently at least not a *replacement* for pathocentric animal ethics but rather an *addition* to classical animal rights approaches. As such, however, they should be taken seriously, and their application should be tested more widely. This is not only important in view of the large number of animals that we use and consume, which we keep under inappropriate and exploitative conditions, or which we threaten with extinction in the wild. This necessity also arises from the fact alone that socio-cognitive abilities are, increasingly, no longer understood as monolithic phenomena that either occur completely or not at all—whether in humans or in nonhuman animals.

Acknowledgments

I would like to thank Mark Kanak for translation, Dr. Susana Monsó for consultancy and proofreading, Evan White for copyediting, Benjamin Suchard for editing, Nidha Jeddy for processing the book through production, Christoph Schachenhofer for indexing, and all reviewers involved in the translation of my work.

Working on this book (or the German version of this book) was partly enabled by grants from the DFG (in terms of a scholarship and publication grant within the Postgraduate Program "Bioethik" at the IZEW, University of Tuebingen), the FWF (P31466-G32), and by translation grants of the Messerli Foundation, the City of Vienna (Wissenschaftsstipendium Magistratsabteilung 7, Kultur, Wissenschafts- und Forschungsförderung) and the Unit of Ethics and Human-Animal Studies of the Messerli Research Institute Vienna.

Chapter 6 (Theory of Mind) has some overlap with the German publication of Benz-Schwarzburg et al. (2008). However, only those sections that were under the sole authorship of Judith Benz-Schwarzburg were used and revised for the present text, see also the explanation by Benz-Schwarzburg et al. (2008) at the end of the publication detailing the distribution of authorship. Parts of the content provided in Chapter 9.1.3. on the situation of great apes in the wild overlaps with Benz & Benz-Schwarzburg (2010) and Benz-Schwarzburg and Benz (2012). A more detailed analysis on great apes as our closest kin and on the kind of loss we face for our identity should these animal get extinct is being provided in Benz-Schwarzburg (2011).

I thank PD Dr. Miriam Haidle for providing me with two of her cognigrams (Fig. 4 and 5) All other illustrations were made by the author, sometimes orienting on photos made by herself, on Open Clipart pictures or other open source clipart. The illustration above the Ambrose Bierce quote stems from The Cosmopolitan 1892, 79 and falls under the ClipArt ETC Paid Commercial License.

Bibliography

AWD (2007). Endangered Mountain Gorillas Shot Dead by Rebels in DRC. Press Release from 17 January 2007. https://www.awf.org/news/endangered-mountain-gorillas-shot-dead-rebels-drc, 2019/05/18.

Allen, C. (1999). Animal concepts revisited: The use of self-monitoring as an empirical approach. Erkenntnis 51: 33–40.

Alpakas für Menschen (2019). Unsere Tiere. http://www.alpakas-fuer-menschen.at/index.php/home-7.html, 2019/05/22.

Alzmann, N. (2009). In: Der ethisch vertretbare Tierversuch: Kriterien und Grenzen (Borchers, D. & Luy, J., eds.). Paderborn: mentis, 141–170.

Ammann, K. (2001). Bushmeat hunting and the great apes. In: Great apes & humans: the ethics of coexistence (Beck, B.B. et al., eds.). Washington; London: Smithsonian Institution Press, 71–85.

AMMPA (a). About the alliance. http://goo.gl/85fB1, 04.08.2012, no longer online available.

AMMPA (b). Our members. http://goo.gl/FTjPc, 04.08.2012, no longer online available.

Ancrenaz, M., Gumal, M., Marshall, A.J., Meijaard, E., Wich, S.A. & Husson, S. (2016). Pongo pygmaeus (errata version published in 2018). The IUCN Red List of Threatened Species 2016: e.T17975A123809220. https://www.iucnredlist.org/species/17975/123809220, 2019/04/13.

Andrews, K. (2005). Chimpanzee theory of mind: looking in all the wrong places? Mind & Language 20 (5), 521–536.

Andrews, K. (2012). Do Apes Read Minds? Toward a New Folk Psychology. Cambridge: The MIT Press.

Andrews, K. (2017). Chimpanzee mind reading: Don't stop believing. Philosophy Compass, 12(1). e12394 https://doi.org/10.1111/phc3.12394.

Andrews, K. & Beck, J. (eds.) (2018): The Routledge Handbook of Philosophy of Animal Minds. London, New York: Routledge.

Andrews et al. (2019): Chimpanzee Rights: The Philosophers' Brief. New York: Routledge.

anonymous (2007a). Lizenzrechte für Produkte: Eisbär Knut beschert Berliner Zoo Zusatzeinnahmen in Millionenhöhe. Wirtschaftswoche 09 June 2007. http://goo.gl/b3Pz7, 2012/07/19.

anonymous (2007b). Zoo lässt »Knut« als Marke schützen (dpa press release). Focus online 25 March 2007. http://goo.gl/9IhsE, 2012/07/19.

anonymous (2008). Briefmarke: Knut gibt's bald zum Abschlecken (dpa press release). Welt Online 27 February 2008. http://goo.gl/G9axR, 2012/07/19.

Aristotle, Ross, W.D., & Brown, L. (2009). The Nicomachean Ethics. Oxford: Oxford University Press.

Arnold, K. & Zuberbühler, K. (2006). Language evolution: semantic combinations in primate calls. Nature 441 (7091), 303.

Arzt, V. & Birmelin, I. (1993). Haben Tiere ein Bewußtsein? Wenn Affen lügen, wenn Katzen denken und Elefanten traurig sind. München: Bertelsmann Verlag.

Auzel, P. & Wilkie, D.S. (2000). Wildlife use in northern Congo: hunting in a commercial logging concession. In: Hunting for sustainability in tropical forests (Robinson, J.G. & Bennett, E., eds.). New York: Columbia University Press, 413–454.

Baber, C. (2003). Cognition and tool use: forms of engagement in human and animal use of tools. London: Taylor & Francis.

Bacon, F. ([1623] 1857). Instauratio magna, De dignitate et augmentis scientiarum, Vol 1, The Works of Francis Bacon. Ed. by J. Spedding, R.L. Ellis and D.D. Heath. London: Longman.

Balda, R.P. & Kamil, A.C. (1992). Long-term spacial memory in Clark's nutcracker, *Nucifraga columbiana*. Animal Behaviour 44 (4), 761–769.

Balluch, M. & Theuer, E. (2007). Trial on personhood for chimp »Hiasl«. Alternatives to Animal Experimentation 24 (4), 335–342.

Barnes, J. (2002). Sea otter foraging and feeding behaviors: a review. Prepared for the US Marine Mammal Commission. 27 pp. http://www.otterproject.org/wp-content/uploads/2012/05/Barnes_2002_Sea_otter_foraging_and_feeding_behaviors_for_MMC.pdf, 2919/05/19.

Baron-Cohen, S. et al. (1985). Does the autistic child have a ›theory of mind‹? Cognition 21 (1), 37–46.

Barta, H. (2007). Online Lehrbuch Zivilrecht: Grundriss und Einführung in das Rechtsdenken. Chap. 4. Wien: WUV Universitätsverlag, 2004. https://www.uibk.ac.at/zivilrecht/buch/pdf/zivilrecht2004_kapitel4.pdf, 2019/05/19.

Bartels, A. (2005). Strukturale Repräsentation. Paderborn: mentis.

Bartsch, K. & Wellmann, H.M. (1995). Children talk about the mind. New York; Oxford: Oxford University Press.

Baverstock, A. & Finlay, F. (2008). Does swimming with dolphins have any health benefits for children with cerebral palsy? Archives of Disease in Childhood 93 (11), 994–995.

Bayertz, K. (1999). Menschenwürde. In: Enzyklopädie Philosophie (Vol 1) (Sandkühler, H.J. et al., eds.). Hamburg: Meiner, 824–826.

BBC (2012). BBC News, Q&A: DR Congo conflict. https://www.bbc.com/news/world-africa-11108589, 2019/05/19.

Beardsworth, A. & Bryman, A.E. (2001). The wild animal in late modernity: the case of the disneyization of zoos. Tourist Studies 1 (1), 83–104.

Beauchamp, T.L. & Childress, J.F. (2008). Principles of biomedical ethics. 6. Aufl. Oxford: Oxford University Press.

Beck, B.B. (1980). Animal tool behavior: the use and manufacture of tools by animals. New York, London: Garland STPM Press.

Beck, B.B. (1982). Chimpocentrism: bias in cognitive ethology. Journal of Human Evolution 11 (1), 3–17.

Beck, B.B. (1986). Tools and intelligence. In: Animal intelligence: insights into the animal mind (Hoage, R.J. & Goldman, L., eds.). Washington D.C.; London: Smithsonian Institution Press, 135–147.

Becker, P.-R. (1993). Werkzeuggebrauch im Tierreich. Stuttgart: Hirzel.

Bekoff, M. (2007a). The emotional lives of animals: a leading scientist explores animal joy, sorrow, and empathy—and why they matter. Novato, CA: New World Press.

Bekoff, M. (ed.) (2007b). Encyclopedia of human-animal relationships: a global exploration of our connections with animals (Vol. 1–4). Westport, Connecticut: Greenwood Press.

Bekoff, M. et al. (eds.) (2002). The cognitive animal: empirical and theoretical perspectives on animal cognition. Cambridge, MA; London: MIT Press.

Benoit-Bird, K. & Au, W. (2009a). Cooperative prey herding by a pelagic dolphin, *Stenella longirostris*. Journal of the Acoustical Society of America 125 (1), 539–546.

Benoit-Bird, K.J. & Au, W.W.L. (2009b). Phonation behavior of cooperatively foraging spinner dolphins. Journal of the Acoustical Society of America 125, 125–137.

Benoit-Bird, K.J. & Au, W.W.L. (2003). Prey dynamics affect foraging by a pelagic predator (*Stenella longirostris*) over a range of spatial and temporal scales. Behavioral Ecology and Sociobiology 53 (6), 364–373.

Bentham, J. ([1789] 1970). An introduction to the principles of morals and legislation, Vol 2, The Collected Works of Jeremy Bentham. Ed. by J.H. Burns, F. Rosen and P. Schofield London: University of London, The Athlon Press.

Benz, J. (2004). Epistemische Ausdrücke in der Kindersprache. Eine linguistische Korpusuntersuchung zum Erstspracherwerb mit quantitativer und qualitativer Auswertung. Unpublished MA thesis (German state exam), submitted to the faculty of linguistics of the University of Tuebingen, Germany.

Benz, S. & Benz-Schwarzburg, J. (2010). Great apes and new wars. Civil Wars 12 (4), 395–430.

Benz-Schwarzburg, J. et al. (2008). Theory of Mind bei Mensch und Tier. In: Kognition und Verhalten: Theory of Mind, Zeit, Imagination, Vergessen, Altruismus. Interdisziplinäre Forschungsarbeiten am Forum Scientiarum (Vol 1) (Evers, D. & Weidtmann, N., eds.). Münster: LIT.

Benz-Schwarzburg, J. (2011). Was fehlt uns, wenn die Großen Menschenaffen aussterben? In: Darwin und die Bioethik (Kovács, L. et al., eds.). Freiburg; München: Alber Verlag, 133–150.

Benz-Schwarzburg, J. & Benz, S. (2012). Driving the Great Apes to Extinction: Perspectives from Conservation Biology, Politics and Bioethics. In: Biopolicy: The Life Sciences and Public Policy (Somit, A. & Peterson, S.A., eds.), Research in Biopolitics Vol. 10. Bingley, U.K: Emerald Books.

Bergl, R.A., et al. (2016). Gorilla ss diehli (errata version published in 2016). The IUCN Red List of Threatened Species 2016: e.T39998A102326240. http://dx.doi.org/10.2305/IUCN.UK.2016-2.RLTS.T39998A17989492.en, 2019/05/21.

Bermejo, M. & German, I. (1999). Tool-set for termite-fishing and honey extraction by wild chimpanzees in the Lossi Forest, Congo. Primates 40 (4), 619–627.

Bermúdez, J.L. (2003). Thinking without words. Oxford: Oxford University Press.

Bernstein, M. (2002). Marginal cases and moral relevance. Journal of Social Philosophy 33 (4), 523–539.

Berwick, R.C. et al. (2010). Songs to syntax: the linguistics of birdsong. Trends in Cognitive Sciences 15 (3), 113–121.

Bieri, P. (1997). Generelle Einführung. In: Analytische Philosophie der Erkenntnis (Bieri, P., ed.). 4., unchanged ed. of the first ed. of 1987, Weinheim: Beltz Athenäum, 9–72.

Birnbacher, D. (2002). Ökologie. In: Enzyklopädie Philosophie (Sandkühler, H.J., ed.), CD-ROM version, Hamburg: Meiner.

Bischof-Köhler, D. (2000). Kinder auf Zeitreise: Theory of Mind, Zeitverständnis und Handlungsorganisation. Bern; Göttingen; Toronto; Seattle: Huber.

Bloom, P. (2000). How children learn the meanings of words. Cambridge, MA: MIT Press.

Bloomsmith, M.A. et al. (1991). Guidelines for developing and managing an environmental enrichment program for non-human primates. Laboratory Animal Science 41 (4), 372–377.

Blumer, K.R. (1997). Tierversuche zum Wohle des Menschen? Ethische Aspekte des Tierversuchs unter besonderer Berücksichtigung transgener Tiere. München: Herbert Utz Verlag.

BMBF (2001). Hightech statt Tiere: Ersatz- und Ergänzungsmethoden zu Tierversuchen. Grundlagen—Ergebnisse—Perspektiven. A joint report prepared by the BMBF, BMELV, BMG and the BMU. https://www.las-interactive.de/fileadmin/user_upload/PDFs/C2/hightech_statt_tiere.pdf, 2919/05/19.

BMELV (2009). Lindemann: BMELV legt Tierversuchszahlen 2008 vor. Press release 252 from 23 October 2009, accessed on 2012/07/19.

BMU (2007). Nationale Strategie zur biologischen Vielfalt. Paderborn: Bonifatius GmbH. https://www.bfn.de/fileadmin/BfN/biologischevielfalt/Dokumente/broschuere_biolog_vielfalt_strategie_bf.pdf, 2019/05/19.

Boedhihartono, A. & Lumpu, M. (2008). Conflict in Goma: what's happening to conservation? The IUCN Forest Conservation Programme Newsletter, ArborVitae 38, 3.

Boesch, C. (1991). Teaching among wild chimpanzees. Animal Behaviour 41 (3), 530–532.

Boesch, C. (1996). The emergence of cultures among wild chimpanzees. In: Evolution of social behaviour patterns in primates and man: a joint discussion meeting of the Royal Society and the British Academy (Runciman, W.G. et al., eds.). Proceedings of the British Academy (Vol. 88), London: The British Academy, 251–268.

Boesch, C. & Boesch, H. (1983). Optimisation of nut-cracking with natural hammers by wild chimpanzees. Behaviour 83 (3–4), 256–286.

Boesch, C. & Boesch, H. (1984). Mental map in wild chimpanzees: an analysis of hammer transports for nut cracking. Primates 25 (2), 160–170.

Boesch, C. & Boesch, H. (1990). Tool use and tool making in wild chimpanzees. Folia Primatologica 54 (1–2), 86–99.

Boesch, C. & Boesch, H. (1993). Different hand postures for pounding nuts with natural hammers by wild chimpanzees. In: Hands of the primates (Preuschoft, H. & Chivers, D., eds.). Wien: Springer, 31–43.

Bohlken, E. (2009). Philosophische Anthropologie („filosofskaja antropologija“). In: Westliche Philosophie der Gegenwart: Ein enzyklopädisches Lexikon („Sovremennaja Sapadnaja Filosofija. Enziklopeditscheskij Slovar“) (Otfried Höffe & Vladimir S. Malachov, eds.). Moskau: Kulturnaja revoljuzija 2009, 112–114.

Bonner, J.T. (1989). The evolution of culture in animals. Princeton; New Jersey: Princeton University Press.

Borchers, D. & Luy, J. (eds.) (2009a). Der ethisch vertretbare Tierversuch: Kriterien und Grenzen. Paderborn: mentis.

Borchers, D. & Luy, J. (2009b). Ethisch vertretbare Tierversuche: Eine Einführung. In: Der ethisch vertretbare Tierversuch: Kriterien und Grenzen (Borchers, D. & Luy, J., eds.). Paderborn: mentis.

Borsche, T. (1990). Was etwas ist: Fragen nach der Wahrheit der Bedeutung bei Platon, Augustin, Nikolaus von Kues und Nietzsche. München: Fink.

Borsche, T. (1995). Sprache (I. Antike). In: Historisches Wörterbuch der Philosophie (Vol. 9) (Ritter, J. & Gründer, K., eds.). Basel: Schwabe & Co AG, 1437–1454.

Bostock, S. (1993). Zoos and animal rights: the ethics of keeping animals. London; New York: Routledge.

Boyd, R. & Richerson, P.J. (1985). Culture and the evolutionary process. Chicago: Chicago University Press.

Boyd, R. & Richerson, P.J. (1996). Why culture is common but cultural evolution is rare. In: Evolution of social behaviour patterns in primates and man: a joint discussion meeting of the Royal Society and the British Academy (Runciman, W.G. et al., eds.). Proceedings of the British Academy (Vol. 88), London: The British Academy, 77–93.

Boyes, S. (2012). Bushmeat: Every Man's Protein until the Forest is Empty... National Geographic Wildlife Blog, https://blog.nationalgeographic.org/2012/02/09/bushmeat-every-mans-protein-until-the-forest-is-empty/, 2019/05/19.

Brakes, P. & Williamson, C. (2008). Delphintherapie: Eine Faktensammlung. Report prepared for the WDCS, transl. into German by Astrid Hackanson and Karsten Brensing. München: WDCS. https://de.whales.org/wp-content/uploads/sites/4/2018/07/delfintherapie_faktensammlung.pdf, 2019/05/19.

Brando, S. (2010). Advances in husbandry training in marine mammal care programs. International Journal of Comparative Psychology 23 (4), 777–791.

Breitenbach, E. et al. (2006). Delfintherapie für Kinder mit Behinderungen: Analyse und Erklärung der Wirksamkeit. Würzburg: Johann Wilhelm Klein-Akademie.

Brensing, K. (2004). Approaches to the behavior of dolphins *Tursiops truncatus* during unstructured swim-with-dolphin programs. Dissertation thesis, submitted to the Faculty of Biology, Chemistry and Pharmacology of the FU Berlin. Berlin: FU Berlin. https://refubium.fu-berlin.de/handle/fub188/13192, 2019/05/19.

Brensing, K. et al. (2005). Impact of different kinds of humans in swim-with-the-dolphin programs in two settings. Anthrozoös 18 (4), 409–429.

Brensing, K. et al. (2003). Can dolphins heal by ultrasound? Journal of Theoretical Biology 225 (1), 99–105.

Breuer, T. et al. (2005). First observation of tool use in wild gorillas. PloS Biology 3 (11), e380. DOI:10.1371/journal.pbio.0030380.

Brook, B.W. et al. (2008). Synergies among extinction drivers under global change. Trends in Ecology and Evolution 23 (8), 453–460.

Broom, D.M. (1999). Animal welfare: the concept of the issues. In: Attitudes to animals (Dolins, F., ed.). Cambridge: Cambridge University Press, 129–142.

Broom, D.M. (2003). The evolution of morality and religion. Cambridge, UK: Cambridge University Press.

Bryman, A. (1999). The disneyization of society. The Sociological Review 47 (1), 25–47.

Buck, C.D. & Schroeder, J.P. (1990). Public health significance of marine mammal diseases. In: Handbook of marine mammal medicine: health, disease, and rehabilitation (Dierauf, L.A., ed.). Cleveland: CRC Press, 163–173.

Bühler, K. (1934). Sprachtheorie: Die Darstellungsfunktion der Sprache, Kap. Das Organonmodell der Sprache. Jena: Verlag Gustav Fischer, 24–33.

Buckner, C. (2013). Morgan's Canon, meet Hume's Dictum: avoiding anthropofabulation in cross-species comparisons. Biology and Philosophy 28 (5): 853–871.

Buckner, C. (2014). The semantic problem(s) with research on animal mindreading. Mind and Language, 29(5), 566–589.

Bugnyar, T. & Heinrich, B. (2005). Ravens, *Corvus corax*, differentiate between knowledgeable and ignorant competitors. Proceedings of the Royal Society of London B: Biological Science 272 (1573), 1641–1646.

Bugnyar, T. & Kotrschal, K. (2002). Observational learning and the raiding of food caches in ravens, *Corvus corax*: is it ›tactical‹ deception? Animal Behaviour 64 (2), 185–195.

Bugnyar, T. & Kotrschal, K. (2004). Leading a conspecific away from food in ravens (*Corvus corax*)? Animal Cognition 7 (2), 69–76.

Buttelmann, D. et al. (2009a). Do great apes use emotional expressions to infer desires? Developmental Science 12 (5), 688–698.

Buttelmann, D. et al. (2009b). Eighteen-month-old infants show false belief understanding in an active helping paradigm. Cognition 112 (2), 337–342.

Buttelmann, D. et al. (2017). Great apes distinguish true from false beliefs in an interactive helping task. PLoS ONE 12(4): e0173793.

Cabib, S. (2006). The neurobiology of stereotypy II: the role of stress. In: Stereotypic animal behaviour: fundamentals and applications to welfare (Mason, G. & Rushen, J., eds.). 2. ed., Wallingford, Oxfordshire; Cambridge, MA: CABI Publishing, 227–255.

Calcutt, S. et al. (2009). Acquisition of nut-cracking behaviors in a group of zoo-living chimpanzees (*Pan troglodytes*). American Journal of Primatology 71 (Supplement 1), 49.

Caldecott, J. & Miles, L. (2005). World atlas of great apes and their conservation. Prepared for the UNEP World Conservation Monitoring Centre. Berkeley: University of California Press.

Caldwell, M.C. & Caldwell, D.K. (1977). Epimeletic (care-giving) behavior in cetacea. In: Whales, dolphins, and porpoises (Norris, K.S., ed.). Berkeley; Los Angeles: University of California Press, 755–788.

Call, J. (2010). Do apes know that they could be wrong? Animal Cognition 3 (5), 689–700.

Call, J. et al. (2004). ›Unwilling‹ versus ›unable‹: chimpanzees' understanding of human intentional action. Developmental Science 7 (4), 488–498.

Call, J. & Tomasello, M. (1999). A nonverbal false belief task: the performance of children and great apes. Child Development 70 (2), 381–395.

Call, J. & Tomasello, M. (2008). Does the Chimpanzee have a theory of mind? 30 years later. Trends in Cognitive Sciences 12 (5), 187–192.

Carey, S. & Bartlett, E. (1978). Acquiring a single new word. Proceedings of the Stanford Child Language Conference 15, 17–29.

Carlstead, K. et al. (1993). Behavioural and adrenocortical responses to environmental changes in leopard cats (*Felis bengalensis*). Zoo Biology 12 (4), 321–331.

Cartmill, E.A. & Byrne, R.W. (2010). Semantics of primate gestures: intentional meanings of orangutan gestures. Animal Cognition 13 (6), 793–804.

Carvalho, S. et al. (2009). Tool-composite reuse in wild chimpanzees (*Pan troglodytes*): archaeologically invisible steps in the technological evolution of early hominins? Animal Cognition 12 (Supplement 1), 103–114.

Cavalieri, P., Singer, P., (1993): The Great Ape Project: Equality beyond humanity. New York: St. Martin's Griffin.

Celli, M.L. et al. (2003). Tool use task as environmental enrichment for captive chimpanzees. Applied Animal Behaviour Science 81 (2), 171–182.

Chaber et al. (2019). The scale of illegal meat importation from Africa to Europe via Paris. Conservation Letters 3 (5), 317–321.

Chamove, A.S. & Anderson, J.R. (1989). Examining environmental enrichment. In: Psychological well-being of primates (Segal, E., ed.). Philadelphia: Noyes Publications, 183–202.

Chojnacki, S. (2005). New List of Wars 1946–2003. Version 1.1., Juli 2005. http://www.polwiss.fu-berlin.de/frieden/pdf/warlist-1.1-7-2005.pdf, 2007/09/15, no longer online available.

Chomsky, N. (1956). Three models for the description of language. IRE Transactions on Information Theory 2 (3), 113–124.

Chomsky, N. (1959). On certain formal properties of grammars. Information and Control 2 (2), 137–167.

Chomsky, N. (2002). On nature and language. Ed. by A. Belletti u. L. Rizzi. Cambridge, UK: Cambridge University Press.

Christiansen, F. et al. (2010). Effects of tourist boats on the behaviour of Indo-Pacific bottlenose dolphins off the south coast of Zanzibar. Endangered Species Research 11 (1), 91–99.

CITES (a). The CITES Appendices. https://www.cites.org/eng/app/index.php, 2019/05/19.

CITES (b). Convention on International Trade in Endangered Species of Wild Fauna and Flora, Appendices I, II and III (valid from 4 October 2017). https://www.cites.org/eng/app/appendices.php, 2019/05/19.

CITES (c). Text of the Convention on International Trade in Endangered Species of Wild Fauna and Flora (entered in force 1975, signed at Washington, D.C., on 3 March 1973, amended at Bonn, on 22 June 1979). https://www.cites.org/sites/default/files/eng/disc/CITES-Convention-EN.pdf, 2019/05/19.

CITES (d). What is CITES? https://www.cites.org/eng/disc/what.php, 2019/05/19.

Clark, C. (2007). Dolphin therapy is all wet, Emory scientists conclude. Emory Report 60 (14). http://www.emory.edu/EMORY_REPORT/erarchive/2007/December/Dec.10/DolphinS&R.htm, 2019/05/19.

Clarke, E. et al. (2006). The syntax and meaning of wild gibbon songs. PLoS ONE 1 (1), e73. DOI:10.1371/journal.pone.0000073.

Clatterbuck, H. (2015). Chimpanzee Mindreading and the Value of Parsimonious Mental Models. Mind & Language, 30(4), 414–436.

Clayton, N.S. & Dickinson, A. (1998). Episodic-like memory during cache recovery by scrub jays. Nature 395 (6699), 272–274.

Clayton, N.S. & Jolliffe, A. (1996). Marsh tits *Parus palustris* use tools to store food. Ibis 138 (3), 554.

Clubb, R. & Mason, G. (1998). Foraging niche and stereotypic behaviour. In: Proceedings of the 32nd Congress of the ISAE (Veissier, I. & Boissy, A., eds.). Clermont-Ferrand: ISAE, 174.

Clubb, R. & Mason, G. (2003). Captivity effects on wide-ranging carnivores. Nature 425 (6957), 473–474.

Clubb, R. et al. (2006). Box 1.1. Motivation and motivational explanations for stereotypies. In: Stereotypic animal behaviour: fundamentals and applications to welfare (Mason, G. & Rushen, J., eds.). 2. ed., Wallingford, Oxfordshire; Cambridge, MA: CABI Publishing, 14–15.

CMS/UNEP (a). Appendices I and II of the Convention on the Conservation of Migratory Species of Wild Animals (CMS) (of the Convention on the Conservation of Migratory Species of Wild Animals (CMS), as amended by the Conference of the Parties in 1985, 1988, 1991, 1994, 1997, 1999, 2002, 2005, 2008, 2011, 2014 and 2017. Effective: 26 January 2018.). https://www.cms.int/en/page/appendix-i-ii-cms, 2018/06/12.

CMS/UNEP (b). Text of the agreement on the conservation of gorillas and their habitats (Gorilla Agreement). Signed in Paris, on 26 October 2007. https://www.cms.int/gorilla/en/documents/agreement, 2019/05/19.

CNN (2008). Dolphin slaughter brings charges from both sides. Published on 11 February 2008.

Coghlan, B. et al. (2007). Mortality in the Democratic Republic of Congo: an ongoing crisis https://www.rescue.org/sites/default/files/document/661/2006-7congomortalitysurvey.pdf, 2019/05/19.

Conard, N.J. (2007). Cultural evolution in Africa and Eurasia during the middle and late pleistocene. In: Handbook of paleoanthropology (Henke, W. & Tattersall, I., eds.). New York: Springer, 2001–2037.

Connor, R.C. & Norris, K.S. (1982). Are dolphins reciprocal altruists? American Naturalist 119 (3), 358–374.

Connor, R.C. & Peterson, D.M. (1994). The lives of whales and dolphins. New York: Henry Holt and Company.

Connor, R.C. et al. (2000). The bottlenose dolphin: social relationships in a fission-fusion society. In: Cetacean societies: field studies of dolphins and whales (Mann, J. et al., eds.) Chicago: Chicago University Press, 91–126.

Cooper, J.J. & Nicol, C.J. (1991). Stereotypic behaviour affects environmental preference in bank voles (*Cletherionomys glareolus*). Animal Behaviour 41 (6), 971–977.

Council Directive 2010/63/EU of the European Parliament and of the Council of 22 September 2010 on the protection of animals used for scientific purposes. [2010] OJ L 276/33, https://eur-lex.europa.eu/LexUriServ/LexUriServ.do?uri=OJ:L:2010:276:0033:0079:en:PDF, 2019/05/19.

Cousteau, J. & Schiefelbein, S. (2008). Der Mensch, die Orchidee und der Oktopus: Mein Leben für die Erforschung und Bewahrung unserer Umwelt. Frankfurt a. M.: Campus.

Crilly, R. (2008). Gorillas in the midst of the Congolese conflict in Virunga National Park. The Irish Times 02. Dezember 2008. https://www.irishtimes.com/news/gorillas-in-the-midst-of-the-congolese-conflict-in-virunga-national-park-1.917734, 2019/05/19.

Darwin, C. ([1879] 2004). The descent of man, and selection in relation to sex. Equals the 2. ed. from 1879, ed. by J. Murray in London; with an introduction by J. Moore und A. Desmond. London: Penguin Books.

Davidson, D. (1982). Rational animals. Dialectica 36 (4), 317–327.

Davidson, D. (1999). The emergence of thought. Erkenntnis 51: 7–17.

de Saussure, F. ([1916] 1967). Grundfragen der allgemeinen Sprachwissenschaft. Transl. Of the French first edition from 1916. Berlin: De Gruyter.

de Waal, F. (2006). Primates and philosophers: how morality evolved. Princeton: Princeton University Press.

de Waal, F.B.M. & Ferrari, P.F. (2010). Towards a bottom-up perspective on animal and human cognition. Trends in Cognitive Sciences 14 (5), 201–207.

DeGrazia, D. (1996a). Taking animals seriously: mental life and moral status, Chap. Back to animal ethics. Cambridge: Cambridge University Press, 258–297.

DeGrazia, D. (1996b). Taking animals seriously: mental life and moral status. Cambridge: Cambridge University Press.

DeGrazia, D. (2005): Regarding the Last Frontier of Bigotry. In: Logos 4(2). http://www.logosjournal.com/issue_4.2/degrazia.htm, 2019/05/19.

Dennett, D. (1981a). Bedingungen der Personalität. In: Analytische Philosophie des Geistes (Bieri, P., ed.). Königstein i. T.: Hain, 303–324.

Dennett, D. (1981b). Intentionale Systeme. In: Analytische Philosophie des Geistes (Bieri, P., ed.). Königstein i. T.: Hain, 162–183.

Dennett, D.C. (1988). Brainstorms: philosophical essays on mind and psychology, Chap. Conditions of personhood. 4. ed., 1. ed. From 1978. Cambridge, MA: MIT Press, 267–285.

Descartes, R. ([1637] 2001). Discours de la méthode / Bericht über die Methode (French/German). Ed. and transl. by H. Ostwald, after the French 1. ed. from 1637. Stuttgart: Reclam.

Desmond, J. (1999). Staging Tourism: Bodies on Display from Waikiki to Sea World. Chicago: University of Chicago Press.

Diamond, J. (1982). Evolution of bowerbirds' bowers: animal origins of the aesthetic sense. Nature 297, 99–102.

Diamond, J. (1986). Animal art: variation in bower decoration style among male bowerbirds *Amblyornis inornatus*. Proceedings of the National Academy of Sciences of the United States of America 83 (9), 3042–3046.

Diamond, J. (1987). Bower building and decoration by the bowerbird *Amblyornis inornatus*. Ethology 74 (3), 177–204.

Diamond, J. (1988). Experimental study of bower decoration by the bowerbird *Amblyornis inornatus*, using colored poker chips. American Naturalist 131 (5), 631–653.

Diamond, J. (2006). Der dritte Schimpanse: Evolution und Zukunft des Menschen. Frankfurt a. M.: Fischer.

Dittrich, L. (ed.) (2007). Zootierhaltung: Tiere in menschlicher Obhut. Grundlagen. 9. corr. ed. of the 1. ed. of 1975 with the title *Wildtiere in Menschenhand*. Frankfurt a. M.: Verlag Harri Deutsch.

Dolphin Research Center (2019). https://dolphins.org/home, 2019/05/19.

Dr. Hadwen Trust for Humane Research (2008). New research reveals 115 million animals used in experiments worldwide. Press release from 13 August 2008, no longer online available.

Elango, N. et al. (2006). Variable molecular clocks in hominoids. Proceedings of the National Academy of Sciences of the United States of America 103 (5), 1370–1375.

Elowson, A.M. et al. (1998). Babbling' and social context in infant monkeys: parallels to human infants. Trends in Cognitive Sciences 2 (1), 31–37.

Elsa Nature Conservancy of Japan et al. (2015): JAZA stops the live capure from the dolphin drives, http://eng.animals-peace.net/entertainment/dolphins/jaza-stops-the-live-capure-from-the-dolphin-drives, 2019/09/13.

Engels, E.-M. (2001). Orientierung an der Natur? Zur Ethik der Mensch-Tier-Beziehung. In: Den Tieren gerecht werden: Zur Ethik und Kultur der Mensch-Tier-Beziehung (Schneider, M., ed.). Witzenhausen: Universität Gesamthochschule Kassel, Fachgebiet Angewandte Nutztierethologie und Artgemäße Tierhaltung, 68–87.

Engels, E.-M. (2007). Charles Darwin. München: C.H. Beck.

Engels, E.-M. (2008). Was und wo ist ein ›naturalistischer Fehlschluss‹? Zur Definition und Identifikation eines Schreckgespenstes der Ethik. In: Wie funktioniert Bioethik? (Brand, C. et al., eds.). Paderborn: mentis, 125–141.

Epstein, R. et al. (1984). ›Insight‹ in the pigeon: antecedents and determinants of an intelligent performance. Nature 308 (5954), 61–62.

Estes, J. et al. (2003). Individual variation in prey selection by sea otters: patterns, causes and implications. Journal of Animal Ecology 72 (1), 144–155.

Fahey, D. (2009). Explaining Uganda's involvement in the DR Congo, 1996–2008. Paper prepared for the 50th International Studies Association Conference in New York City, February 2009. http://citation.allacademic.com/meta/p312862_index.html, 2019/05/19.

FAO (2005). Global Forest Resources Assessment 2005—15 Key Findings. http://foris.fao.org/static/data/fra2005/kf/common/GlobalForestA4-ENsmall.pdf, 2019/05/19.

Farm Animal Welfare Council, UK (1992). FAWC updates the five freedoms. The Veterinary Record 131, 357.

Ferrari, A. (ed.) (2008). Genmaus & Co: Gentechnisch veränderte Tiere in der Biomedizin. Erlangen: Harald Fischer.

Ferrari, A. (2009). Gentechnisch veränderte Tiere: ein Sonderfall? In: Der ethisch vertretbare Tierversuch: Kriterien und Grenzen (Borchers, D. & Luy, J., eds.). Paderborn: mentis, 265–295.

Fitch, W.T. & Hauser, M.D. (2004). Computational constraints on syntactic processing in a nonhuman primate. Science 303 (5656), 377–380.

Flavell, J.H. et al. (1983). Development of the appearance-reality distinction. Cognitive Psychology 15 (1), 95–120.

Fletcher, J. (1979). Humanhood: essays in biomedical ethics. Buffalo: Prometheus Books.

Forster, M.N. (2007). Menschen und andere Tiere: Über das Verhältnis von Mensch und Tier bei Tomasello. Deutsche Zeitschrift für Philosophie 55 (5), 761–767.

Fragaszy, D.M. & Perry, S. (eds.) (2003). The biology of traditions: models and evidence. New York: Cambridge University Press.

Frankena, W.K. (1979). Ethics and the environment. In: Ethics and problems of the 21st century (Goodpaster, K.E. & Sayre, K.M., eds.). London: University of Notre Dame Press, 3–20.

Freud, S. (1917). Eine Schwierigkeit der Psychoanalyse. Imago. Zeitschrift für Anwendung der Psychoanalyse auf die Geisteswissenschaften V (1), 1–7. Ed. 1919 by S. Freud, redacted by O. Rank and H. Sachs. Leipzig; Wien: Internationaler Psychoanalytischer Verlag.

Frith, C. & Frith, U. (2005). Theory of mind. Current Biology 15 (17), R644–R646.

Frohoff, T.G. (2000). Behavioral indicators of stress in odontocetes during interactions with humans: a preliminary review and discussion. IWC Scientific Committee, SC/52/WW2.

Frohoff, T.G. & Packard, J.M. (1995). Interactions between humans and free-ranging and captive bottlenose dolphins. Anthrozoös 8 (1), 44–54.

Fruth, B. et al. 2016. Pan paniscus (errata version published in 2016). The IUCN Red List of Threatened Species 2016: e.T15932A102331567. http://dx.doi.org/10.2305/IUCN.UK.2016-2.RLTS.T15932A17964305.en, 2019/04/13.

Galef, B.G. & Laland, K.N. (2005). Social learning in animals: empirical studies and theoretical models. Bioscience 55 (6), 489–499.

Ganslandt, H.R. (1995). Menschenrechte. In: Enzyklopädie Philosophie und Wissenschaftstheorie (Vol. 2) (Mittelstraß, J., ed.). Stuttgart; Weimar: Metzler, 846–847.

Gardner, H. (1999). Intelligence reframed: multiple intelligences for the 21st century. New York: Basic Books.

Garner, J.P. (2005). Stereotypies and other abnormal repetitive behaviors: potential impact on validity, reliability, and replicability of scientific outcomes. ILAR Jounal 46 (2), 106–117.

Garner, J.P. (2006). Perseveration and stereotypy—system-level insights from clinical psychology. In: Stereotypic animal behaviour: fundamentals and applications to welfare (Mason, G. & Rushen, J., eds.). 2. ed., Wallingford, Oxfordshire; Cambridge, MA: CABI Publishing, 121–152.

Gehlen, A. ([1940] 1993). Der Mensch: Seine Natur und Stellung in der Welt. In: Arnold Gehlen: Gesamtausgabe (Vol. 3), critical edition based on the entire text body of the 1. ed. from 1940, Issue 1 (Rehberg, K.-S., ed.). Frankfurt a. M.: Vittorio Klostermann.

Gehlen, A. ([1942] 1998). Ein Bild vom Menschen. In: Anthropologie (Gebauer, G., ed.). Leipzig: Reclam.

Gigon, O. (2002). Die philosophische Ethik der Griechen (Einführung III). In: Aristoteles: Die Nikomachische Ethik (Gigon, O., ed.). 5. ed. of the entire ed. from 1991, transl. and commented by O. Gigon, München: dtv, 53–75.

Gilligan, C. (1999). Die andere Stimme: Lebenskonflikte und Moral der Frau. 5. ed. of the 1. ed. from 1988. München: Piper Verlag.

Global Whitness (2003). Briefing document: recommendations and situational update presented to the UN Security Council. Pressemeldung vom 20.06.2003. https://www.globalwitness.org/en/archive/briefing-document-recommendations-and-situational-update-presented-un-security-council/, 2019/05/19.

Glock, H.J. (2000). Animals, thoughts, and concepts. Synthese 123: 35–64.

Goodall, J. (1964). Tool-using and aimed throwing in a community of free-living chimpanzees. Nature 201, 1264–266.

Gopnik, A. et al. (1994). Changing your views: how understanding visual perception can lead to a new theory of the mind. In: Children's early understanding of mind: origins and development (Lewis, C. & Mitchell, P., eds). Hillsdale, NJ: Erlbaum, 157–81.

Gorilla Diary (August 2007—März 2009). Gorilla Diary August 2007—March 2009: a weekly diary by Virunga rangers Diddy and Innocent. Published by BBC News, page last updated 24 March 2009. http://news.bbc.co.uk/2/hi/science/nature/7549444.stm, 2019/05/19.

Gorke, M. (1999). Artensterben: Von der ökologischen Theorie zum Eigenwert der Natur. Stuttgart: Klett-Cotta.

Gorke, M. (2000). Was spricht für eine holistische Umweltethik? Natur und Kultur 1 (2), 86–105.

Gorke, M. (2010). Eigenwert der Natur: Ethische Begründung und Konsequenzen. Stuttgart: Hirzel.

Gould, J.L. & Grant Gould, C. (1994). The Animal Mind. New York: Scientific American Library.

GRD (2019). Delphinarien. "Leben" in Gefangenschaft. https://www.delphinschutz.org/delfine/delfinarien/, 2019/05/31.

Greenpeace (2005). Die Danzer-Gruppe und der Urwald. Published on the Greenpeace Germany Homepage on 14 February 2005. https://www.greenpeace.de/themen/waelder/die-danzer-gruppe-und-der-urwald, 2019/05/19.

Gregg, J.D. et al. (2007). Do dolphins eavesdrop on the echolocation signals of conspecifics? International Journal of Comparative Psychology 20 (1), 65–88.

Grewendorf, G. et al. (1999). Sprachliches Wissen: Eine Einführung in moderne Theorien der grammatischen Beschreibung. 11. ed. of the 1. ed. from 1987. Frankfurt a. M.: Suhrkamp.

Griesser, M. (2008). Referential calls signal predator behavior in a group-living bird species. Current Biology 18, 69–73.

Griffin, A.S. (2004). Social learning about predators: a review and prospectus. Learning Behavior 32 (1), 131–40.

Griffin, D.R. (1985). Wie Tiere denken: Ein Vorstoß ins Bewußtsein der Tiere. München; Wien; Zürich: BLV Verlagsgesellschaft.

Griffin, D.R. (1991). Progress toward a cognitive ethology. In: Cognitive Ethology (Ristau, C.A., ed.). Hillsdale: Lawrence Erlbaum, 3–17.

Griffin, D.R. (2001a). Animal minds: beyond cognition to consciousness. Revised and extended edition. Chicago; London: The University of Chicago Press.

Griffin, D.R. (2001b). Animal minds: Beyond cognition to consciousness, Chap. Tools and special devices. Revised and extended edition. Chicago; London: The University of Chicago Press, 113–26.

Grignon, F. (2009). Rape as a weapon of war in Congo. Spiegel Online 11. Juni 2009. https://www.spiegel.de/international/world/opinion-rape-as-a-weapon-of-war-in-congo-a-629885.html, 2019/05/19.

Grimm, D. (2015): Has U.S. biomedical research on chimpanzees come to an end? Published on Science online 18 August 2015, doi:10.1126/science.aad1635, https://www.sciencemag.org/news/2015/08/has-us-biomedical-research-chimpanzees-come-end, 2019/04/14.

Güntürkün, O. & Bugnyar, T. (2016): Cognition without Cortex. Trends in Cognitive Sciences 20 (4), 291–303.

Habermas, J. (1997). Die Herausforderung der ökologischen Ethik für eine anthropozentrisch ansetzende Konzeption. In: Naturethik: Grundtexte der gegenwärtigen tier- und ökoethischen Diskussion (Krebs, A., ed.). Frankfurt a. M.: Suhrkamp, 92–99.

Haidle, M.N. (2012). How to think tools? A comparison of cognitive aspects in tool behavior of animals and during human evolution. Habilitation thesis published online by TOBIAS-lib. Tübingen: University of Tübingen. http://hdl.handle.net/10900/49627, 2019/05/19.

Halina, M. (2015). There Is No Special Problem of Mindreading in Nonhuman Animals. Philosophy of Science, 82(3), 473–90.

Hall, K.R.L. & Schaller, G.B. (1964). Tool-using behavior of the California sea otter. Journal of Mammalogy 45 (2), 287–98.

Hampton, R.R. (2001). Rhesus monkeys know when they remember. Proceedings of the National Academy of Sciences of the United States of America 98 (9), 5359–362.

Hampton, R.R. et al. (2004). Rhesus monkeys (*Macaca mulatta*) discriminate between knowing and not knowing and collect information as needed before acting. Animal Cognition 7 (4), 239–46.

Hardus, M. et al. (2009). A description of the orangutan's vocal and sound repertoire, with a focus on geographic variation. In: Orangutans: geographic variation in behavioral ecology and conservation (Wich, S.A. et al., eds.). Oxford University Press, 49–64.

Hare, B. et al. (2000). Chimpanzees know what conspecifics do and do not see. Animal Behaviour 59 (4), 771–85.

Hare, B. et al. (2001). Do chimpanzees know what conspecifics know? Animal Behaviour 61 (1), 139–51.

Hare, B. et al. (2006). Chimpanzees deceive a human competitor by hiding. Cognition 101 (3), 495–514.

Hare, B. & Tomasello, M. (2004). Chimpanzees are more skillful in competitive than cooperative cognitive tasks. Animal Behaviour 68 (3), 571–81.

Harnell, B. (2008). Secret film will show slaughter to the world: covert operation finally exposes Taiji's annual dolphin horror. The Japan Times 30. März 2008. No longer online available on Japan Times but copy available here: http://relespider.blog.sohu.com/83388086.html, 2019/05/20.

Hart, T. (2008): Bushmeat 3: The History of Hunting in TL2. http://www.bonoboincongo.com/2008/02/26/bushmeat-3-the-history-of-hunting-in-tl2/, 2018/07/22.

Hauser, M. et al. (2002a). The faculty of language: what is it, who has it, and how did it evolve? Science 298 (5598), 1569–579.

Hauser, M. et al. (2002b). Rule learning by cotton-top tamarins. Cognition 86, B15–B22.

Hauser, M.D. (2001). Wild minds: what animals really think, Chap. Gossip on the ark. London: Penguin Books, 215–59.

Hauser, M.D. et al. (2001). Segmentation of the speech stream in a nonhuman primate: statistical learning in cotton-top tamarins. Cognition 78, B53–B64.

Hediger, H. (1942). Wildtiere in Gefangenschaft: Ein Grundriss der Tiergartenbiologie. Basel: Verlag Benno Schwabe.

Hediger, H. (1965). Mensch und Tier im Zoo: Tiergartenbiologie. Rüschlikon-Zürich; Stuttgart; Wien: Albert Müller Verlag.

Hegel, G.W.F. ([1837] 1970). Vorlesungen über die Philosophie der Geschichte (Works Vol. XII). Frankfurt a. M.: Suhrkamp.

Heidegger, M. ([1927] 1967). § 34: Da-sein und Rede. Die Sprache. In: Sein und Zeit. 11., unchanged ed. of the 1. ed. from 1927, Tübingen: Max Niemeyer Verlag, 160–66.

Heidegger, M. ([1946] 1976). Brief über den »Humanismus«. Gesamtausgabe I. Abteilung: Veröffentlichte Schriften 1910–1976 (Vol. 9 Wegmarken 1919–1961). Frankfurt a. M.: Verlag Vittorio Klostermann, 313–64.

Heinrich, B. (1995). An experimental investigation of insight in the common raven *Corvus corax*. The Auk 112 (4), 994–1003.

Heinrich, B. (1999). The mind of the raven: investigations and adventures with wolf-birds. New York: Cliff Street Books.

Helweg, D.A. et al. (1992). Humpback whale song: Our current understanding. In: Sensory Processes of Marine Mammals (Thomas, J.R. et al., eds.). New York: Plenum, 459–83.

Herder, J.G. ([1772] 2002). Abhandlung über den Ursprung der Sprache. Ed. by H.D. Irmscher. Bibliographically extended edition. Stuttgart: Reclam.

Herman, L.M. (2002). Exploring the cognitive world of the bottlenosed dolphin. In: The cognitive animal: empirical and theoretical perspectives on animal cognition (Bekoff, M. et al., eds.). Cambridge, MA; London: MIT Press, 275–83.

Herman, L.M. (2009). Can dolphins understand language? In: LACUS Forum XXXIV: Speech and beyond (Sutcliffe, P. et al., eds.). Houston, TX: LACUS, 3–20.

Herman, L.M. et al. (1993). Responses to anomalous gestural sequences by a language-trained dolphin: Evidence for processing of semantic relations and syntactic information. Journal of Experimental Psychology: General 122 (2), 184–94.

Herman, L.M. et al. (1984). Comprehension of sentences by bottlenosed dolphins. Cognition 16, 129–219.

Herman, L.M. & Uyeyama, R.K. (1999). The dolphin's grammatical competency: comments on Kako (1998). Animal Learning & Behavior 27 (1), 18–23.

Hernandez-Aguilar, R.A. et al. (2007). Savanna chimpanzees use tools to harvest the underground storage organs of plants. Proceedings of the National Academy of Sciences of the United States of America 104 (49), 19210–9213.

Herrnstein, R.J. (1979). Acquisition, generalization, and discrimination reversal of a natural concept. Journal of Experimental Psychology: Animal Behavior Processes 5, 116–29.

Herrnstein, R.J. & De Villiers, P.A. (1980). Fish as a natural category for people and pigeons. In: The psychology of learning and motivation (Vol. 14) (Bower, G.H., ed.). New York: Academic Press, 59–95.

Heyes, C. (1994). Social learning in animals: categories and mechanisms. Biological Review 69 (2), 207–31.

Heyes, C. & Galef, B.J. (1996). Social learning and the roots of culture. San Diego: Academic Press.

Hickey, J.R., et al. (2018). Gorilla beringei ssp. beringei. The IUCN Red List of Threatened Species 2018: e.T39999A17989719. http://dx.doi.org/10.2305/IUCN.UK.2018-2.RLTS.T39999A17989719.en, 2019/04/13.

Hickman, C.P. et al. (2008). Zoologie. 13., revised edition. München: Pearson.

Hilpert, K. (1998). Menschenrechte. In: Lexikon der Bioethik (Vol. 2) (Korff, W. et al., eds.). Gütersloh: Gütersloher Verlagshaus, 670–79.

Hirata, S. (2006). Chimpanzee learning and transmission of tool use to fish for honey. In: Cognitive development in chimpanzees (Matsuzawa, T. et al., eds.). Tokyo: Springer, 201–13.

Hirata, S. et al. (2009). How to crack nuts: acquisition process in captive chimpanzees (*Pan troglodytes*) observing a model. Animal Cognition 12 (Supplement 1), 87–101.

Hobbes, T. ([1655] 1967). Vom Körper. Selected and tranl. by M. Frischeisen-Köhler, unchanged print of the version from 1949, 2. ed. with references and register. Hamburg: Meiner.

Höffe, O. (1997). Lexikon der Ethik (Entry on *Person*). 5., newly ed. and extend. edition. München: C.H. Beck, 226–28.

Hoffmann, L. (2007). Der Mensch und seine Sprache—eine anthropologische Skizze. In: Diskurse und Texte: Festschrift für Konrad Ehlich zum 65. Geburtstag (Redder, A., ed.). Tübingen: Stauffenburg, 21–37.

Holder, H.D. et al. (1993). A bottlenosed dolphin's responses to anomalous gestural sequences expressed within an artificial gestural language. In: Language and communication: comparative perspectives (Roitblat, H.R. et al., eds.). Hillsdale, N.Y.: Erlbaum, 199–308.

Hopkins, W.D. et al. (2007). Chimpanzees differentially produce novel vocalizations to capture the attention of a human. Animal Behaviour 73 (2), 281–86.

Horner, V. & Whiten, A. (2005). Causal knowledge and imitation/emulation switching in chimpanzees (*Pan troglodytes*) and children. Animal Cognition 8 (3), 164–81.

Houk, J.L. & Geibel, J.J. (1974). Observation of underwater tool use by the sea otter *Enhydra lutris Linnaeus*. California Fish and Game 60 (4), 207–08.

Hülsken, C. (2001). Training in der Theory of Mind-Forschung: Die Rolle von Kohärenz und Feedback in der Entwicklung einer naiven Alltagspsychologie. Aachen: Shaker.

Huffman, M.A. (2003). Animal self-medication and ethno-medicine: exploration and exploitation of the medicinal properties of plants. Proceedings of the Nutrition Society 62 (2), 371–81.

Huffman, M.A. et al. (1997). Seasonal trends in intestinal nematode infection and medicinal plant use among chimpanzees in the Mahale Mountains, Tanzania. Primates 38 (2), 111–25.

Huffman, M.A. & Wrangham, R.W. (1993). Diversity of medical plant use by chimpanzees in the wild. In: Chimpanzee cultures (Wrangham, R.W. et al., eds.). Cambridge, MA: Harvard University Press, 129–48.

Hughes, B.O. & Duncan, I.J.H. (1988). The notion of ethological need, models of motivation and animal-welfare. Animal Behaviour 36 (6), 1696–707.

Hume (1751, 1777). An Enquiry concerning the Principles of Morals. Hume Texts Online, ed. by Peter Millican and Amyas Merivale. https://davidhume.org/texts/m/, 2019/05/16.

Humle, T. (2016a). Pan troglodytes (errata version published in 2018). The IUCN Red List of Threatened Species 2016: e.T15933A129038584. http://dx.doi.org/10.2305/IUCN.UK.2016-2.RLTS.T15933A17964454.en, 2019/04/13.

Humle, T., et al. (2016b). Pan troglodytes ssp. verus (errata version published in 2016). The IUCN Red List of Threatened Species 2016: e.T15935A102327574. http://dx.doi.org/10.2305/IUCN.UK.2016-2.RLTS.T15935A17989872.en, 2019/05/20.

Humphries, T.L. (2003). Effectiveness of dolphin-assisted therapy as a behavioral intervention for young children with disabilities. Bridges: Practice-Based Research Synthesis 1 (6), 1–9.

Hunt, G.R. & Gray, R.D. (2003). Diversification and cumulative evolution in New Caledonian crow tool manufacture. Proceedings of the Royal Society of London B: Biological Science 270 (1517), 867–74.

Hunt, T.D. et al. (2008). Health risks for marine mammal workers. Diseases of Aquatic Organisms 81 (1), 81–92.

Hutchins, M. et al. (eds.) (2003). Grzimek's animal life encyclopedia (Vol. 12, Mammals I). 2. edition. Farmington Hills, MI: Gale Group.

ICG (2011). Conflict in Congo. Last updated 27.01.2011, no longer online available. For the ICG's crisis watch on DRC see: https://www.crisisgroup.org/africa/central-africa/democratic-republic-congo, 2019/05/20.

Ingensiep, H.W. (1999). Mensch und Menschenaffe: Die besondere Beziehung. In: Tiere und Menschen: Geschichte und Aktualiät eines prekären Verhältnisses (Münch, P., ed.). Paderborn; Munchen; Wien; Zürich: Schöningh, 429–46.

Inglis, I.R. & Fergusson, N.J.K. (1986). Starlings search for food rather than eat freely-available, identical food. Animal Behaviour 34 (2), 614–17.

Inglis, I.R. et al. (2001). An information primacy model of exploratory and foraging behaviour. Animal Behaviour 62 (3), 543–57.

Irrgang, B. (1998). Pathozentrik. In: Lexikon der Bioethik (Vol. 2) (Korff, W. et al., eds.). Gütersloh: Gütersloher Verlagshaus, 834–35.

IUCN (2019). The IUCN Red List of Threatened Species. Version 2019–1. http://www.iucnredlist.org, 2019/04/13.

IUCN (2001). IUCN Red List Categories and Criteria: Version 3.1. Published by the IUCN Species Survival Commission. Gland, Switzerland; Cambridge, UK. https://www.iucnredlist.org/resources/categories-and-criteria.

Jantschke, F. (1997). Zoo- und Zirkustiere. In: Das Buch vom Tierschutz (Sambraus, H.H. & Steiger, A., eds.). Stuttgart: Ferdinand Enke, 402–23.

Jarvis, E.D. et al. (2005). Avian brains and a new understanding of vertebrate brain evolution. Nature Reviews Neuroscience 6 (2), 151–59.

Jenkins, M. (2008). Wer ermordete die Virunga-Gorillas? National Geographic Deutschland 8 (1 August 2008), 26–55.

Jerison, H.J. (1986). The perceptual world of dolphins. In: Dolphin cognition and behavior: A comparative view (Thomas, J.A. & Wood, F.G., eds.). Hillsdale, NJ: Lawrence Erlbaum Associates.

Jha, A. (2006). Questions raised over ban on research using great apes. The Guardian (3 June 2006). https://www.theguardian.com/science/2006/jun/03/animalrights.research, 2019/05/20.

Johnson, M.H. & Morton, J. (1991). Biology and cognitive development: The case of face recognition. Oxford: Blackwell.

Johnson, S. (2007). Congo's Gorilla Tragedy. Newsweek (8 May 2007). https://www.newsweek.com/congos-gorilla-tragedy-99553, 2019/05/20.

Kaldor, M. (1999). New & Old Wars: organized violence in a global era. Stanford: Stanford University Press.

Kaldor, M. (2007). New & Old Wars: organized violence in a global era. 2. ed. Stanford: Stanford University Press.

Kamawar, D. & Olson, D.R. (1998). Theory of mind in young human primates: does Heyes's task measure it? Behavioral & Brain Sciences 21 (1), 122–23.

Kaminski, J. et al. (2004). Word learning in a domestic dog: evidence for »Fast-Mapping«. Science 304 (5677), 1682–683.

Kaminski, J. et al. (2008). Chimpanzees know, what others know, but not what they believe. Cognition 109 (2), 224–34.

Kant, I. ([1797] 2017). The Metaphysics of Morals. Revised edition, ed. by L. Denis, transl. by M. Gregor. New York: Cambridge UP.

Kant, I. ([1785] 2011): Groundwork of the Metaphysics of Morals: A German-English Edition, ed. and transl. by Mary Gregor and Jens Timmermann. Cambridge: Cambridge University Press.

Kasten, H. (2005). 4–6 Jahre: Entwicklungspsychologische Grundlagen. Weinheim; Basel: Beltz.

Kawai, M. (1963). On the newly-acquired behaviors of the natural troop of Japanese monkeys on Koshima Island. Primates 4 (1), 113–15.

Kenward, B. et al. (2006). Development of tool use in New Caledonian crows: Inherited action patterns and social influence. Animal Behaviour 72 (6), 1329–343.

Kenward, B. et al. (2005). Behavioural ecology: tool manufacture by naïve juvenile crows. Nature 433 (7022), 121.

Kestin, S. (2004). Captive mammals can net big profits for exhibitors (Part 3: Park business of marine attractions. Below the surface). Sun-Sentinel 17. Mai 2007. https://www.tribpub.com/gdpr/sun-sentinel.com/, 2019/05/21.

King, B.J. & Shanker, S.G. (2003). How can we know the dancer from the dance? The dynamic nature of African great ape social communication. Anthropological Theory 3 (1), 5–26.

Klann-Delius, G. (1999). Spracherwerb. Stuttgart; Weimar: Metzler.

Klenner, H. (1990). Menschenrechte. In: Europäische Enzyklopädie zu Philosophie und Wissenschaften (Vol. 3) (Sandkühler, H.J., ed.). Hamburg: Meiner, 366–72.

Kluge, H.-G. (ed.) (2002). Tierschutzgesetz: Kommentar. Stuttgart: Kohlhammer.

Knight, A. (2007). The poor contribution of chimpanzee experiments to biomedical progress. Journal of Applied Animal Welfare Science 10 (4), 281–308.

Knight, A. (2008a). 127 million non-human vertebrates used worldwide for scientific purposes in 2005. Alternatives to Laboratory Animals 36 (5), 494–96.

Knight, A. (2008b). The beginning of the end for chimpanzee experiments? Philosophy, Ethics & Humanities in Medicine 3 (16), 1–14.

Knörnschild, M. et al. (2006). Babbling behavior in the sac-winged bat (*Saccopteryx bilineata*). Naturwissenschaften 93 (9), 451–54.

Köhler, W. ([1921] 1963). Intelligenzprüfungen an Menschenaffen. Unchanged version of the 2. Ed. with the title *Intelligenzprüfungen an Anthropoiden I*, Abhandlungen der Königlich-Preußischen Akademie der Wissenschaften (1917), Physikal.-Mathemat. Klasse I. Berlin: Springer.

Koshimizu, K. et al. (2003). Use of *Vernonia amygdalina* by wild chimpanzee: possible roles of ist bitter and related constituents. Physiology & Behavior 56 (6), 1209–216.

Krebs, A. (1993). Haben wir moralische Pflichten gegenüber Tieren? Deutsche Zeitschrift für Philosophie 41 (6), 995–1008.

Krebs, A. (1999): Ethics of Nature. A Map. Berlin; New York: De Gruyter.

Kreger, M.D. et al. (1998). Context, ethics, and environmental enrichment in zoos and aquariums. In: Second nature: environmental enrichment for captive animals (Shepherdson, D.J. et al., eds.). Washington; London: Smithsonian Institution Press, 59–87.

Kroeber, A.L. & Kluckhohn, C. (1978). Culture: a critical review of concepts and definitions, Vol. XLVII No. 1 of the Papers of the Peabody Museum of American Archaeology and Technology, Harvard University. Published by the museum in 1952. Millwood, NY: Kraus Reprint CO.

Krutzen, M. et al. (2005). Cultural transmission of tool use in bottlenosed dolphins. Proceedings of the National Academy of Sciences of the United States of America 102 (25), 8939–943.

Krupenye, C. (2016). Great apes anticipate that other individuals will act according to false beliefs. Science 354 (6308), 110–14.

Kruuk, H. (2006). Otters: ecology, behaviour and conservation. Oxford; New York: Oxford University Press.

Kulke, L. et al. (2018). How robust are anticipatory looking measures of Theory of Mind? Replication attempts across the life span. Cognitive Development 46 (April–June 2018), 97–111.

Kuper, A. (1999). Culture: the anthropologist's account. Cambridge MA; London: Harvard University Press.

Laakso, L. & Hiekkanen, H. (2002). Africa's First World War in the Democratic Republic of the Congo. In: Agents, motives and instruments: an atlas of African conflicts (Kivimäki, T. & Laakso, L., eds.). Finland: CTS.

Lauer, D. (2010). Sinn und Präsenz: Über Transparenz und Opazität in der Sprache. In: Hide and Seek: Das Spiel von Transparenz und Opazität (Rautzenberg, M. & Wolfsteiner, A., eds.). München: Fink, 311–24.

Lazari-Radek, K., Singer, P. (2014): The Point of View of the Universe. Oxford: Oxford University Press.

Leaky, R.E. & Lewin, R. (1996). Wie der Mensch zum Menschen wurde: Neue Erkenntnisse über den Ursprung und die Zukunft des Menschen. Hamburg: Hoffmann und Campe.

Leavens, D.A. et al. (2010). Multimodal communication by captive chimpanzees (*Pan troglodytes*). Animal Cognition 13 (1), 33–40.

Lecky, W.E.H. ([1862] 1869). The history of European morals from Augustinus to Charlemagne (Vol. 1). 2. edition. London: Longmans, Green and Co.

Lee, Jane J. (2019). Military whales and dolphins: What do they do and who uses them? National Geographic 3 May 2019. https://news.nationalgeographic.com/news/2014/03/140328-navy-dolphin-sea-lion-combat-ocean-animal-science/, 2019/05/30.

Lewis, M.H. et al. (2006). The neurobiology of stereotypy I: environmental complexity. In: Stereotypic animal behaviour: fundamentals and applications to welfare (Mason, G. & Rushen, J., eds.). 2. ed., Wallingford, Oxfordshire; Cambridge, MA: CABI Publishing, 190–226.

Limbaugh, C. (1961). Observations on the California sea otter. Journal of Mammalogy 42 (2), 271–73.

Lindl, T. et al. (2005). Tierversuche in der biomedizinischen Forschung: Eine Bestandsaufnahme der klinischen Relevanz von genehmigten Tierversuchsvorhaben. Nach 10 Jahren keine Umsetzung in der Humanmedizin nachweisbar. Alternatives to Animal Experimentation 3, 143–51.

Locke, J. ([1690] 1975). An essay concerning human understanding. The Clarendon edition of the works of John Locke. Ed. with an introduction, a critical commentary and a clossary by P.H. Nidditch. Oxford: The Clarendon Press.

Lonsdorf, E.V. (2006). What is the role of mothers in the acquisition of termite-fishing behaviors in wild chimpanzees (*Pan troglodytes schweinfurthii*)? Animal Cognition 9 (1), 36–46.

Loreau, M. et al. (2006). Diversity without representation. Nature 442 (7100), 245–46.

Lorenz, K. (1935). Der Kumpan in der Umwelt des Vogels. Journal of Ornithology 83 (2), 137–213.

Louv, R. (2008). Last child in the woods: saving our children from nature-deficit disorder. Chapel Hill, NC; New York: Algonquin Books of Chapel Hill / Workman Publishing.

Lozano, G.A. (1998). Parasitic stress and self-medication in wild animals. Advances in the Study of Behavior 27, 291–317.

Lurz, R.W. (ed.) (2009). The philosophy of animal minds. New York: Cambridge University Press.

Lufthansa Cargo (2019). FAQ's About Live. https://lufthansa-cargo.com/documents/20184/35769/20190524_Live_FAQs_EN.pdf/9f816fe5-971b-1a58-c09c-71385a92444a?t=1562074967000, 2019/09/14.

MacIntyre, A. (1999). Dependant Rational Animals. Why Human Beings Need the Virtues. Chicago and La Salle, IL: Open Court.

MacWhinney, B. (2000). The CHILDES Project: Tools for analyzing talk. 3. edition. Mahwah, NJ: Lawrence Erlbaum Associates.

Maehle, A.-H. (1999). Vom rechtlosen Geschöpf zum Träger von Rechten? Die historischen Wurzeln unserer Tierethik im Denken der Aufklärung. In: Tiere ohne Rechte? (Joerden, J.C. & Busch, B., eds.). Berlin; Heidelberg; New York: Springer, 1–11.

Maisels, F., Bergl, R.A. & Williamson, E.A. (2018). Gorilla gorilla (amended version of 2016 assessment). The IUCN Red List of Threatened Species 2018: e.T9404A136250858. http://dx.doi.org/10.2305/IUCN.UK.2018-2.RLTS.T9404A136250858.en, 2019/04/13.

Majetschak, S. (1995). Sprache (III. Neuzeit). In: Historisches Wörterbuch der Philosophie (Vol. 9) (Ritter, J. & Gründer, K., eds.). Basel: Schwabe & Co AG, 1468–495.

Manser, M.B. (2001). The acoustic structure of suricates' alarm calls varies with predator type and the level of response urgency. Proceedings of the Royal Society of London B: Biological Sciences 268 (1483), 2315–324.

Manser, M.B. et al. (2001). The information that receivers extract from alarm calls in suricates. Proceedings of the Royal Society of London B: Biological Sciences 268 (1484), 2485–491.

Manser, M.B. et al. (2002). Suricate alarm calls signal predator class and urgency. Trends in Cognitive Sciences 6 (2), 55–57.

Maple, T.L. & Perkin, L.A. (1998). Enclosure furnishings and structural environmental enrichment. In: Wild mammals in captivity (Kleiman, D.G. et al., eds.). Chicago: University of Chicago Press, 212–22.

Marino, L. & Lilienfeldt, S. (1998). Dolphin-assisted therapy: flawed data, flawed conclusions. Anthrozoös 11 (4), 194–200.

Marino, L. & Lilienfeldt, S. (2006). Dolphin-assisted therapy: a discussion paper. https://de.scribd.com/document/49004225/Marino-Lilenfeld-Discussion-paper-on-DAT, 2019/05/21.

Marino, L. & Lilienfeldt, S.O. (2007). Dolphin-assisted therapy: more flawed data and more flawed conclusions. Anthrozoös 20 (3), 239–49.

Marks, J. (2000). Scientific and folk ideas about heredity. In: The Human Genome Project and minority communities: ethical, social and political dilemmas (Zilinskas,

R.A. & Balint, P.J., eds.). Westport, CT: Praeger Frederick, 53–66. http://www.pbs.org/race/000_About/002_04-background-01-12.htm, 2019/05/21.

Marler, P. & Slabbekoorn, H.W. (eds.) (2004). Nature's music: the science of birdsong. San Diego, CA; London, UK: Elsevier.

Mason, G. & Rushen, J. (eds.) (2006). Stereotypic animal behaviour: fundamentals and applications to welfare. 2. edition. Wallingford, Oxfordshire; Cambridge, MA: CABI Publishing.

Mason, G.J. (1991). Stereotypies: a critical review. Animal Behaviour 41 (6), 1015–037.

Mason, G.J. et al. (2001). Frustrations of fur-farmed mink. Nature 410 (6824), 35–36.

Mason, G.J. & Latham, N.R. (2004). Can't stop, won't stop: is stereotypy a reliable animal welfare indicator? Animal Welfare 13 (Supplement 1), 57–69.

Matsuzawa, T. (1996). Chimpanzee intelligence in nature and in captivity: isomorphism of symbol use and tool use. In: Great ape societies (McGrew, W.C. et al., eds.). Cambridge: Cambridge University Press, 196–209.

Matsuzawa, T. (1999). Communication and tool use in chimpanzees: cultural and social context. In: The design of animal communication (Hauser, M. & Konishi, M., eds.). Cambridge, MA: MIT Press, 645–71.

Matsuzawa, T. (2003). Koshima monkeys and Bossou chimpanzees: long-term research on culture in nonhuman primates. In: Animal social complexity: intelligence, culture, and individualized societies (de Waal, F.B.M. & Tyack, P.L., eds.). Cambridge, MA; London: Harvard University Press, 374–87.

Matsuzawa, T. et al. (2001). Emergence of culture in wild chimpanzees: education by masterapprenticeship. In: Primate origins of human cognition and behaviour (Matsuzawa, T., ed.). Tokyo: Springer, 557–74.

Mayr, P. (2003). Das pathozentrische Argument als Grundlage einer Tierethik. Münster: Monsenstein & Vannerdat.

Mazet, J.A.K. et al. (2004). Assessment of the risk of zoonotic disease transmission to marine mammal workers and the public: survey of occupational risks. Final report (Research Agreement Number K005486-01) prepared for the US Marine Mammal Commission by the Wildlife Health Center, School of Veterinary Medicine, University of California, Davis, CA. https://swfsc.noaa.gov/uploadedFiles/Divisions/PRD/Programs/Photogrammetry/Marine_Mammal_Zoonoses_Final_Report-2.pdf, 2019/04/15.

McFarland, D. (2006). Oxford dictionary of animal behaviour. Oxford: Oxford University Press.

McGrew, W.C. (1994). Tools compared: the material of culture. In: Chimpanzee cultures (Wrangham, R.W. et al., eds.). Harvard: Harvard University Press.

McGrew, W.C. & Tutin, C.E.G. (1978). Evidence for a social custom in wild chimpanzees? Man 13 (2), 234–51.

McMahan, E. (1982). Bait-and-capture strategy of a termite-eating assassin bug. Insectes Sociaux 29 (2), 346–51.

Melis, A.P. et al. (2006). Chimpanzees (*Pan troglodytes*) conceal visual and auditory information from others. Journal of Comparative Psychology 120 (2), 154–62.

Mench, J.A. (1998). Environmental enrichment and the importance of exploratory behavior. In: Second nature: environmental enrichment for captive animals (Shepherdson, D.J. et al., eds.). Washington; London: Smithsonian Institution Press, 30–46.

Mercader, J. et al. (2007). 4300-year-old chimpanzee sites and the origins of percussive stone technology. Proceedings of the National Academny of Science of the United States of America 104 (9), 3043–048.

Mezzina, M.J. (2018): Proposed brief by Amici Curiae philosophers in support of the petitioner-appellant (in re Nonhuman Rights Project, Inc., on behalf of Tommy, against Patrick C. Lavery). Court of Appeals, State of New York. New York County Clerk's Index Nos. 162358/15 and 150149/16, https://www.nonhumanrights.org/content/uploads/In-re-Nonhuman-Rights-v.-Lavery-Proposed-Brief-by-PHILOSOPHERS-74435.pdf, 2018/05/25.

Milner-Gulland, E.J. et al. (2003). Wild meat: the bigger picture. Trends in Ecology and Evolution 18 (7), 351–57.

Monen, J. et al. (1998). What does a pigeon see in a picture? Journal of the Experimental Analysis of Behavior 69 (2), 223–26.

Monsó, S. (2017): Morality without Mindreading. In: Mind & Language 32 (3): 338–57.

Münkler, H. (2002). Über den Krieg: Stationen der Kriegsgeschichte im Spiegel ihrer theoretischen Reflexion. Weilerswist: Velbrück Wissenschaft.

Münkler, H. (2006). Was ist neu an den neuen Kriegen? Eine Erwiderung auf die Kritiker. In: Den Krieg überdenken (Geis, A., ed.). Baden-Baden: Nomos, 133–50.

Mulcahy, N.J. & Call, J. (2006). Apes save tools for future use. Science 312 (5776), 1038–040.

Naess, A. (1973). The shallow and the deep, long-range ecology-movements: a summary. Inquiry 16, 95–100.

Naess, A. (2008). Ecology of wisdom: writings by Arne Naess. Ed. by A. Drengson und B. Devall. Berkeley, CA: Counterpoint.

Nagel, T. (1974). What is it like to be a bat? The Philosophical Review 83 (4), 435–50.

Nathanson, D.E. (1998). Long-term effectiveness of dolphin-assisted therapy for children with severe disabilities. Anthrozoös 11 (1), 22–32.

Nathanson, D.E. et al. (1997). Effectiveness of short-term dolphin-assisted therapy for children with severe disabilities. Anthrozoös 10 (2–3), 90–100.

Nellemann, C. et al. (eds.) (2007). The last stand of the orangutan—state of emergency: illegal logging, fire and palm oil in Indonesias national parks. Arendal, Norway: UNEP/GRID-Arendal. http://wedocs.unep.org/handle/20.500.11822/7524, 2019/05/21.

Newen, A. & Bartels, A. (2007). Animal minds and the possession of concepts. Philosophical Psychology 20 (3), 283–308.

Newsweek Staff (2007). The editor's desk. Newsweek 8 May 2007. https://www.newsweek.com/editors-desk-99583, 2019/05/21.

Nogge, G. (1999). Über den Umgang mit Tieren im Zoo: Tier- und Artenschutzaspekte. In: Tiere und Menschen: Geschichte und Aktualiät eines prekären Verhältnisses (Münch, P., ed.). Paderborn; München; Wien; Zürich: Schöningh, 447–57.

Norris, K.S. (1991). Dolphin days: the life and times of the spinner dolphin. New York: W.W. Norton.

Norris, K.S. (1994). Comparative view of cetacean social ecology, culture, and evolution. In: The Hawaiian spinner dolphin (Norris, K.S. et al., eds.). Berkeley, CA: University of California Press, 301–44.

Nowak, M.G. et al. (2017). Pongo tapanuliensis. The IUCN Red List of Threatened Species 2017: e.T120588639A120588662. http://dx.doi.org/10.2305/IUCN.UK.2017-3.RLTS.T120588639A120588662.en, 2019/04/13.

Oates, J.F. (1996). Habitat alteration, hunting and the conservation of folivorous primates in African forests. Australian Journal of Ecology 21 (1), 1–9.

Ogden, C.K. & Richards, I.A. (1923). The meaning of meaning: a study of the influence of language upon thought and of the science of symbolism. With essays by B. Malinowski and F.G. Crookshank. London: Kegan Paul, Trench, Trubner & Co.

Onishi, K.H. & Baillargeon, R. (2005). Do 15-month-old infants understand false beliefs? Science 308 (5719), 255–58.

Ouattara, K. et al. (2009). Campbell's monkeys concatenate vocalizations into context-specific call sequences. Proceedings of the National Academy of Sciences of the United States of America 106 (51), 22026–2031.

Pack, A.A. & Herman, L.M. (2006). Dolphin social cognition and joint attention: our current understanding. Aquatic Mammals 32 (4), 443–60.

Parker, S.T. & Gibson, K.R. (1977). Object manipulation, tool use and sensorimotor intelligence as feeding adaptations in cebus monkeys and great apes. Journal of Human Evolution 6 (7), 623–41.

Pepperberg, I. (2002). The Alex studies: cognitive and communicative abilities of grey parrots. Cambridge, MA; London: Harvard University Press.

Pepperberg, I. (2008). Alex & Me. New York: Harper Collins.

Perler, D. & Wild, M. (2005). Der Geist der Tiere: Philosophische Texte zu einer aktuellen Diskussion. Frankfurt a. M.: Suhrkamp.

Perner, J. et al. (1987). Three-year-olds' difficulty with false belief: the case for a conceptual deficit. British Journal of Developmental Psychology 5 (2), 125–37.

Plato ([after 399 v. Chr.] 1986). Kratylos. German transl. by F.D.E. Schleiermacher from 1807. Berlin: Akademie Verlag.

Plumptre, A., Robbins, M.M. & Williamson, E.A. (2019). Gorilla beringei. The IUCN Red List of Threatened Species 2019: e.T39994A115576640. http://dx.doi.org/10.2305/IUCN.UK.2019-1.RLTS.T39994A115576640.en, 2019/05/21.

Poole, T. (1997). Happy animals make good science. Laboratory Animals 31 (2), 116–24.

Poole, T.B. (1992). The nature of evolution of behavioural needs in mammals. Animal Welfare 1 (3), 203–20.

Povinelli, D.J. & Barth, J. (2005). Reinterpreting behavior: a human specialization? Behavioral and Brain Sciences 28 (5), 712–13.

Povinelli, D.J. & Eddy, T.J. (1996). What young chimpanzees know about seeing. Monographs of the Society for Research in Child Development 61 (3), 1–152.

Povinelli, D.J. et al. (2000). Toward a science of other minds: escaping the argument by analogy. Cognitive Science 24 (3), 509–41.

Povinelli, D.J. & Vonk, J. (2003). Chimpanzee minds: suspiciously human? Trends in Cognitive Sciences 7 (4), 157–60.

Povinelli, D.J. & Vonk, J. (2004). We don't need a microscope to explore the chimpanzee's mind. Mind & Language 19 (1), 1–28.

Precht, R.D. (2000). Noahs Erbe: Vom Recht der Tiere und den Grenzen des Menschen, Chap. *Treuhand Zoo: Oder vom Nachteil und Nutzen der Tiergärten für das Tierleben*. 1. ed. from 1997. Reinbek bei Hamburg: Rowohlt, 316–49.

Premack, D. & Woodruff, G. (1978). Does the chimpanzee have a theory of mind? Behavioral and Brain Sciences 1 (4), 515–26.

Primatt, H. (1778). Ueber Barmherzigkeit und Grausamkeit gegen die thierische Schöpfung. German transl. of the Engl. 1. edition *On the Duty of Mercy and Sin of Cruelty to Brute Animals* v. 1776. Halle: J.J. Gebauer.

Projekt Anahata (2008). Bototerapia—therapia assistida por Golfinhos de Rio & Integracao Estrutural através do Rolfing. No longer available online.

Pruetz, J.D. & Bertolani, P. (2007). Savanna chimpanzees, *Pan troglodytes verus*, hunt with tools. Current Biology 17 (5), 412–17.

Pylyshyn, Z.W. (1990). Computation and cognition: issues in the foundations of cognitive science. In: Foundations of cognitive science: the essential readings (Garfield, J.L., ed.). New York: Paragon House, 18–74.

Rat der Europäischen Union (1999). Council Directive 1999/22/ECof 29 March 1999 relating to the keeping of wild animals in zoos. Official Journal of the European Communities L 94/24 (9 April 1999), 24–26. https://eur-lex.europa.eu/legal-content/EN/TXT/PDF/?uri=CELEX:31999L0022&from=DE, 2019/05/21.

Regan, T. (2004). The case for animal rights. Updated ed. of the 1. Ed. of 1983. Berkeley; Los Angeles: University of California Press.

Rendell, L. & Whitehead, H. (2001). Culture in whales and dolphins. The Behavioral and Brain Sciences 24 (2), 309–24.

Rentsch, T. (1999). Mensch. In: Enzyklopädie Philosophie (Vol. 1) (Sandkühler, H.J. et al., eds.). Hamburg: Meiner, 814–18.

Reynolds III, J.E. et al. (2000). The bottlenose dolphin: biology and conservation. Gainsville, FL: University Press of Florida.

Richard, B.P. (1967). Le déterminisme de la construction des barrages chez le castor du Rhone. La Terre et la Vie 114 (4), 338–470.

Richard, P.B. (1980). Les castors. Poitiers: Balland.

Riedman, M.L. & Estes, J.A. (1990). The sea otter (*Enhydra lutris*): behavior, ecology, and natural history. U.S. Fish and Wildlife Service Biological Report 90 (14). Washington, D.C.: US Department of the Interior, Fish and Wildlife Service.

Rippe, K.-P. (2011). Die Würde des Tieres aus rechtsphilosophischer Sicht. TIERethik 3, 8–31.

Ritzer, G. (1993). The McDonaldization of society. Thousand Oaks, CA: Pine Forge Press.

Ritzer, G. (1998). The McDonaldization thesis. London: Sage.

Roberts, G.J. (1980). Apparent baiting behaviour by a black kite. Emu 82 (1), 53–54.

Roberts, M. (2008). Spanish parliament to extend rights to apes. Reuters press release from 25 June 2008. https://www.reuters.com/article/us-spain-apes/spanish-parliament-to-extend-rights-to-apes-idUSL256586320080625, 2019/05/21.

Robinson, M.H. (1998). Enriching the lives of zoo animals, and their welfare: where research can be fundamental. Animal Welfare 7 (2), 151–75.

Rogers, L.J. & Kaplan, G. (2004). All animals are not equal: the interface between scientific knowledge and legislation for animal rights. In: Animal rights: current debates and new directions (Sunstein, C.R. & Nussbaum, M.C., eds.). Oxford: Oxford University Press, 175–202.

Rogoff, B. (2003). The cultural nature of human development. Oxford: Oxford University Press.

Rose, A.L. (2002). Conservation must pursue human-nature biosynergy in the era of social chaos and bushmeat commerce. In: Primates face to face: conservation implications of human and nonhuman primate interconnections (Fuentes, A. & Wolfe, L.D., eds.). Cambridge, UK: Cambridge University Press, 208–39.

Rose, N.A. et al. (2009). The case against marine mammals in captivity, Chap. Risks to human health. 4. ed. of a report of the HSUS and the WSPA. Washington, DC; Boston, MA: HSUS and WSPA, 29–32.

Rowlands, M. (2012): Can Animals Be Moral? New York: Oxford University Press.

Rozak, T. (1992). The voice of the earth: an exploration of ecopsychology. New York: Simon & Schuster.

Russel, W.M.S. & Burch, R.L. (1992). The principles of humane experimental technique. 1. ed. of 1959. Wheathampstead, UK: Universities Federation for Animal Welfare.

Ryden, H. (1989). Lily pond. New York: William Morrow.

Samuels, A. & Spradlin, T. (1995). Quantitative behavioral study of bottlenose dolphins in swim-with-dolphin programs in the United States. Marine Mammal Science 11 (4), 520–44.

Sanz, C. et al. (2009). Design complexity in termite-fishing tools of chimpanzee (*Pan troglodytes*). Biology Letters 5 (3), 293–96.

Sanz, C. et al. (2004). New insights into chimpanzees, tools and termites from the Congo Basin. The American Naturalist 164 (5), 567–81.

Sanz, C.M. et al. (2010). Chimpanzees prey on army ants with specialized tool set. American Journal of Primatology 72 (1), 17–24.

Sapir, E. ([1921] 2006). Language: an Introduction to the study of speech. Based on the French 1. Edition from 1921. Teddington, UK: The Echo Library.

Sargeant, B.L. & Mann, J. (2008). The question of animal culture. In: From social learning to culture: intrapopulation variation in bottlenose dolphins (Galef Jr., B.G. & Laland, K.N., eds.). Cambridge, MA: Harvard University Press, 152–73.

Sargeant, B.L. & Mann, J. (2009). Developmental evidence for foraging traditions in wild bottlenose dolphins. Animal Behaviour 78 (3), 715–21.

Sargeant, B.L. et al. (2005). Specialization and development of beach hunting, a rare foraging behavior, by wild bottlenose dolphins (*Tursiops sp.*). Canadian Journal of Zoology 83 (11), 1400–410.

Savage-Rumbaugh, S. et al. (1998). Apes, language, and the human mind. New York; Oxford: Oxford University Press.

Schäffter, O. (1991). Das Fremde: Erfahrungsmöglichkeiten zwischen Faszination und Bedrohung. Chap. Modi des Fremderlebens: Deutungsmuster im Umgang mit Fremdheit. Opladen: Westdeutscher Verlag, 11–42.

Schaler, J.A. (ed.) (2009). Peter Singer under fire: the moral iconoclast faces his critics. Chicago; La Salle, IL: Open Court.

Schertler, G. (2012). Schimpansenshow: Tierschutz geht gegen Schwabenpark vor. Stuttgarter Nachrichten.de 7 August 2012. https://www.stuttgarter-nachrichten.de/inhalt.schimpansenshow-tierschutz-geht-gegen-schwabenpark-vor.a34afeca-dde0-4214-9e75-e3165fb27801.html, 2019/05/21.

Schicktanz, S. (2002). Organlieferant Tier? Medizin- und tierethische Probleme der Xenotransplantation. Frankfurt a. M.: Campus Verlag.

Schmidt, K. (2008). Blinde Hühner als Testfall tierethischer Theorien. Zeitschrift für philosophische Forschung 62 (4), 537–561.

Schmidt, M. (2010). Primates, Primaten, Herrentiere. In: Spezielle Zoologie (Teil 2: Wirbel- und Schädeltiere) (Westheide, W. & Rieger, R., eds.). 2. Edition. Heidelberg: Spektrum Akademischer Verlag, 565–592.

Schmoeckel, M. et al. (eds.) (2003). Historisch-kritischer Kommentar zum BGB (Vol I: Allgemeiner Teil: §§1-240). Tübingen: Mohr Siebeck.

Schneider, J.H.J. (1995). Sprache (II. Mittelalter, Renaissance, Humanismus). In: Historisches Wörterbuch der Philosophie (Vol. 9) (Ritter, J. & Gründer, K., eds.). Basel: Schwabe & Co AG, 1454–1468.

Schopenhauer, A. ([1841] 2007). Über die Grundlage der Moral. Ed. by P. Welsen, 1. ed. with an introduction, comments and register. Hamburg: Meiner.

Schramm, S., Etzold, S., Schuh H. (2007). Leidet der Eisbär? Ein Streitgespräch zwischen Günther Nogge und Hanno Würbel. Die Zeit 18 (26 April 2007), 38–39.

Schütt, H.P. (ed.) (1990). Die Vernunft der Tiere. Frankfurt a. M.: Keip.

Schuster, G. et al. (2007). Die Denker des Dschungels: Der Orangutan-Report. Bilder, Fakten, Hintergründe. Tandem Verlag.

Schweitzer, A. ([1923] 1971). Kultur und Ethik, Vol. 2 from selected works in five volumes. Berlin: Union Verlag.

SeaWorld Parks & Entertainment (2016). SeaWorld announces last generation of orcas in its care. Press Release from 21 March 2016. https://www.seaworldcalifornia.co.uk/latest-news/seaworld-announces-last-generation-of-orcas-in-its-care, 2019/05/31.

Seidensticker, J. & Doherty, J.G. (1998). Integrating animal behavior and exhibit design. In: Wild mammals in captivity (Kleiman, D.G. et al., eds.). Chicago: University of Chicago Press, 180–190.

Seidensticker, J. & Fortman, D.L. (1998). Evolution and ecology in zoo animal enrichment.In: Second nature: environmental enrichment for captive animals (Shepherdson, D.J. et al., eds.). Washington; London: Smithsonian Institution Press, 15–29.

Semaw, S. (2000). The world's oldest stone artefacts from Gona Ethiopia: their implications for understanding stone technology and patterns of human evolution between 2.6–1.5 million years ago. Journal of Archaeological Science 27 (12), 1197–1214.

Seyfarth, R. et al. (1980). Monkey responses to three different alarm calls: evidence of predator classification and semantic communication. Science 210 (4471), 801–803.

Seyfarth, R.M. & Cheney, D.L. (1993). Wie Affen sich verstehen. Spektrum der Wissenschaft February 1993, 88–95.

Shatz, M. et al. (1983). The acquisition of mental verbs: a systematic investigation of the first reference to mental state. Cognition 14 (3), 301–321.

Shearer, D. (1999). Africa's great war. Survival 41 (2), 89–106.

Shepherdson, D.J. (1998). Tracing the path of environmental enrichment in zoos. In: Second nature: environmental enrichment for captive animals (Shepherdson, D.J. et al., eds.). Washington; London: Smithsonian Institution Press.

Shepherdson, D.J. et al. (eds.) (1998). Second nature: environmental enrichment for captive animals. Washington; London: Smithsonian Institution Press.

Shriver, A. (2009). Knocking Out Pain in Livestock: Can Technology Succeed Where Morality Has Stalled? Neuroethics 2 (3): 115–24.

Shriver, A., and McConnachie, E. (2018). Genetically Modifying Livestock for Improved Welfare: A Path Forward. Journal of Agricultural and Environmental Ethics 31 (2): 161–80.

Simmel, G. (1958). Soziologie: Untersuchungen über die Formen der Vergesellschaftung. Berlin: Duncker & Humblot.

Singer, P. (1976). Animal liberation: a new ethics for our treatment of animals. 1. ed. from 1975. London: J. Cape.

Singer, P. (1981). The expanding circle: ethics and sociobiology. Oxford: Oxford University Press.

Singer, P. (1997). The drowning child and the expanding circle. The New Internationalist 1 April 1997. https://www.utilitarian.net/singer/by/199704--.htm, 2019/05/21.

Singer, P. (2006). The great ape debate unfolds in Europe. The Japan Times Online 22. May 2006. https://www.japantimes.co.jp/opinion/2006/05/22/commentary/world-commentary/the-great-ape-debate-unfolds-in-europe/#.XOPoJaRCSuU, 2019/05/21.

Singer, P. (2008). Practical Ethics. 3. ed. of the 1. ed. from first edition from 1980. New York: Cambridge University Press.

Singer, P. (2009a). Animal liberation: the definitive classic of the animal movement. Revised and extended edition of the first edition from 1975. New York; London; Toronto; Sydney; New Delhi; Auckland: Harper Perennial.

Singer, P. (2009b). An intellectual autobiography. In: Peter Singer under fire: the moral iconoclast faces his critics (Schaler, J.A., ed.). Chicago; La Salle, IL: Open Court, 1–74.

Singer, P. (2009c). Reply to Don Marquis: *Singer on abortion and infanticide*. In: Peter Singer under fire: the moral iconoclast faces his critics (Schaler, J.A., ed.). Chicago; La Salle, IL: Open Court, 153–162.

Singer, P. (2009d). Reply to Harry J. Gensler: *Singer's unsanctity of human life. A critique*. In: Peter Singer under fire: the moral iconoclast faces his critics (Schaler, J.A., ed.). Chicago; La Salle, IL: Open Court, 185–194.

Singleton, I., et al. (2017). Pongo abelii. The IUCN Red List of Threatened Species 2017: e.T121097935A115575085. http://dx.doi.org/10.2305/IUCN.UK.2017-3.RLTS.T121097935A115575085.en, 2019/05/21.

SJD (a). Frequently asked questions. http://savedolphins.eii.org/news/entry/frequently-asked-questions-save-japan-dolphins-campaign, 2019/05/21.

Smolker, R. (2001). To touch a wild dolphin. New York: Doubleday.

Sommer, V. (2003). Geistlose Affen oder äffische Geistwesen? Eine Exkursion durch die mentale Welt unserer Mitprimaten. In: Gene, Meme und Gehirne: Geist und Gesellschaft als Natur. Eine Debatte (Becker, A. et al., eds.). Frankfurt a. M.: Suhrkamp, 112–136.

Sommer, V. (2007). Darwinisch Denken: Horizonte der Evolutionsbiologie. Stuttgart: Hirzel.

Sorenson, J. (2009). Ape. London: Reaktion Books.

Spaemann, R. (1984). Tierschutz und Menschenwürde. In: Tierschutz: Testfall unserer Menschlichkeit (Händel, U.M., ed.). Frankfurt a. M.: Fischer Taschenbuch, 71–81.

Špinka, M. & Wemelsfelder, F. (2011). Environmental challenge and animal agency. Animal welfare. Wallingford, UK: CAB International, 27–44.

Stiftung für das Tier im Recht (2019). Rechtsauskünfte / Tiere sind keine Sachen / Haben Tiere Rechte? https://www.tierimrecht.org/de/recht/rechtsauskunfte/tiere-sind-keine-sachen/haben-tiere-rechte/, 2019/05/22.

Streck, B. (2000a). Kulturanthropologie. In: Wörterbuch der Ethnologie. 2., ext. edition. Wuppertal: Peter Hammer, 141–144.

Streck, B. (2000b).Kulturökologie. In: Wörterbuch der Ethnologie. 2., ext. edition. Wuppertal: Peter Hammer, 144–146.

Struhsaker, T. (1967). Auditory communication among vervet monkeys (*Certopithecus aethiops*). In: Social communication among primates (Altmann, S.A., ed.). Chicago: University of Chicago Press, 281–324.

Sturma, D. (1997). Philosophie der Person: Die Selbstverhältnisse von Subjektivität und Moralität. Paderborn; Wien; Zürich: Schöningh.

Sturma, D. (2002). Person. In: Enzyklopädie Philosophie (Sandkühler, H.J., ed.). CD-ROM version. Hamburg: Meiner.

Su, S. et al. (2008). East learns from west: Asiatic honeybees can understand dance language of European honeybees. PLoS ONE 3 (6), e2365. DOI:10.1371/journal.pone.0002365.

Suzuki, S. et al. (1995). Tool-set for termite-fishing by chimpanzees in the Ndoki Forest, Congo. Behaviour 132 (3–4), 219–235.

Suzuki, R. et al. (2006). Information entropy of humpback whale songs. Journal of the Acoustical Society of America 119 (3), 1849–1866.

Szagun, G. (2000). Sprachentwicklung beim Kind. Unchanged version of the 6., completely rev. ed. from 1996. Weinheim; Basel; Berlin: Beltz.

Taylor, A.H. et al. (2007). Spontanuous metatool use by New Caledonian crows. Current Biology 17 (17), 1504–1507.

Taylor, K. & Gordon, N. (2008). Authors response to Andrew Knights article *127 million non human vertebrates used worldwide for scientific purposes in 2005*. Alternatives to Laboratory Animals 36 (5), 496–497.

Taylor, K. et al. (2008). Estimates for worldwide laboratory animal use in 2005. Alternatives to Laboratory Animals 36 (3), 327–342.

Taylor, P.W. (1986). Respect for nature: a theory of environmental ethics. Princeton, New Jersey: Princeton University Press.

Tennie, C. et al. (2009). Ratcheting up the ratchet: on the evolution of cumulative culture. Philosophical Transactions of the Royal Society B: Biological Sciences 364 (1528), 2405–2415.

The Animal Fund (2019): The Taiji Dolphin Drives. https://theanimalfund.net/en/taiji-dolphin-drives/, 2019/05/21.

Terrace, H.S. et al. (2003). Serial expertise of rhesus macaques. Psychological Science 14, 66–73.

Tierschutzbund, Deutscher (2019). Bürgerliches Recht. Tiere sind keine Sachen. https://www.tierschutzbund.de/information/hintergrund/recht/buergerliches-recht/, 2019/05/18.

Tiergarten Nürnberg (2010). Eisbären schützen. Press release, no longer online available.

Tomasello, M. (1999). The cultural origins of human cognition. Cambridge, MA: Harvard University Press.

Tomasello, M. & Call, J. (2006). Do chimpanzees know what others see—or only what they are looking at? In: Rational Animals? (Hurley, S. & Nudds, M., eds.). Oxford: Oxford University Press, 371–384.

Tomasello, M. et al. (2003a). Chimpanzees understand psychological states—the question is which ones and to what extent. Trends in Cognitive Sciences 7 (4), 153–156.

Tomasello, M. et al. (2003b). Chimpanzees versus humans: it's not that simple. Trends in Cognitive Sciences 7 (4), 239–240.

Tudge, C. (1993). Letzte Zuflucht Zoo: Über die Erhaltung bedrohter Arten in den Zoologischen Gärten. Berlin; Heidelberg; Oxford: Spektrum Akademischer Verlag.

Tugendhat, E. (1994). Vorlesungen über Ethik. Frankfurt a. M.: Suhrkamp.

Tyack, P. (2000). Functional aspects of cetacean communication. In: Cetacean societies: field studies of dolphins and whales (Mann, J. et al., eds.). Chicago: Chicago University Press, 270–307.

Tylor, E. (1871): Primitive Culture: Research into the Development of Mythology, Philosophy, Religion, Art, and Custum. Vol. 1. London: John Murray.

UK Parliamentary Office of Science and Technology (2005). The Bushmeat Trade. Postnote of the Parliamentary Office of Science and Technology (UK) 236 (February 2005). https://www.parliament.uk/documents/post/postpn236.pdf, 2019/05/22.

UNEP (2009). From conflict to peacebuilding: the role of natural resources and the environment. https://postconflict.unep.ch/publications/pcdmb_policy_01.pdf, 2019/05/22.

UNESCO/WHC (2008a). Violent clashes in southern sector of Virunga National Park. Meldung vom 10. Oktober 2008. http://whc.unesco.org/en/news/461, 2019/05/22.

UNESCO/WHC (2008b). Virunga National Park: Mountain gorillas remain target for armed groups. Press release 6 June 2008. http://whc.unesco.org/en/news/351, 2019/05/22.

Universität Würzburg, Abteilung Rehabilitationspsychologie (2015). Forschung der Abteilung Rehabilitationspsychologie / aktuelle Forschungsprojekte / Delphintherapie. Last updated 9 December 2015. https://www.reha.hu-berlin.de/de/lehrgebiete/rhp/forschung/delfintherapie?set_language=de, 2019/05/22.

van Lawick-Goodall, J. (1970). Tool-using in primates and other vertebrates. In: Advances in the study of behavior (Vol. 3) (Lehrman, D.S. et al., eds.). New York: Academic Press, 195–249.

van Leeuwen, E.J.C., Cronin, K.A. & Haun, D.B.M. (2014): A group-specific arbitrary tradition in chimpanzees (*Pan troglodytes*). Animal Cognition 17 (6), S. 1421–1425.

van Schaik, C.P. et al. (2003). Orangutan cultures and the evolution of material culture. Science 299 (5603), 102–105.

van Schaik, C.P. et al. (2006). Innovation in wild Bornean orangutans (*Pongo pygmaeus wurmbii*). Behaviour 143 (7), 839–876.

VandeBerg, J.L. et al. (2005). A unique biomedical resource at risk. Nature 437 (7055), 30–32.

Varner, G.E. (2012): Personhood, Ethics, and Animal Cognition: Situating Animals in Hare's Two Level Utilitarianism. New York: Oxford University Press.

Warneken, F. & Tomasello, M. (2006). Altruistic helping in human infants and young chimpanzees. Science 311 (5765), 1301–1303.

Watanabe, S. et al. (1995). Pigeons' discrimination of paintings by Monet and Picasso. Journal of Experimental Analysis of Bahavior 63 (2), 165–174.

WAZA (a). Committing to Conservation. The World Zoo and aquarium Conservation Strategy. https://www.waza.org/wp-content/uploads/2019/03/WAZA-Conservation-Strategy-2015_Landscape.pdf, 2019/05/22.

WAZA (b). How to Join. Downloaded 19 July 2012. No longer online available as of May 2019, possibly because of ongoing restructuring of the WAZA homepage.

WAZA (c). WAZA Code of Ethics and Animal Welfare. https://www.waza.org/wp-content/uploads/2019/05/WAZA-Code-of-Ethics.pdf, 24/08/19.

WAZA (d). WAZA Ethics and Animal Welfare Committee briefing on general principles and practice with particular reference to dolphin capture developments, 2010. Downloaded on 19 July 2012. No longer online available.

WDC (a). Delphinarien in Deutschland. https://de.whales.org/wdc-ziele/delfinarien-schliessen/deutschland/, 2019/05/19.

WDC (b). Swimming with dolphins. https://us.whales.org/swimming-with-dolphins/, 2019/05/31.

WDCS (a). Kritische Betrachtung zum Abschlussbericht des Forschungsprojekts Delfintherapie der Uni Würzburg und des Nürnberger Delfinariums. https://de.whales.org/wp-content/uploads/sites/4/2018/07/delfintherapie_stellungnahme.pdf, 2019/04/15.

WDCS (b). WDCS Stellungnahme zur Haltung von Walen und Delfinen in Gefangenschaft. Downloader 19 July 2012. No longer online available.

Webb, N.L. & Drummond, P.D. (2001). The effect of swimming with dolphins on human well-being and anxiety. Anthrozoös 14 (2), 81–85.

Wehner, R. & Gehring, W. (2007). Zoologie. 24., entirely rev. ed. of the 1. ed. from 1922. Stuttgart: Georg Thieme Verlag.

Weir, A.A.S. et al. (2002). Shaping of hooks in New Caledonian crows. Science 297 (5583), 981.

Weiss, J.M. (1972). Psychological factors in stress and disease. Scientific American 226 (6), 104–113.

Wemelsfelder, F. (1994). Animal boredom—A model of chronic suffering in captive animals and its consequences for environmental enrichment. Humane Innovations and Alternatives 8, 587–591. Downloaded on 19 July 2012 from the Society & Animals Forum (Download link no longer available).

Wemelsfelder, F. (2005). Animal Boredom: Understanding the Tedium of Confined Lives. In: Mental Health and Well-Being in Animals (McMillan, F.D. ed.). Oxford; Victoria: Blackwell Publishing, 79–91.

White, T. (2007). In defense of dolphins: the new moral frontier. Malden (MA); Oxford; Victoria: Blackwell Publishing.

Whiten, A. (2005). The second inheritance system of chimpanzees and humans. Nature 437 (7055), 52–55.

Whiten, A. et al. (1999). Cultures in chimpanzees. Nature 399 (6737), 682–685.

Whiten, A. et al. (2005). Conformity to cultural norms of tool use in chimpanzees. Nature 437 (7059), 737–740.

Whiten, A. et al. (2007). Transmission of multiple traditions within and between chimpanzee groups. Current Biology 17 (12), 1038–1043.

Wich, S.A. et al. (2009). A case of spontaneous acquisition of a human sound by an orangutan. Primates 50 (1), 56–64.

Wiedemann, R. (1999). Die Fremdheit der Tiere. In: Tiere und Menschen: Geschichte und Aktualiät eines prekären Verhältnisses (Münch, P., ed.). Paderborn; Munchen; Wien; Zürich: Schöningh, 351–381.

Wiesing, U. (2000). Der Hippokratische Eid: Einführung. In: Ethik in der Medizin: Ein Reader (Nida-Rümelin, J., ed.). In collaboration with J.S. Ach, M. Bormuth and G. Marckmann, Stuttgart: Reclam, 21–25.

Wild, M. (2008). Tierphilosophie: Zur Einführung. Hamburg: Junius.

Wildman, D.E. et al. (2003). Implications of natural selection in shaping 99.4 % nonsynonymous DNA identity between humans and chimpanzees: enlarging genus Homo. Proceedings of the National Academy of Sciences of the United States of America 100 (12), 7181–7188.

Wilkie, D.S. (2001). Bushmeat hunting in the Congo Basin. In: Great Apes & humans: the ethics of coexistence (Beck, B.B. et al., eds.). Washington; London: Smithsonian Institution Press, 86–109.

Wilkie, D.S. & Carpenter, J.F. (1999). Bushmeat hunting in the Congo Basin: an assessment of impacts and options for mitigation. Biodiversity and Conservation 8 (7), 927–955.

Williams, B.G. et al. (1996). The effect of a moving bait on the behaviour of captive cheetahs (*Acinonyx jubatus*). Animal Welfare 5 (3), 271–281.

Williams, T.D., Bengtsson, J.O. (2018): Personalism. In: Edward N. Zalta (ed.): The Stanford Encyclopedia of Philosophy (Winter 2018 Edition), https://plato.stanford.edu/entries/personalism/, 2019/04/23.

Williamson, C. (2018). Why whale and dolphin captivity is cruel—the science. Summary provided for the WDCS US. https://us.whales.org/2018/07/25/why-whale-and-dolphin-captivity-is-cruel-the-science/, 2019/05/31.

Wilson, E.O. (2003). Biophilia: the human bond with other species. 12. ed. of the 1. ed. from 1984. Cambridge, MA: Harvard University Press.

Wilson, L. (1971). Observations and experiments on the ethology of the European beaver (*Castor fiber L.*): a study in the development of phylogenetically adapted behaviour in a highly specialized mammal. Viltrevy: Swedish Wildlife 8 (3), 115–266.

Wimmer, H. & Perner, J. (1983). Beliefs about beliefs: representation and constraining function of wrong beliefs in young children's understanding of deception. Cognition 13 (1), 103–128.

Wimmer, R. (1999). Ethische Aspekte des Personbegriffs. In: Biologie und Ethik (Engels, E.-M., ed.). Stuttgart: Reclam, 327–345.

Wimpenny, J.H. et al. (2009). Cognitive processes associated with sequential tool use in New Caledonian crows. PLoS ONE 4 (8), e6471. DOI:10.1371/journal.pone.0006471.

Wise, S.M. (2000). Rattling the cage: toward legal rights for animals. Cambridge, MA: Perseus Books.

Wittgenstein, L. (1922). Tractatus Logico-Philosophicus. Correct. Version of the 1. ed. from 1921, with an introd. By B. Russell, transl. by C.K. Ogden. London: Kegan Paul, Trench, Trubner & Co.

Wolf, J.-C. (1992). Tierethik: Neue Perspektiven für Menschen und Tiere. Freiburg, CH: Paulusverlag.

Woolfenden, J. (1985). The California sea otter: saved or doomed? Pacific Grove, CA: Boxwood Press.

Workman, L. & Reader, W. (2004). Evolutionary psychology: an introduction. Cambridge: Cambridge University Press.

Wrangham, R.W. et al. (2008). The Great Ape World Heritage Species Project. In: Conservation in the 21st Century: Gorillas as a Case Study (Stoinski, T.S. et al., eds.). New York: Springer, 282–295.

Würbel, H. (2001). Ideal homes? Housing effects on rodent brain and behavior. Trends in Neuroscience 24 (4), 207–211.

Würbel, H. (2007). Biologische Grundlagen zum ethischen Tierschutz. In: Tierrechte: Eine interdisziplinäre Herausforderung (Interdisziplinäre Arbeitsgemeinschaft Tierethik (IAT), ed.). Erlangen: Harald Fischer, 11–30.

Würbel, H. et al. (2006). Box 1.3. The coping hypothesis of stereotypic behaviour. In: Stereotypic animal behaviour: fundamentals and applications to welfare (Mason, G. & Rushen, J., eds.). 2. ed., Wallingford, Oxfordshire; Cambridge, MA: CABI Publishing, 14–15.

Würbel, H. & Garner, J.P. (2007). Refinement of rodent research through environmental enrichment and systematic randomization. NC3Rs Newsletter 9, 1–9. https://

www.nc3rs.org.uk/sites/default/files/documents/Refinementenvironmentalenrichmentandsystematicrandomization.pdf, 2019/05/31.

Würsig, B. (1983). The question of animal awareness approached through studies in nature. Cetus 5 (1), 4–7.

Würsig, B. et al. (1994). A spinner dolphin's day. In: The Hawaiian spinner dolphin (Norris, K.S. et al., eds.). Berkeley, CA: University of California Press, 65–102.

Yerkes, R. (1925). Almost human. New York: Century.

Young, R.J. (2003). Environmental enrichment for captive animals. UFAW Animal Welfare Series. Oxford: Blackwell Publishing.

Zerres, T. (2005). Bürgerliches Recht: Ein einführendes Lehrbuch in das Zivil- und Zivilprozessrecht. 5. rev. and ext. ed. from 2005. Berlin; Heidelberg; New York: Springer.

Zuberbühler, K. et al. (1999). Conceptual semantics in a nonhuman primate. Journal of Comparative Psychology 113 (1), 33–42.

Zuberbühler, K. et al. (1997). Diana monkey long-distance calls: messages for conspecifics and predators. Animal Behaviour 53 (3), 589–604.

Index